VOLUME ONE

ADVANCES IN STEM CELLS AND THEIR NICHES

Hematopoietic Stem Cell Niche

VOLUME ONE

ADVANCES IN STEM CELLS AND THEIR NICHES

Hematopoietic Stem Cell Niche

Edited by

DOMINIQUE BONNET
The Francis Crick Institute
London, United Kingdom

ACADEMIC PRESS
An imprint of Elsevier

Academic Press is an imprint of Elsevier
50 Hampshire Street, 5th Floor, Cambridge, MA 02139, United States
525 B Street, Suite 1800, San Diego, CA 92101-4495, United States
The Boulevard, Langford Lane, Kidlington, Oxford OX5 1GB, United Kingdom
125 London Wall, London EC2Y 5AS, United Kingdom

First edition 2017

ISBN: 978-0-12-811375-2
ISSN: 2468-5097

For information on all Academic Press publications
visit our website at https://www.elsevier.com/books-and-journals

Publisher: Zoe Kruze
Acquisition Editor: Zoe Kruze
Senior Editorial Project Manager: Helene Kabes
Production Project Manager: Surya Narayanan Jayachandran
Senior Cover Designer: Miles Hitchen

Typeset by SPi Global, India

CONTENTS

CONTRIBUTORS

P. Charbord
Sorbonne Universités, UPMC Univ Paris 06, IBPS, CNRS UMR7622, Inserm U 1156, Laboratoire de Biologie du Développement, Paris, France

D. Duarte
Imperial College London; The Francis Crick Institute, London, United Kingdom

C. Durand
Sorbonne Universités, UPMC Univ Paris 06, IBPS, CNRS UMR7622, Inserm U 1156, Laboratoire de Biologie du Développement, Paris, France

A. García-García
Wellcome Trust-Medical Research Council Cambridge Stem Cell Institute, University of Cambridge; National Health Service Blood and Transplant, Cambridge, United Kingdom; Centro Nacional de Investigaciones Cardiovasculares Carlos III (CNIC), Madrid, Spain

K. Golan
Weizmann Institute of Science, Rehovot, Israel

L. Han
The University of Texas MD Anderson Cancer Center, Houston, TX, United States

E. Khatib-Massalha
Weizmann Institute of Science, Rehovot, Israel

O. Kollet
Weizmann Institute of Science, Rehovot, Israel

M. Konopleva
The University of Texas MD Anderson Cancer Center, Houston, TX, United States

D.S. Krause
Georg-Speyer-Haus, Institute for Tumor Biology and Experimental Therapy, Frankfurt, Germany

A. Kumari
Weizmann Institute of Science, Rehovot, Israel

T. Lapidot
Weizmann Institute of Science, Rehovot, Israel

L. Li
Stowers Institute for Medical Research, Kansas City, MO; University of Kansas Medical Center, Kansas City, KS, United States

C. Lo Celso
Imperial College London; The Francis Crick Institute, London, United Kingdom

S. Méndez-Ferrer
Wellcome Trust-Medical Research Council Cambridge Stem Cell Institute, University of Cambridge; National Health Service Blood and Transplant, Cambridge, United Kingdom; Centro Nacional de Investigaciones Cardiovasculares Carlos III (CNIC), Madrid, Spain

G. Stik
Sorbonne Universités, UPMC Univ Paris 06, IBPS, CNRS UMR7622, Inserm U 1156, Laboratoire de Biologie du Développement, Paris, France

D. Verma
Georg-Speyer-Haus, Institute for Tumor Biology and Experimental Therapy, Frankfurt, Germany

M. Zhao
Zhongshan School of Medicine; Key Laboratory for Stem Cells and Tissue Engineering, Ministry of Education, Sun Yat-Sen University, Guangzhou, Guangdong, China

B.O. Zhou
State Key Laboratory of Cell Biology, CAS Center for Excellence in Molecular Cell Science, Shanghai Institute of Biochemistry and Cell Biology, Chinese Academy of Sciences, University of Chinese Academy of Sciences, Shanghai, China

PREFACE

It is essential to understand how stem cells interact with their microenvironment, the so-called stem cell niche to establish and maintain their function.

In this new series "Advances in Stem Cells and their niches" each volume will focus on a specific organ looking at how the niche components could regulate stem cells in both normal and disease conditions.

The first volume is dedicated to the *Hematopoietic Stem Cell Niche*. It contains seven contributions aiming at providing the latest reviews on our current understanding of the role of the hematopoietic stem cell (HSC) niche in the regulation of both normal and malignant hematopoiesis as well as new technological advances in in vivo imaging of HSC in their niche as well as new development on how to improve homing or mobilization of HSC in and out of the bone marrow niche.

In the first contribution, Dr. Stik et al. discuss our current knowledge on the cellular and molecular identity of the HSC niches during ontogeny and explore the molecular requirements for the HSC-supportive capacity of the niche cells and their cross talk with HSCs.

In the second chapter, Dr. Zhou et al. discuss the origin of the HSC niche concept, and examine the complexity of the bone marrow environment by describing the different players being involved. In the third chapter, Dr. García-García and Méndez-Ferrer discuss our current knowledge on the interplay between the nervous and the hematopoietic systems, with particular emphasis on the regulation of HSCs.

The next chapter by Duarte and Celso describes the use of intravital imaging to visualize and follow the dynamic behavior of HSC in their bone marrow niche allowing direct live observation of HSC–niche interactions in vivo. The following chapter by Kumari et al. addresses the mechanisms responsible for the homing and mobilization of the HSC.

The next chapter by Han and Konopleva discusses the recent insights into the role of altered niche function in the setting of myeloid malignancies, and the last chapter by Verma and Krause addresses the potential approaches to restore normal hematopoiesis through targeting of malignant niches.

This book should thus be of interest for academic researchers, research scientists, and graduate students in universities, industry, and government.

D. Bonnet
The Francis Crick Institute, London, United Kingdom

CHAPTER ONE

Stroma Cell Niche Regulation During HSC Development

G. Stik*[,1], P. Charbord*, C. Durand*[,2]
*Sorbonne Universités, UPMC Univ Paris 06, IBPS, CNRS UMR7622, Inserm U 1156, Laboratoire de Biologie du Développement, Paris, France
[2]Corresponding author: e-mail address: charles.durand@upmc.fr

Contents

1. INTRODUCTION

Hematopoietic stem cells (HSCs) constitute a rare population of cells at the foundation of the adult hematopoietic system. Through their ability to self-renew and to differentiate, HSCs are responsible for the continuous production of mature blood cells throughout life. HSCs thus represent a unique model to study stem cell biology and have also major implications in the field of regenerative medicine (Wagers, 2012). Hematopoietic stem and progenitor cells (HSPCs) reside in close association with stromal cells that provide a supportive microenvironment (also called niche) for HSPCs. In the adult, HSPCs are located in the bone marrow (BM). The combination of transgenic mouse models with imaging and functional assays revealed that HSPCs are closely associated with blood vessels (arterioles and sinusoids) in contact with mesenchymal stem/stromal cells (MSCs), endothelial cells, and pericytes (Ding & Morrison, 2013; Ding, Saunders, Enikolopov, &

[1] Present address: Centre for Genomic Regulation (CRG), C/Dr. Aiguader, 88, PRBB Building, 08003 Barcelona, Spain.

Advances in Stem Cells and their Niches, Volume 1
ISSN 2468-5097
http://dx.doi.org/10.1016/bs.asn.2016.12.001

Morrison, 2012; Greenbaum et al., 2013; Kiel, Yilmaz, Iwashita, Terhorst, & Morrison, 2005; Kunisaki et al., 2013; Mendez-Ferrer et al., 2010). The BM HSPC niche appears to be highly complex, since Schwann cells, osteoblasts, osteoclasts, megakaryocytes, and macrophages have all been found to play an active role in the regulation of HSPCs (Mendez-Ferrer, Scadden, & Sanchez-Aguilera, 2015; Morrison & Scadden, 2014). At the molecular level, niche cells are known to regulate HSPC biology through the production extrinsic factors (such as growth factors and morphogens) and the regulation of cell adhesion and extracellular matrix remodeling cues. On top of these noncell autonomous mechanisms, systemic factors involving for instance the sympathetic nervous system and the circadian clock have also been shown to play an important role (Mendez-Ferrer, Lucas, Battista, & Frenette, 2008).

Importantly, HSCs are not produced in the BM, but early during embryonic life. The first adult-type HSCs are autonomously generated in the aorta-gonad-mesonephros (AGM) region at midgestation (Fig. 1; Cumano, Dieterlen-Lievre, & Godin, 1996; Medvinsky & Dzierzak, 1996; Muller, Medvinsky, Strouboulis, Grosveld, & Dzierzak, 1994). The hematopoietic activity in the AGM is characterized by the production of intraaortic hematopoietic clusters (IAHCs) closely attached to the ventral side of the dorsal aorta (Cumano & Godin, 2007; Dzierzak & Speck, 2008; Medvinsky, Rybtsov, & Taoudi, 2011). In the mouse, IAHCs are also observed on the dorsal side of the aorta. Interestingly, if both dorsal and ventral IAHCs harbor clonogenic potential, the HSC repopulating activity is preferentially assigned to ventral IAHCs (Souilhol et al., 2016; Taoudi & Medvinsky, 2007). Taken together, these observations demonstrate that the AGM hematopoietic activity is polarized along the dorsoventral axis and that the embryonic tissues underneath the dorsal aorta may serve as an important niche for the first emerging HSCs. The process of endothelial to hematopoietic transition by which HSPCs are produced from hemogenic endothelial cells lining the floor of the dorsal aorta is well conserved during evolution and has been documented in multiple vertebrate models (de Bruijn et al., 2002; Jaffredo, Gautier, Eichmann, & Dieterlen-Lievre, 1998; Oberlin, Tavian, Blazsek, & Peault, 2002). Moreover, these findings have been recently confirmed through in vivo intravital live imaging in the zebrafish and mouse embryos (Bertrand et al., 2010; Boisset et al., 2010; Kissa & Herbomel, 2010). Following their emergence in the dorsal aorta, HSPCs then colonize the fetal liver (FL) where they massively amplify. By the end of gestation, HSPCs migrate to the BM where they will be

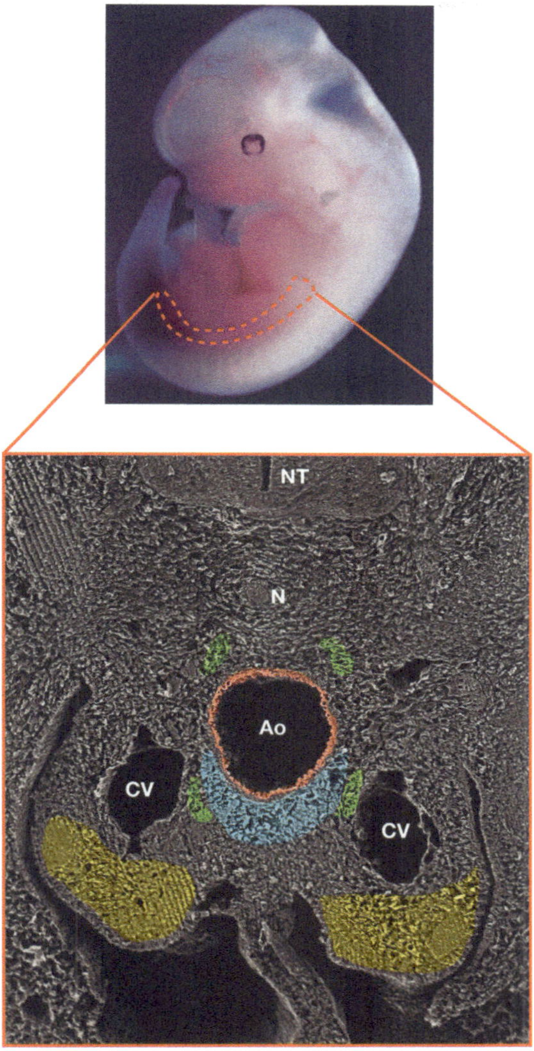

Fig. 1 The AGM hematopoietic microenvironment. *Top panel*: E11.5 mouse embryo. The AGM is indicated with a *red-dotted line*. *Bottom panel*: Electron micrograph showing different cellular compartments of the AGM HSPC microenvironment: endothelial cells (*red*), vascular smooth muscle cells, and subaortic mesenchyme (*blue*), sympathetic nervous system (*green*), and urogenital ridges (*yellow*). *Ao*, dorsal aorta; *CV*, cardinal vein; *NT*, neural tube; *N*, notochord. *Courtesy of Thierry Jaffredo.*

maintained throughout life. Thus, during their ontogeny, HSPCs are found in different niches that successively support their emergence, amplification, and maintenance. As emphasized by recent studies on the reprogramming of somatic cells to HSPCs (Batta, Florkowska, Kouskoff, & Lacaud, 2014; Pereira et al., 2013; Riddell et al., 2014; Sandler et al., 2014), identifying the similarities and differences between these distinct developmental HSPC niches is then of critical importance for our basic understanding of HSPC biology and also for their efficient ex vivo amplification and production for clinical purposes.

2. THE AGM HEMATOPOIETIC MICROENVIRONMENT

To better understand the role of the AGM hematopoietic microenvironment, a large panel of stromal cell lines was established not only from the subcompartments of the AGM, the Ao and its surrounding mesenchyme, and the urogenital ridges (UG) but also from the embryonic liver (EL) and the gastrointestinal tract of midgestation mouse embryos (Oostendorp, Medvinsky, et al., 2002). Coculture experiments with embryonic or adult HSPCs followed by colony forming cell or in vivo transplantation assays revealed that some embryonic stromal lines (including UG26-1B6 or EL08-1D2) provide a potent support for HSPCs, whereas other lines are less efficient in maintaining HSPCs ex vivo (Buckley et al., 2011; Oostendorp, Harvey, et al., 2002; Oostendorp et al., 2005). Interestingly, some of these stromal lines have also been shown to support the hematopoietic differentiation of mouse and human embryonic stem cells (Krassowska et al., 2006; Ledran et al., 2008). These AGM stromal cells have a mesenchymal origin and exhibit a myofibroblastic morphology. They are committed to the vascular smooth muscle cell (VSMC) differentiation pathway (Charbord et al., 2002) but can still give rise to other mesenchymal derivatives (such as osteoblasts and adipocytes) when cultured under appropriate in vitro conditions (Durand, Robin, & Dzierzak, 2006). The existence in the AGM of primary mesenchymal progenitors with adipogenic, osteogenic, and chondrogenic potentials has been documented at time of HSC production in the AGM (Mendes, Robin, & Dzierzak, 2005). Mesenchymal progenitor cells are then detected in the FL, neonatal, and adult BM, suggesting a developmental association between HSPCs and primitive mesenchymal cells. Existence of mesenchymal stem/progenitor cells in the AGM was also shown in human embryo (Wang et al., 2008). In vivo, the process of

HSPC production in the AGM occurs in a very short developmental window and is finely influenced by the surrounding embryonic tissues. Understanding how the AGM microenvironment is shaped in the developing embryo and how the adjacent tissues of mesodermal, endodermal, and ectodermal origins cooperate to initiate and control AGM HSPC production thus constitutes a major challenge in the field (Fig. 1). Experimental embryology approaches combined with gene expression profiling and functional studies have started to address this important question. In the chick embryo, preventing by ligature the migration and the installation of the subaortic mesenchyme was shown to inhibit the production of IAHCs and the expression of *Runx1* encoding a major transcription factor involved in the establishment of definitive hematopoiesis (Richard et al., 2013). In the mouse, explant cultures of early E10 AGM alone or associated with dorsal (neural tube) or ventral (gut) tissues followed by in vivo transplantation revealed that ventral tissue induce and increase HSC activity whereas dorsal tissues have a negative effect (Peeters et al., 2009). More recently, by taking advantage of an ex vivo culture system based on dissociation and reaggregation of embryonic tissues (Taoudi et al., 2008), it was shown that AGM HSC activity is strongly influenced by signals emanating from the ventral and dorsal domains of the Ao as well as from the adjacent UG (Souilhol et al., 2016). In the zebrafish, experimental evidence suggested that the somites together with the noncanonical Wnt signaling play an important role in the specification of HSCs, possibly via the recruitment of VSMC precursors (Clements et al., 2011).

At the molecular level, few cytokines, such as IL-3, Tpo, and Scf (Petit-Cocault, Volle-Challier, Fleury, Peault, & Souyri, 2007; Robin et al., 2006; Souilhol et al., 2016); the cell cycle regulator p57Kip2 (Mascarenhas, Parker, Dzierzak, & Ottersbach, 2009); and multiple signaling pathways, including Notch (Burns, Traver, Mayhall, Shepard, & Zon, 2005; Kumano et al., 2003; Robert-Moreno, Espinosa, de la Pompa, & Bigas, 2005; Robert-Moreno et al., 2008; Yoon et al., 2008), Hedgehog (Peeters et al., 2009; Souilhol et al., 2016; Wilkinson et al., 2009), bone morphogenetic protein (Bmp) (Crisan et al., 2015; Durand et al., 2007; Marshall, Kinnon, & Thrasher, 2000; Souilhol et al., 2016; Wilkinson et al., 2009), transforming growth factor (Monteiro et al., 2016), fibroblast growth factor (Lee et al., 2014; Pouget et al., 2014), and insulin growth factor (Mascarenhas et al., 2009), have all been shown to be active in the AGM and to play a role in the process of HSPC specification, maturation, or maintenance (Clements & Traver, 2013). How these pathways are regulated in time and space and how are they eventually interconnected is also a critical issue. For example, studies in the zebrafish revealed a functional link between the Shh, Vegfa, and Notch signaling (Lawson, Vogel, & Weinstein, 2002). Shh is produced by dorsal tissues (the floor plate, the notochord, and the hypochord) and regulates the expression of Vegfa in the somites. Vegfa in turn induces Notch receptor expression in aortic endothelial cells. Bmp4 is also an interesting regulator since its expression at the mRNA and protein levels is preferentially located in the subaortic mesenchyme (Durand et al., 2007; Marshall et al., 2000; Pimanda et al., 2007; Souilhol et al., 2016; Wilkinson et al., 2009) and together with the Scl transcriptional network the Bmp/Smad signaling was shown to regulate Runx1 activity (Pimanda et al., 2007). Using a Bmp responsive element green fluorescent protein transgenic mouse line, it was recently reported that all E11 HSCs were in the GFP^+ fraction (Crisan et al., 2015). Functional studies in the zebrafish and mouse embryos suggested that the Bmp pathway is required for the emergence and maintenance of HSCs (Durand et al., 2007; Wilkinson et al., 2009). However, a recent study revealed that the Bmp signaling is downregulated during the maturation of HSCs (Souilhol et al., 2016), suggesting a complex role for the Bmp pathway in the production, maturation, and maintenance of AGM HSCs. Interestingly, loss- and gain-of-function experiments in the zebrafish indicated that the Fgf signaling negatively regulates the formation of HSCs in the dorsal aorta by repressing Bmp activity in the subaortic mesenchyme (Pouget et al., 2014).

On top of these local signals, systemic factors have recently emerged as critical regulators of AGM HSPCs. Among them, the blood flow and associated mechanical forces (Adamo et al., 2009; Diaz et al., 2015; North et al., 2009), the inflammatory pathway (Espin-Palazon et al., 2014; Li et al., 2014; Sawamiphak, Kontarakis, & Stainier, 2014), the sympathetic nervous system via the production of catecholamines (Fitch et al., 2012), and the central nervous system via stress-responsive glucocorticoid receptor signaling (Kwan et al., 2016) play an important role in the biology of AGM HSPCs. Collectively, these observations strongly support that the specification and subsequent maturation of HSPCs in the dorsal aorta is a complex event controlled in time and space by cell autonomous and noncell autonomous mechanisms. These latters operate both locally in the AGM and also at the systemic level to connect the development of the definitive hematopoietic system with other embryonic tissues and with the needs of the developing embryo.

3. THE FL AND PLACENTA HEMATOPOIETIC MICROENVIRONMENTS

Several tissues harbor HSC activity in the embryo including the AGM, umbilical and vitelline vessels, yolk sac, and placenta (Dzierzak & Speck, 2008; Medvinsky et al., 2011). Following their emergence, HSCs then massively amplify in the FL and in the placenta. So far, very little is known regarding the niches that support HSC expansion. As for the AGM, the establishment of a large collection of stromal lines was instrumental for exploring the role of the FL hematopoietic microenvironment (Moore, Ema, & Lemischka, 1997). Moreover, the comparison of the gene expression profiles of FL stromal lines with differing capacity to support HSPCs ex vivo leads to a better understanding of the molecular signature of the HSPC niche and to the identification of novel HSPC regulators (Charbord et al., 2014; Hackney et al., 2002). More recently, by taking advantage of transgenic mice, $Nestin^{+}NG2^{+}$ pericytes were shown to be associated with portal vessels in the E14.5 FL and to promote HSC expansion (Khan et al., 2016). Interestingly, BM and FL $Nestin^{+}NG2^{+}$ cells exhibit similar gene expression, although FL cells express preferentially genes involved in cell cycle and metabolism, suggesting that HSCs and their niche cells may actively proliferate in the FL (Khan et al., 2016). Based on a coculture system, it has been reported that DLK^{+} FL hepatic progenitors expressing the α-fetoprotein and albumin as well as hematopoietic cytokines such as Scf, Cxcl12 and Tpo support, and expand HSPCs in vitro (Chou, Flygare, & Lodish, 2013). These observations thus suggest that hepatic progenitors may also constitute an important constituent of the HSPC niche in the FL (Chou & Lodish, 2010).

The placenta was recently shown to harbor hematopoietic progenitors (Alvarez-Silva, Belo-Diabangouaya, Salaun, & Dieterlen-Lievre, 2003) and stem cells with long-term, high-level, multilineage hematopoietic repopulating potential (Gekas, Dieterlen-Lievre, Orkin, & Mikkola, 2005; Ottersbach & Dzierzak, 2005). Importantly, HSC activity in the placenta starts to be detected at E10.5–E11, dramatically increases between E11.5–E12.5 and then declines from E13.5, whereas the pool of HSCs continues to expand in the FL (Gekas et al., 2005). By utilizing Ly6A-GFP transgenic mice, where GFP expression is under the control of the regulatory elements of the HSC marker Sca-1, Ottersbach, and Dzierzak reported that all HSCs in the placenta (as in the AGM and adult BM) are GFP^{+} and are associated with the vasculature of the placenta (Ottersbach & Dzierzak, 2005). In human, HSPCs are also detected in the placenta. However, in contrast to the transient hematopoietic potential harbored by the murine placenta, HSCs with the ability to repopulate the hematopoietic system of NOD-SCID recipients are still present in the full term human placenta (Robin et al., 2009). In the same study, stromal lines were generated from various developmental stages (3–38 weeks of gestation) and characterized for cell morphology, cell surface phenotype, and hematopoietic supporting capacity. Some of these stromal lines exhibit osteogenic, adipogenic, and eventually endothelial differentiation potential when cultured in vitro under appropriate conditions and express the pericyte markers NG2 and CD146. Coculture experiments with $CD34^{+}$ cord blood cells revealed that the placental stromal lines support the expansion of $CD34^{+}$ cells and hematopoietic progenitors. Together with the identification of pericytes stained for CD146, NG2 and smooth alpha actin and associated with endothelial cells on placental cryosections, these results suggest that perivascular cells may play an important role in the HSC niche in the placenta (Robin et al., 2009).

4. EXTRACTING THE MOLECULAR CORE OF THE HSPC NICHE

Although important extrinsic regulators (such as cytokines, morphogens, and components of the extracellular matrix) have been identified and shown to be active in the HSC niche throughout development, a comprehensive and integrated understanding of the molecular framework of the HSPC niche still remained elusive. To address this critical question, we have recently designed a systems biology approach based on the comparison of the gene expression profiling of murine stromal lines established from the AGM, FL, and adult BM and differing in their capacity support HSPCs ex vivo (Charbord et al., 2014). Our working hypothesis was that important genes for the stromal function should be conserved in at least two out of the three potent supportive lines generated from the developmental hematopoietic tissues. By combining bioinformatics and statistical analyses, we identified a set of 481 mRNAs and 17 microRNAs that was representative and predictive of the ability of stromal cells (including stromal lines and primary stromal cells) to support HSPCs. Using Weighted Gene Correlation Network Analysis (WGCNA) (Langfelder & Horvath, 2008), we showed that the mRNA gene set is organized into a network containing subnetworks or modules positively and negatively correlated to the factor "HSPC support" and containing highly connected genes or hubs. Importantly, the set of genes contains most of the known regulators of HSPCs including Scf, Cxcl12, Kirrel3, Pdgfra, and Angptl4, but also a number of unexpected ones as shown by functional validation in the zebrafish embryos (Charbord et al., 2014). More recently, by using single cell analysis and by comparing the transcriptomes of BM osteolineage cells located at a proximal or distal distance from HSPCs, Silberstein et al. further explored the molecular signature of BM niche cells and also identified novel HSPC regulators such as the secreted RNAse angiogenin, the adhesion molecule Embigin, and the cytokine IL-18 (Silberstein et al., 2016).

Interestingly, we and others have observed that BM HSPCs influence the transcriptome of stromal cells (Charbord et al., 2014; Istvanffy et al., 2015). Indeed, by taking advantage of two FL stromal lines (AFT024 and BFC012) that strongly differ in their capacity to support HSPCs (AFT024 being a potent supporter of HSPCs whereas BFC012 is nonsupportive) and by performing short-term coculture experiments with BM lineage $^{-}$Sca-1^{+}c-kit^{+} (KLS) cells, we showed that the stromal function dedicated to the HSPC support exists prior contact with KLS cells and is extended after contact. This function and the associated biological activities and gene network are not found in nonsupportive stromal cells, even after contact with KLS cells (Charbord et al., 2014). In agreement with this notion, MSCs derived from patients having myelodysplastic syndromes (MDS) have been shown to play an important role in the propagation of MDS-initiating cells. They are different at the functional and molecular levels from healthy MSCs and, importantly, MDS hematopoietic cells may also directly influence gene expression in healthy MSCs and "reprogram" them into MDS MSC-like cells (Medyouf et al., 2014). In addition, in a mouse model of myeloproliferative neoplasia, it has been reported that leukemic myeloid cells stimulate the production in the BM of osteoblastic cells from MSCs and remodel the endosteal niche. Through the deregulation of genes implicated in the maintenance of HSPCs (such as *Scf*, *Cxcl12*, *Angpt1*, and *Slit2*), in the inflammatory milieu and myelofibrosis, altered osteoblastic cells appear to efficiently support the development of the leukemia to the detriment of normal HSPCs (Schepers et al., 2013).

Taken together, these data strongly support the hypothesis that the dialogue between HSPCs and their niche cells is a bidirectional process that is playing a major role in the development and homeostasis of the hematopoietic system as well as in its deregulation during hematological disorders.

5. CONCLUSION

Although major advances have been made regarding the cellular composition of the BM HSPC niche, very little is known about the hematopoietic microenvironments that drive HSPC production in the embryo and their subsequent expansion during fetal life. In the AGM, endothelial cells, VSMCs, mesenchymal cells located in the subaortic mesenchyme, as well as derivatives of the neural crest cells and the sympathetic nervous system are all playing an important role in the formation and regulation of the first emerging HSPCs. Understanding how the adjacent embryonic tissues and systemic factors also control these processes constitutes a major challenge in the field of developmental hematopoiesis. In the FL and placenta, pericytes and endothelial cells as well as hepatoblasts (in the FL) are likely to participate to the HSPC niche. However, a comprehensive understanding of the molecular requirement of these developmental HSPC niches still needs to be addressed. Moreover, the complexity and the dynamics of the molecular crosstalk between HSPCs and their niche cells are still poorly understood. We are confident that this knowledge is of critical importance for efficiently amplifying HSPCs ex vivo and their production from precursor cells, and also for a better understanding of HSPC biology in physiological and pathological situations.

ACKNOWLEDGMENTS

We thank Drs. Marion Roques and Thierry Jaffredo (UMR CNRS 7622) for kindly providing us the electron microscopy image of the AGM and Sophie Gournet (UMR CNRS 7622) for drawing assistance.

REFERENCES

Adamo, L., Naveiras, O., Wenzel, P. L., McKinney-Freeman, S., Mack, P. J., Gracia-Sancho, J., et al. (2009). Biomechanical forces promote embryonic haematopoiesis. *Nature*, *459*(7250), 1131–1135. http://dx.doi.org/10.1038/nature08073. nature08073 [pii].

Alvarez-Silva, M., Belo-Diabangouaya, P., Salaun, J., & Dieterlen-Lievre, F. (2003). Mouse placenta is a major hematopoietic organ. *Development*, *130*(22), 5437–5444. http://dx.doi.org/10.1242/dev.00755. dev.00755 [pii].

Batta, K., Florkowska, M., Kouskoff, V., & Lacaud, G. (2014). Direct reprogramming of murine fibroblasts to hematopoietic progenitor cells. *Cell Reports*, *9*(5), 1871–1884. http://dx.doi.org/10.1016/j.celrep.2014.11.002. S2211-1247(14)00956-5 [pii].

Bertrand, J. Y., Chi, N. C., Santoso, B., Teng, S., Stainier, D. Y., & Traver, D. (2010). Haematopoietic stem cells derive directly from aortic endothelium during development. *Nature*, *464*(7285), 108–111. http://dx.doi.org/10.1038/nature08738. nature08738 [pii].

Boisset, J. C., van Cappellen, W., Andrieu-Soler, C., Galjart, N., Dzierzak, E., & Robin, C. (2010). In vivo imaging of haematopoietic cells emerging from the mouse aortic endothelium. *Nature, 464*(7285), 116–120. http://dx.doi.org/10.1038/nature08764. nature08764 [pii].

Buckley, S. M., Ulloa-Montoya, F., Abts, D., Oostendorp, R. A., Dzierzak, E., Ekker, S. C., et al. (2011). Maintenance of hsc by wnt5a secreting agm-derived stromal cell line. *Experimental Hematology, 39*(1). http://dx.doi.org/10.1016/j.exphem.2010.09.010. 114–123, e111-115, S0301-472X(10)00491-1 [pii].

Burns, C. E., Traver, D., Mayhall, E., Shepard, J. L., & Zon, L. I. (2005). Hematopoietic stem cell fate is established by the notch-runx pathway. *Genes & Development, 19*(19), 2331–2342. http://dx.doi.org/10.1101/gad.1337005. gad.1337005 [pii].

Charbord, P., Oostendorp, R., Pang, W., Herault, O., Noel, F., Tsuji, T., et al. (2002). Comparative study of stromal cell lines derived from embryonic, fetal, and postnatal mouse blood-forming tissues. *Experimental Hematology, 30*(10), 1202–1210. http://dx.doi.org/10.1016/S0301-472X(02)00895-0.

Charbord, P., Pouget, C., Binder, H., Dumont, F., Stik, G., Levy, P., et al. (2014). A systems biology approach for defining the molecular framework of the hematopoietic stem cell niche. *Cell Stem Cell, 15*(3), 376–391. http://dx.doi.org/10.1016/j.stem.2014.06.005. S1934-5909(14)00253-7 [pii].

Chou, S., Flygare, J., & Lodish, H. F. (2013). Fetal hepatic progenitors support long-term expansion of hematopoietic stem cells. *Experimental Hematology, 41*(5), 479–490, e474. http://dx.doi.org/10.1016/j.exphem.2013.02.003. S0301-472X(13)00016-7 [pii].

Chou, S., & Lodish, H. F. (2010). Fetal liver hepatic progenitors are supportive stromal cells for hematopoietic stem cells. *Proceedings of the National Academy of Sciences of the United States of America, 107*(17), 7799–7804. http://dx.doi.org/10.1073/pnas.1003586107. 1003586107 [pii].

Clements, W. K., Kim, A. D., Ong, K. G., Moore, J. C., Lawson, N. D., & Traver, D. (2011). A somitic wnt16/notch pathway specifies haematopoietic stem cells. *Nature, 474*(7350), 220–224. http://dx.doi.org/10.1038/nature10107. nature10107 [pii].

Clements, W. K., & Traver, D. (2013). Signalling pathways that control vertebrate haematopoietic stem cell specification. *Nature Reviews Immunology, 13*(5), 336–348. http://dx.doi.org/10.1038/nri3443. nri3443 [pii].

Crisan, M., Kartalaei, P. S., Vink, C. S., Yamada-Inagawa, T., Bollerot, K., van, IJcken, W., et al. (2015). Bmp signalling differentially regulates distinct haematopoietic stem cell types. *Nature Communications, 6*, 8040. http://dx.doi.org/10.1038/ncomms9040. ncomms 9040 [pii].

Cumano, A., Dieterlen-Lievre, F., & Godin, I. (1996). Lymphoid potential, probed before circulation in mouse, is restricted to caudal intraembryonic splanchnopleura. *Cell, 86*(6), 907–916. http://dx.doi.org/10.1016/S0092-8674(00)80166-X.

Cumano, A., & Godin, I. (2007). Ontogeny of the hematopoietic system. *Annual Review of Immunology, 25*, 745–785. http://dx.doi.org/10.1146/annurev.immunol.25.022106.141538.

de Bruijn, M. F., Ma, X., Robin, C., Ottersbach, K., Sanchez, M. J., & Dzierzak, E. (2002). Hematopoietic stem cells localize to the endothelial cell layer in the midgestation mouse aorta. *Immunity, 16*(5), 673–683. http://dx.doi.org/10.1016/S1074-7613(02)00313-8.

Diaz, M. F., Li, N., Lee, H. J., Adamo, L., Evans, S. M., Willey, H. E., et al. (2015). Biomechanical forces promote blood development through prostaglandin e2 and the camp-pka signaling axis. *The Journal of Experimental Medicine, 212*(5), 665–680. http://dx.doi.org/10.1084/jem.20142235. jem.20142235 [pii].

Ding, L., & Morrison, S. J. (2013). Haematopoietic stem cells and early lymphoid progenitors occupy distinct bone marrow niches. *Nature, 495*(7440), 231–235. http://dx.doi.org/10.1038/nature11885. nature11885 [pii].

Ding, L., Saunders, T. L., Enikolopov, G., & Morrison, S. J. (2012). Endothelial and perivascular cells maintain haematopoietic stem cells. *Nature*, *481*(7382), 457–462. http://dx.doi.org/10.1038/nature10783. nature10783 [pii].

Durand, C., Robin, C., Bollerot, K., Baron, M. H., Ottersbach, K., & Dzierzak, E. (2007). Embryonic stromal clones reveal developmental regulators of definitive hematopoietic stem cells. *Proceedings of the National Academy of Sciences of the United States of America*, *104*(52), 20838–20843. http://dx.doi.org/10.1073/pnas.0706923105. 0706923105 [pii].

Durand, C., Robin, C., & Dzierzak, E. (2006). Mesenchymal lineage potentials of aorta-gonad-mesonephros stromal clones. *Haematologica*, *91*(9), 1172–1179.

Dzierzak, E., & Speck, N. A. (2008). Of lineage and legacy: The development of mammalian hematopoietic stem cells. *Nature Immunology*, *9*(2), 129–136. http://dx.doi.org/10.1038/ni1560. ni1560 [pii].

Espin-Palazon, R., Stachura, D. L., Campbell, C. A., Garcia-Moreno, D., Del Cid, N., Kim, A. D., et al. (2014). Proinflammatory signaling regulates hematopoietic stem cell emergence. *Cell*, *159*(5), 1070–1085. http://dx.doi.org/10.1016/j.cell.2014.10.031. S0092-8674(14)01319-1 [pii].

Fitch, S. R., Kimber, G. M., Wilson, N. K., Parker, A., Mirshekar-Syahkal, B., Gottgens, B., et al. (2012). Signaling from the sympathetic nervous system regulates hematopoietic stem cell emergence during embryogenesis. *Cell Stem Cell*, *11*(4), 554–566. http://dx.doi.org/10.1016/j.stem.2012.07.002. S1934-5909(12)00417-1 [pii].

Gekas, C., Dieterlen-Lievre, F., Orkin, S. H., & Mikkola, H. K. (2005). The placenta is a niche for hematopoietic stem cells. *Developmental Cell*, *8*(3), 365–375. http://dx.doi.org/10.1016/j.devcel.2004.12.016. S1534-5807(04)00468-X [pii].

Greenbaum, A., Hsu, Y. M., Day, R. B., Schuettpelz, L. G., Christopher, M. J., Borgerding, J. N., et al. (2013). Cxcl12 in early mesenchymal progenitors is required for haematopoietic stem-cell maintenance. *Nature*, *495*(7440), 227–230. http://dx.doi.org/10.1038/nature11926. nature11926 [pii].

Hackney, J. A., Charbord, P., Brunk, B. P., Stoeckert, C. J., Lemischka, I. R., & Moore, K. A. (2002). A molecular profile of a hematopoietic stem cell niche. *Proceedings of the National Academy of Sciences of the United States of America*, *99*(20), 13061–13066. http://dx.doi.org/10.1073/pnas.192124499. 192124499 [pii].

Istvanffy, R., Vilne, B., Schreck, C., Ruf, F., Pagel, C., Grziwok, S., et al. (2015). Stroma-derived connective tissue growth factor maintains cell cycle progression and repopulation activity of hematopoietic stem cells in vitro. *Stem Cell Reports*, *5*(5), 702–715. http://dx.doi.org/10.1016/j.stemcr.2015.09.018. S2213-6711(15)00280-5 [pii].

Jaffredo, T., Gautier, R., Eichmann, A., & Dieterlen-Lievre, F. (1998). Intraaortic hemopoietic cells are derived from endothelial cells during ontogeny. *Development*, *125*(22), 4575–4583.

Khan, J. A., Mendelson, A., Kunisaki, Y., Birbrair, A., Kou, Y., Arnal-Estape, A., et al. (2016). Fetal liver hematopoietic stem cell niches associate with portal vessels. *Science*, *351*(6269), 176–180. http://dx.doi.org/10.1126/science.aad0084. science.aad0084 [pii].

Kiel, M. J., Yilmaz, O. H., Iwashita, T., Terhorst, C., & Morrison, S. J. (2005). Slam family receptors distinguish hematopoietic stem and progenitor cells and reveal endothelial niches for stem cells. *Cell*, *121*(7), 1109–1121. http://dx.doi.org/10.1016/j.cell.2005.05.026. S0092-8674(05)00540-4 [pii].

Kissa, K., & Herbomel, P. (2010). Blood stem cells emerge from aortic endothelium by a novel type of cell transition. *Nature*, *464*(7285), 112–115. http://dx.doi.org/10.1038/nature08761. nature08761 [pii].

Krassowska, A., Gordon-Keylock, S., Samuel, K., Gilchrist, D., Dzierzak, E., Oostendorp, R., et al. (2006). Promotion of haematopoietic activity in embryonic stem cells by the aorta-gonad-mesonephros microenvironment. *Experimental Cell Research*,

312(18), 3595–3603. http://dx.doi.org/10.1016/j.yexcr.2006.08.001. S0014-4827(06)00313-2 [pii].
Kumano, K., Chiba, S., Kunisato, A., Sata, M., Saito, T., Nakagami-Yamaguchi, E., et al. (2003). Notch1 but not notch2 is essential for generating hematopoietic stem cells from endothelial cells. *Immunity*, *18*(5), 699–711. http://dx.doi.org/10.1016/S1074-7613(03)00117-1.
Kunisaki, Y., Bruns, I., Scheiermann, C., Ahmed, J., Pinho, S., Zhang, D., et al. (2013). Arteriolar niches maintain haematopoietic stem cell quiescence. *Nature*, *502*(7473), 637–643. http://dx.doi.org/10.1038/nature12612. nature12612 [pii].
Kwan, W., Cortes, M., Frost, I., Esain, V., Theodore, L. N., Liu, S. Y., et al. (2016). The central nervous system regulates embryonic hspc production via stress-responsive glucocorticoid receptor signaling. *Cell Stem Cell*, *19*(3), 370–382. http://dx.doi.org/10.1016/j.stem.2016.06.004. S1934-5909(16)30154-0 [pii].
Langfelder, P., & Horvath, S. (2008). Wgcna: An r package for weighted correlation network analysis. *BMC Bioinformatics*, *9*, 559. http://dx.doi.org/10.1186/1471-2105-9-559. 1471-2105-9-559 [pii].
Lawson, N. D., Vogel, A. M., & Weinstein, B. M. (2002). Sonic hedgehog and vascular endothelial growth factor act upstream of the notch pathway during arterial endothelial differentiation. *Developmental Cell*, *3*(1), 127–136. http://dx.doi.org/10.1016/S1534-5807(02)00198-3.
Ledran, M. H., Krassowska, A., Armstrong, L., Dimmick, I., Renstrom, J., Lang, R., et al. (2008). Efficient hematopoietic differentiation of human embryonic stem cells on stromal cells derived from hematopoietic niches. *Cell Stem Cell*, *3*(1), 85–98. http://dx.doi.org/10.1016/j.stem.2008.06.001. S1934-5909(08)00280-4 [pii].
Lee, Y., Manegold, J. E., Kim, A. D., Pouget, C., Stachura, D. L., Clements, W. K., et al. (2014). Fgf signalling specifies haematopoietic stem cells through its regulation of somitic notch signalling. *Nature Communications*, *5*, 5583. http://dx.doi.org/10.1038/ncomms6583. ncomms6583 [pii].
Li, Y., Esain, V., Teng, L., Xu, J., Kwan, W., Frost, I. M., et al. (2014). Inflammatory signaling regulates embryonic hematopoietic stem and progenitor cell production. *Genes & Development*, *28*(23), 2597–2612. http://dx.doi.org/10.1101/gad.253302.114. gad.253302.114 [pii].
Marshall, C. J., Kinnon, C., & Thrasher, A. J. (2000). Polarized expression of bone morphogenetic protein-4 in the human aorta-gonad-mesonephros region. *Blood*, *96*(4), 1591–1593.
Mascarenhas, M. I., Parker, A., Dzierzak, E., & Ottersbach, K. (2009). Identification of novel regulators of hematopoietic stem cell development through refinement of stem cell localization and expression profiling. *Blood*, *114*(21), 4645–4653. http://dx.doi.org/10.1182/blood-2009-06-230037. blood-2009-06-230037 [pii].
Medvinsky, A., & Dzierzak, E. (1996). Definitive hematopoiesis is autonomously initiated by the agm region. *Cell*, *86*(6), 897–906. http://dx.doi.org/10.1016/S0092-8674(00)80165-8.
Medvinsky, A., Rybtsov, S., & Taoudi, S. (2011). Embryonic origin of the adult hematopoietic system: Advances and questions. *Development*, *138*(6), 1017–1031. http://dx.doi.org/10.1242/dev.040998. 138/6/1017 [pii].
Medyouf, H., Mossner, M., Jann, J. C., Nolte, F., Raffel, S., Herrmann, C., et al. (2014). Myelodysplastic cells in patients reprogram mesenchymal stromal cells to establish a transplantable stem cell niche disease unit. *Cell Stem Cell*, *14*(6), 824–837. http://dx.doi.org/10.1016/j.stem.2014.02.014. S1934-5909(14)00078-2 [pii].
Mendes, S. C., Robin, C., & Dzierzak, E. (2005). Mesenchymal progenitor cells localize within hematopoietic sites throughout ontogeny. *Development*, *132*(5), 1127–1136. http://dx.doi.org/10.1242/dev.01615. dev.01615 [pii].
Mendez-Ferrer, S., Lucas, D., Battista, M., & Frenette, P. S. (2008). Haematopoietic stem cell release is regulated by circadian oscillations. *Nature*, *452*(7186), 442–447. http://dx.doi.org/10.1038/nature06685. nature06685 [pii].

Mendez-Ferrer, S., Michurina, T. V., Ferraro, F., Mazloom, A. R., Macarthur, B. D., Lira, S. A., et al. (2010). Mesenchymal and haematopoietic stem cells form a unique bone marrow niche. *Nature, 466*(7308), 829–834. http://dx.doi.org/10.1038/nature09262. nature 09262 [pii].

Mendez-Ferrer, S., Scadden, D. T., & Sanchez-Aguilera, A. (2015). Bone marrow stem cells: Current and emerging concepts. *Annals of the New York Academy of Sciences, 1335*, 32–44. http://dx.doi.org/10.1111/nyas.12641.

Monteiro, R., Pinheiro, P., Joseph, N., Peterkin, T., Koth, J., Repapi, E., et al. (2016). Transforming growth factor beta drives hemogenic endothelium programming and the transition to hematopoietic stem cells. *Developmental Cell, 38*(4), 358–370. http://dx.doi.org/10.1016/j.devcel.2016.06.024. S1534-5807(16)30427-0 [pii].

Moore, K. A., Ema, H., & Lemischka, I. R. (1997). In vitro maintenance of highly purified, transplantable hematopoietic stem cells. *Blood, 89*(12), 4337–4347.

Morrison, S. J., & Scadden, D. T. (2014). The bone marrow niche for haematopoietic stem cells. *Nature, 505*(7483), 327–334. http://dx.doi.org/10.1038/nature12984. nature 12984 [pii].

Muller, A. M., Medvinsky, A., Strouboulis, J., Grosveld, F., & Dzierzak, E. (1994). Development of hematopoietic stem cell activity in the mouse embryo. *Immunity, 1*(4), 291–301. http://dx.doi.org/10.1016/1074-7613(94)90081-7.

North, T. E., Goessling, W., Peeters, M., Li, P., Ceol, C., Lord, A. M., et al. (2009). Hematopoietic stem cell development is dependent on blood flow. *Cell, 137*(4), 736–748. http://dx.doi.org/10.1016/j.cell.2009.04.023. S0092-8674(09)00448-6 [pii].

Oberlin, E., Tavian, M., Blazsek, I., & Peault, B. (2002). Blood-forming potential of vascular endothelium in the human embryo. *Development, 129*(17), 4147–4157.

Oostendorp, R. A., Harvey, K. N., Kusadasi, N., de Bruijn, M. F., Saris, C., Ploemacher, R. E., et al. (2002). Stromal cell lines from mouse aorta-gonads-mesonephros subregions are potent supporters of hematopoietic stem cell activity. *Blood, 99*(4), 1183–1189.

Oostendorp, R. A., Medvinsky, A. J., Kusadasi, N., Nakayama, N., Harvey, K., Orelio, C., et al. (2002). Embryonal subregion-derived stromal cell lines from novel temperature-sensitive sv40 t antigen transgenic mice support hematopoiesis. *Journal of Cell Science, 115*(Pt. 10), 2099–2108.

Oostendorp, R. A., Robin, C., Steinhoff, C., Marz, S., Brauer, R., Nuber, U. A., et al. (2005). Long-term maintenance of hematopoietic stem cells does not require contact with embryo-derived stromal cells in cocultures. *Stem Cells, 23*(6), 842–851. http://dx.doi.org/10.1634/stemcells.2004-0120. 23/6/842 [pii].

Ottersbach, K., & Dzierzak, E. (2005). The murine placenta contains hematopoietic stem cells within the vascular labyrinth region. *Developmental Cell, 8*(3), 377–387. http://dx.doi.org/10.1016/j.devcel.2005.02.001. S1534-5807(05)00048-1 [pii].

Peeters, M., Ottersbach, K., Bollerot, K., Orelio, C., de Bruijn, M., Wijgerde, M., et al. (2009). Ventral embryonic tissues and hedgehog proteins induce early agm hematopoietic stem cell development. *Development, 136*(15), 2613–2621. http://dx.doi.org/10.1242/dev.034728. dev.034728 [pii].

Pereira, C. F., Chang, B., Qiu, J., Niu, X., Papatsenko, D., Hendry, C. E., et al. (2013). Induction of a hemogenic program in mouse fibroblasts. *Cell Stem Cell, 13*(2), 205–218. http://dx.doi.org/10.1016/j.stem.2013.05.024. S1934-5909(13)00217-8 [pii].

Petit-Cocault, L., Volle-Challier, C., Fleury, M., Peault, B., & Souyri, M. (2007). Dual role of mpl receptor during the establishment of definitive hematopoiesis. *Development, 134*(16), 3031–3040. http://dx.doi.org/10.1242/dev.001818. dev. 001818 [pii].

Pimanda, J. E., Donaldson, I. J., de Bruijn, M. F., Kinston, S., Knezevic, K., Huckle, L., et al. (2007). The scl transcriptional network and bmp signaling pathway interact to regulate

runx1 activity. *Proceedings of the National Academy of Sciences of the United States of America*, *104*(3), 840–845. http://dx.doi.org/10.1073/pnas.0607196104. 0607196104 [pii].
Pouget, C., Peterkin, T., Simoes, F. C., Lee, Y., Traver, D., & Patient, R. (2014). Fgf signalling restricts haematopoietic stem cell specification via modulation of the bmp pathway. *Nature Communications*, *5*, 5588. http://dx.doi.org/10.1038/ncomms6588. ncomms 6588 [pii].
Richard, C., Drevon, C., Canto, P. Y., Villain, G., Bollerot, K., Lempereur, A., et al. (2013). Endothelio-mesenchymal interaction controls runx1 expression and modulates the notch pathway to initiate aortic hematopoiesis. *Developmental Cell*, *24*(6), 600–611. http://dx.doi.org/10.1016/j.devcel.2013.02.011. S1534-5807(13)00123-8 [pii].
Riddell, J., Gazit, R., Garrison, B. S., Guo, G., Saadatpour, A., Mandal, P. K., et al. (2014). Reprogramming committed murine blood cells to induced hematopoietic stem cells with defined factors. *Cell*, *157*(3), 549–564. http://dx.doi.org/10.1016/j.cell.2014.04.006. S0092-8674(14)00472-3 [pii].
Robert-Moreno, A., Espinosa, L., de la Pompa, J. L., & Bigas, A. (2005). Rbpjkappa-dependent notch function regulates gata2 and is essential for the formation of intra-embryonic hematopoietic cells. *Development*, *132*(5), 1117–1126. http://dx.doi.org/10.1242/dev.01660. dev.01660 [pii].
Robert-Moreno, A., Guiu, J., Ruiz-Herguido, C., Lopez, M. E., Ingles-Esteve, J., Riera, L., et al. (2008). Impaired embryonic haematopoiesis yet normal arterial development in the absence of the notch ligand jagged1. *The EMBO Journal*, *27*(13), 1886–1895. http://dx.doi.org/10.1038/emboj.2008.113. emboj2008113 [pii].
Robin, C., Bollerot, K., Mendes, S., Haak, E., Crisan, M., Cerisoli, F., et al. (2009). Human placenta is a potent hematopoietic niche containing hematopoietic stem and progenitor cells throughout development. *Cell Stem Cell*, *5*(4), 385–395. http://dx.doi.org/10.1016/j.stem.2009.08.020. S1934-5909(09)00444-5 [pii].
Robin, C., Ottersbach, K., Durand, C., Peeters, M., Vanes, L., Tybulewicz, V., et al. (2006). An unexpected role for il-3 in the embryonic development of hematopoietic stem cells. *Developmental Cell*, *11*(2), 171–180. http://dx.doi.org/10.1016/j.devcel.2006.07.002. S1534-5807(06)00303-0 [pii].
Sandler, V. M., Lis, R., Liu, Y., Kedem, A., James, D., Elemento, O., et al. (2014). Reprogramming human endothelial cells to haematopoietic cells requires vascular induction. *Nature*, *511*(7509), 312–318. http://dx.doi.org/10.1038/nature13547. nature13547 [pii].
Sawamiphak, S., Kontarakis, Z., & Stainier, D. Y. (2014). Interferon gamma signaling positively regulates hematopoietic stem cell emergence. *Developmental Cell*, *31*(5), 640–653. http://dx.doi.org/10.1016/j.devcel.2014.11.007. S1534-5807(14)00724-2 [pii].
Schepers, K., Pietras, E. M., Reynaud, D., Flach, J., Binnewies, M., Garg, T., et al. (2013). Myeloproliferative neoplasia remodels the endosteal bone marrow niche into a self-reinforcing leukemic niche. *Cell Stem Cell*, *13*(3), 285–299. http://dx.doi.org/10.1016/j.stem.2013.06.009. S1934-5909(13)00267-1 [pii].
Silberstein, L., Goncalves, K. A., Kharchenko, P. V., Turcotte, R., Kfoury, Y., Mercier, F., et al. (2016). Proximity-based differential single-cell analysis of the niche to identify stem/progenitor cell regulators. *Cell Stem Cell*, *19*(4), 530–543. http://dx.doi.org/10.1016/j.stem.2016.07.004. S1934-5909(16)30199-0 [pii].
Souilhol, C., Gonneau, C., Lendinez, J. G., Batsivari, A., Rybtsov, S., Wilson, H., et al. (2016). Inductive interactions mediated by interplay of asymmetric signalling underlie development of adult haematopoietic stem cells. *Nature Communications*, 7, 10784. http://dx.doi.org/10.1038/ncomms10784. ncomms10784 [pii].
Taoudi, S., Gonneau, C., Moore, K., Sheridan, J. M., Blackburn, C. C., Taylor, E., et al. (2008). Extensive hematopoietic stem cell generation in the agm region via maturation of ve-cadherin + cd45 + pre-definitive hscs. *Cell Stem Cell*, *3*(1), 99–108. http://dx.doi.org/10.1016/j.stem.2008.06.004. S1934-5909(08)00283-X [pii].

Taoudi, S., & Medvinsky, A. (2007). Functional identification of the hematopoietic stem cell niche in the ventral domain of the embryonic dorsal aorta. *Proceedings of the National Academy of Sciences of the United States of America*, *104*(22), 9399–9403. http://dx.doi.org/10.1073/pnas.0700984104. 0700984104 [pii].

Wagers, A. J. (2012). The stem cell niche in regenerative medicine. *Cell Stem Cell*, *10*(4), 362–369. http://dx.doi.org/10.1016/j.stem.2012.02.018. S1934-5909(12)00077-X [pii].

Wang, X. Y., Lan, Y., He, W. Y., Zhang, L., Yao, H. Y., Hou, C. M., et al. (2008). Identification of mesenchymal stem cells in aorta-gonad-mesonephros and yolk sac of human embryos. *Blood*, *111*(4), 2436–2443. http://dx.doi.org/10.1182/blood-2007-07-099333. blood-2007-07-099333 [pii].

Wilkinson, R. N., Pouget, C., Gering, M., Russell, A. J., Davies, S. G., Kimelman, D., et al. (2009). Hedgehog and bmp polarize hematopoietic stem cell emergence in the zebrafish dorsal aorta. *Developmental Cell*, *16*(6), 909–916. http://dx.doi.org/10.1016/j.devcel.2009.04.014. S1534-5807(09)00176-2 [pii].

Yoon, M. J., Koo, B. K., Song, R., Jeong, H. W., Shin, J., Kim, Y. W., et al. (2008). Mind bomb-1 is essential for intraembryonic hematopoiesis in the aortic endothelium and the subaortic patches. *Molecular and Cellular Biology*, *28*(15), 4794–4804. http://dx.doi.org/10.1128/MCB.00436-08. MCB.00436-08 [pii].

CHAPTER TWO

The Evolvement of Hematopoietic Stem Cell Niches

B.O. Zhou*, L. Li†,‡, M. Zhao§,¶,[1]

*State Key Laboratory of Cell Biology, CAS Center for Excellence in Molecular Cell Science, Shanghai Institute of Biochemistry and Cell Biology, Chinese Academy of Sciences, University of Chinese Academy of Sciences, Shanghai, China
†Stowers Institute for Medical Research, Kansas City, MO, United States
‡University of Kansas Medical Center, Kansas City, KS, United States
§Zhongshan School of Medicine, Sun Yat-Sen University, Guangzhou, Guangdong, China
¶Key Laboratory for Stem Cells and Tissue Engineering, Ministry of Education, Sun Yat-Sen University, Guangzhou, Guangdong, China
[1]Corresponding author: e-mail address: zhaom38@mail.sysu.edu.cn

Contents

1. THE ORIGIN OF HSC NICHE CONCEPT

The stem cells can divide through mitosis to produce more stem cells (self-renew) and can also differentiate into multiple specialized lineage cells. Till and McCulloch used in vivo spleen colony-forming (CFU-S) assay to provide the first experimental evidence for stem cells. They showed that single cells could yield multilineages and preserve multipotency in CFU-S assay (Siminovitch, McCulloch, & Till, 1963; Till, McCulloch, & Siminovitch, 1964). This assay also led Schofield to propose the niche hypothesis in 1978. He noticed that CFU-S "stem cells" have less robust reconstitution capacity in irradiated mice as compared to cells from the bone marrow (BM), which led to his hypothesis that BM holds a specialized niche to preserve the

Advances in Stem Cells and their Niches, Volume 1
ISSN 2468-5097
http://dx.doi.org/10.1016/bs.asn.2017.01.001

hematopoietic stem cell (HSC) pool (Schofield, 1978). Subsequently, several cytokines from BM stromal cells were shown to support HSCs by in vitro culture assays (Dexter, Allen, & Lajtha, 1977; Taichman & Emerson, 1994). However, the existence of in vivo HSC niche was not proved till 2003 when mouse genetic engineering technology was much developed. Studies from two independent groups found that bone-lining osteoblastic cells contribute to HSC maintenance (Calvi et al., 2003; Zhang et al., 2003). This is the first evidence to show that a microenvironment, or niche, can contribute to stem cell regulation in mammals. Furthermore, genetic perturbation of osteoprogenitors impaired normal hematopoiesis while promoting the formation of myelodysplasia and secondary leukemia in mouse models, which supports the concept of niche-induced oncogenesis (Raaijmakers et al., 2010). Subsequent studies showed that vascular endothelial cells and mesenchymal stem cells (MSCs) also contribute to HSC regulation in vivo (Kiel, Yilmaz, Iwashita, Terhorst, & Morrison, 2005; Mendez-Ferrer et al., 2010). These studies reveal the complexity of BM microenvironment and the heterogeneity of HSCs.

2. RELOCATION OF HSCs DURING DEVELOPMENT

The development of HSCs involves multiple anatomical sites. HSCs are specified from mesoderm, which contributes to both hematopoietic and vascular fates. Primitive hematopoiesis starts from blood islands of the yolk sac at embryonic day 7.5 (E7.5) in mice and at 30 days postconception in humans. Aorta-gonadmesonephros (AGM) region, which consists of the dorsal aorta, surrounding mesenchyme, and the urogenital ridges, is a source of pre-HSCs at E10.5 in mice and at 4 weeks postconception (wpc) in humans (Godin & Cumano, 2002; Medvinsky & Dzierzak, 1996). These pre-HSCs have unique features in their transcriptional machinery, metabolism state, and signaling pathways compared to definitive HSCs (Zhou et al., 2016). They move from AGM to the placenta and fetal liver at E11.5 in mice and at 5 wpc in humans. Fetal liver serves as the main organ for HSC expansion and maturation until BM hematopoiesis is established (Mikkola & Orkin, 2006). Most fetal HSCs are cycling while retaining self-renewal ability, which means that they undergo quickly symmetric cell division to expand the HSC pool (Lessard, Faubert, & Sauvageau, 2004). HSCs move to BM at E18 in mice and at 12 wpc in humans, where they reside permanently in lifetime. Adult HSCs in BM are quiescent and can switch from dormancy to self-renewal to maintain the appropriate HSC pool size during homeostasis and repair (Wilson & Trumpp, 2006). The migration of HSCs during development indicates the importance of defined cellular niches in supporting HSCs at each developmental stage. In this chapter, more information about the role of the different HSC niche during development and the molecular mechanisms involved in HSC regulation are being discussed.

3. THE IDENTIFICATION OF HSC NICHES

Stem cell niche refers to an anatomic location where stem cells reside and is composed by cellular components that functionally contribute to stem cell regulation. These two characteristics result in two separate strategies to identify the HSC niche. One is a "proximity principle" in which researchers are looking for the HSC-interacting cells. The other one is a "taking one out" strategy, in which researchers are searching for the HSC regulatory cellular components.

To approach the "proximity principle" strategy, researchers suffered a long time to identify the rare and indistinguishable HSCs in the BM. Because HSCs are a functionally defined population with the properties of long-term self-renewal and the ability to reconstitute all hematopoietic lineages. Although researchers can cooperate multiple cell surface markers to identify functional HSCs using fluorescence-activated cell sorting (FACS) technology, the multiple fluorochrome dye combination is too complex for microscopic detection, especially for in vivo HSC imaging (Mendez-Ferrer, Scadden, & Sanchez-Aguilera, 2015; Morrison & Scadden, 2014). Therefore, multiple alternative approaches have been used to localize HSCs in the BM.

As HSCs are well characterized for being in a quiescent state which is correlated with their functionality (Cheng et al., 2000; Ogawa, 1993; Wilson et al., 2008), the distribution of quiescent cells potentially reflects HSC localization in the BM. The quiescent cells are slow-cycling cells that tend to retain labeled DNA and are resistant to DNA damage. Various methods such as bromodeoxyuridine (BrdU) label retaining, GFP-labeled histone (H2B-GFP) retention, and chemotherapy (e.g., 5-fluorouracil, 5-FU)-resistance cells are studied in the BM. Interestingly, the quiescent populations were often found to be close to the endosteum (the interface between bone and BM) of the trabecular bone, albeit by different approaches (Arai et al., 2004; Foudi et al., 2009; Sugimura et al., 2012; Zhang et al., 2003), suggesting that the endosteum forms a niche to maintain quiescent HSC subpopulations. However, this strategy is often challenged by the specificity of the observed populations, given that some mature cells, such as memory lymphocytes, can also retain DNA labeling due to their long life span (Sprent & Tough, 1994). Although some HSC-positive markers such as Sca-1 were included in the immunostaining of BM to improve the accuracy of observed cells, the nonspecificity issue cannot be solved. One group used a combination of HSC-positive marker c-kit with negative markers ($CD48^-CD41^-Lineage^-$) and found that the observed cells preferentially localized in endosteal zones where microvessels were enriched (Nombela-Arrieta et al., 2013). However, this study suffered from a major technical challenge that the combined markers recognized more progenitor cells than HSCs, which renders their observation rather nonspecific. To overcome this issue, two groups independently reported that immunophenotypic highly purified HSCs tended to locate at the endosteum in postirradiated mice, but were randomly distributed and unstable in non-irradiated mice (Lo Celso et al., 2009; Xie et al., 2009). This provided additional evidence for the endosteum zone as the HSC niche, though how such severely stressed transplantation model reflects the distribution of endogenous HSC in homeostasis needs to be further explored.

To solve these technique issues, researchers simplified the complicated HSC-FACS immunofluorescent markers into a two-color stain, which enable them to pinpoint endogenous HSCs under microscopy. The recently developed powerful imaging cytometry platform has also made possible comprehensive quantitative analysis of HSC distribution. Combination using positive staining for CD150 and negative markers for lineage, CD41, and CD48 detected a highly enriched HSC population in the two-color staining (Kiel et al., 2005). In this scenario, HSCs were found more likely to be adjacent to sinusoids and further confirmed by in vivo deep imaging studies (Acar et al., 2015; Chen et al., 2016; Kiel et al., 2005). Furthermore, the HSC heterogeneity studies revealed that most $Ki67^-$ quiescent HSCs located at periarteriolar regions instead of perisinusoidal regions in a mathematic-based random distribution model (Kunisaki et al., 2013). Consistent with this, dormant HSCs, in a low reactive oxygen species (ROS) state, were found at regions around less permeable arteriolar blood vessels, whereas active HSCs, in a ROS-high state, were located around more permeable sinusoids (Itkin et al., 2016). Only a fraction of quiescent HSCs, in low ROS state, were found by sinusoids and adjacent to megakaryocytes (MKs), a type of mature blood cells, that produces multiple factors to regulate HSC quiescence (Bruns et al., 2014; Zhao et al., 2014).

To facilitate locating endogenous HSCs by confocal imaging, scientists have been trying to develop reporter mice that can distinguish primitive HSCs from all downstream lineages, including multipotent progenitors. This would require identification of genes that are uniquely expressed in HSCs rather than other blood cells or stromal cells in BM. Three mouse lines have been developed to meet this criterion including *Fgd5*-mCherry (Gazit et al., 2014), *α-catulin*-GFP (Acar et al., 2015), and *Hoxb5*-mCherry knock-in mice (Chen et al., 2016), in all of which virtually all long-term repopulating HSCs were found to be included in the reporter-expressing cells. Whole-tissue clearing and three-dimensional confocal imaging techniques were then applied to these HSC reporter mice to systematically measure the distance of the reported HSCs to different cell types in the BM or in the spleen that underwent extramedullary hematopoiesis (EMH). In both *α-catulin*-GFP and *Hoxb5*-mCherry mice, most reporter-expressing cells, no matter dividing or nondividing, were found to reside at regions close to, or in direct contact with, the sinusoidal blood vessels in the BM or in the spleen (Acar et al., 2015; Chen et al., 2016; Inra et al., 2015). However, because both BM and spleen contain a dense network of sinusoidal blood vessels, over 60% of all hematopoietic cells were within 5 μm to the sinusoids (Acar et al., 2015). HSCs were actually not significantly more close to sinusoids relative to randomly selected cells (Acar et al., 2015). Besides, it needs to be taken into account that tissue clearing is always associated with severe tissue swelling or shrinkage of vessels, while bone-lining cells are more stable under the tissue clearing treatment, which may influence the interpretation of relative distance measurement.

Altogether, these anatomy studies indicated the coexistence of different niche cells that independently contribute to the heterogeneous HSC regulation in the BM and highlighted possible limitations of each test methods.

The HSC niche refers to cellular components that functionally contribute to HSC maintenance in which HSCs are protected from depletion and overexuberant proliferation (Scadden, 2006). The HSC niche is initially evidenced by the in vivo study showing that the increase of trabecular bone and the associated osteoblastic cells led to an increase of HSC number (Calvi et al., 2003; Zhang et al., 2003), which indicated the role of osteoblastic cells in regulating HSCs. Later studies proposed that a perivascular niche (endothelial cells and pericytes) might also produce factors that contribute to HSC regulation. To directly study how stromal cells contribute to HSC regulation through those niche factors, a "taking one out" strategy was performed in which niche factors were specifically deleted from each individual niche cells and its influence on HSC function was assessed. Several key HSC regulation factors have been studied using this strategy.

Stem cell factor (SCF, also known as kit ligand, Kitl) is a growth factor that binds to the c-kit receptor (CD117), a receptor tyrosine kinase that expresses in HSCs (Kent et al., 2008). SCF is noncell autonomously required for HSC proliferation, survival, and function in vivo (Broudy, 1997). Mice that do not express SCF die in utero from severe anemia (Broudy, 1997). To investigate the functional source of SCF that contributes to HSC regulation, a genetic approach of deleting *Scf* from each individual niche component was performed. Researchers found that HSC frequency and function were not affected when *Scf* was conditionally deleted from hematopoietic cells using *Vav-1-Cre*, from osteoblasts using *Col2.3-Cre*, or from Nestin-expressing cells using *Nestin-Cre* or *Nestin-Cre/ERT2*. However, HSC number was significantly reduced in the BM when *Scf* was deleted from *Tie-2-Cre*-marked endothelial cells or leptin receptor (Lepr)-expressing perivascular stromal cells (Ding, Saunders, Enikolopov, & Morrison, 2012; Oguro, Ding, & Morrison, 2013). This study proposed that endothelial and Lepr-expressing perivascular cells are the critical source of SCF for HSC maintenance in the BM.

The stromal cell-derived factor 1 (SDF-1, which is encoded by C-X-C motif chemokine 12 gene, *Cxcl12*) is a chemokine that plays an important role in leukocyte and hematopoietic precursor migration (Ma et al., 1998) along with G-CSF-induced HSC mobilization (Levesque, Hendy, Takamatsu, Simmons, & Bendall, 2003). Deletion of CXCR4, a receptor for CXCL12, in HSCs severely impaired HSC maintenance; thus, the cells abundantly expressing CXCL12 were identified as CXCL12-abundant reticular (CAR) cells, a critical HSC-regulating stromal cell in the BM. CAR cells, as adipo-osteogenic progenitors, were found surrounding sinusoidal endothelial cells or located near the endosteum (Omatsu, Seike, Sugiyama, Kume, & Nagasawa, 2014; Sugiyama, Kohara, Noda, & Nagasawa, 2006). A genetic approach to delete *Cxcl12* from different niches revealed that both $Prx1^{+}$ early mesenchymal progenitors and $Tie2^{+}$ endothelial cells serve as the functional CXCL12 sources for HSC maintenance. Conditional deletion of *Cxcl12* from hematopoietic cells or Nestin-expressing cells had little or no effect on HSCs or restricted progenitors. However, deletion of *Cxcl12* from osteoblasts depleted certain early lymphoid progenitors (Ding & Morrison, 2013; Greenbaum et al., 2013). This suggests that CXCL12 secreted from distinct cellular niches in BM might have different functions.

Transforming growth factor beta 1 (TGFβ1) is a secreted protein that performs many cellular functions, including the control of cell growth, cell proliferation, cell differentiation, and apoptosis (Massague, 2012). TGFβ1 signaling has been identified as a critical signaling to keep HSCs in their quiescent state (Yamazaki et al., 2009). TGFβ receptor (*Tgfβr1* or *Tgfβr2*)-deficient HSCs have increased proliferation but impaired long-term self-renewal potential (Larsson et al., 2003; Yamazaki et al., 2011). Recently, MKs have been demonstrated to constitute a critical niche for HSC quiescence maintenance during homeostasis and HSC regeneration under stress by supplying TGFβ1 in BM. Deleting TGFβ1 from MKs led to HSC proliferation (Zhao et al., 2014). Nonmyelinating Schwann cells (glial cells ensheathed autonomic nerves) have been identified as a critical TGFβ signal regulator which contributes to TGFβ activation and HSC maintenance in the BM (Yamazaki et al., 2011). This suggests that MKs and nonmyelinating Schwann cells form an integrated niche for HSC quiescence maintenance.

The osteoblast niche in the endosteal region was the first identified HSC niche in the BM. Subsequent studies have suggested that osteoblasts express many factors that promote HSC maintenance, including SCF, CXCL12, angiopoietin-1, and thrombopoietin (Arai et al., 2004; Kollet et al., 2006; Wilson & Trumpp, 2006; Yoshihara et al., 2007). However, deleting either *Scf*, *Cxcl12*, or angiopoietin-1 using *Col2.3-Cre* did not affect HSC function in mice (Ding & Morrison, 2013; Ding et al., 2012; Greenbaum et al., 2013; Zhou, Ding, & Morrison, 2015). Consistently, *Col2.3-Cre*-induced osteoblast ablation in mice did not affect the overall HSC numbers but impaired the long-term self-renewal of HSCs (Bowers et al., 2015; Zhao & Li, 2015), which suggests that *Col2.3-Cre*-marked osteoblastic cells might produce unknown factors for HSC maintenance. These studies can also be explained by the inefficient recombination efficiency of *Col2.3-Cre* in endosteal niche, which also is composed by osteoblast cells and MSCs. Further studies need to be performed to test the potential contributions of thrombopoietin (Yoshihara et al., 2007), Jagged-1 (Guezguez et al., 2013), and noncanonical Wnt signals (Sugimura et al., 2012) from the osteoblastic niche in the endosteal region for HSC regulation.

Conditionally deleting critical HSC regulation factors from the perivascular niche dramatically reduced HSC abundance in the BM; however, a significant portion of the functional HSC still retained (Ding & Morrison, 2013; Ding et al., 2012; Greenbaum et al., 2013). Although uneven recombination efficiencies of Cre recombinase among cell types and transgenic strains explain variability in current studies, it also suggests that some important niche components may have been missed in the "taking one off" studies due to lack of appropriate genetic tools or the redundancy of each individual niches in the BM.

Among recent studies, BM MSCs (also known as skeletal stem cells) are proved to be critical for forming the HSC niche in the BM. MSCs are self-renewing cells that can differentiate into bone, fat, cartilage, and stromal cells of the BM. The first approach using Nestin-GFP reporter mouse line found that $Nestin^+$ MSCs are spatially associated with HSCs and highly express HSC maintenance factors, suggesting that MSCs and HSCs form a unique BM niche (Mendez-Ferrer et al., 2010). Furthermore, different genetic approach studies demonstrated that MSCs supply SCF and CXCL12 for HSC regulation by using *Lepr-Cre*, *Prx1-Cre*, and *NG2-Cre* lines (Greenbaum et al., 2013; Kunisaki et al., 2013; Zhou, Yue, Murphy, Peyer, & Morrison, 2014). All $Lepr^+$, $Nestin^+$, $Prx1^+$, and $NG2^+$ cells could behave as MSCs, which indicated that the heterogeneity of MSCs needs to be further explored.

4. BM NICHES FOR HSC REGENERATION

Under myeloablation or hemorrhage stress when homeostatic hematopoiesis is damaged, HSCs quickly switch from a quiescent state to a proliferating state to expand stem and progenitor cell pool for recovery. During this period, several extrinsic signals from BM niche play a critical role in supporting HSC regeneration.

Wnt signaling has been demonstrated to regulate the development of various tissues. Initial β-catenin (Ctnnb1, an important component of Wnt pathway) gain-of-function studies demonstrated that activation of β-catenin in HSCs promotes HSC expansion in vitro and hematopoietic reconstitution in vivo upon transplantation, especially in cooperation with the activation of AKT pathway (Perry et al., 2011; Reya et al., 2003; Willert et al., 2003). Genetically deleting β-catenin from HSCs did not affect their self-renewal in transplantation model (Cobas et al., 2004; Jeannet et al., 2008; Koch et al., 2008) but impaired HSC regeneration postinjury through enhanced oxidative stress (Lento et al., 2014). This could be partially explained by the switch between noncanonical Wnt signaling and canonical Wnt signaling during homeostasis and under stress. During homeostasis, LT-HSCs were maintained by noncanonical Wnt signaling, via Frizzled 8 (Fzd8) receptor and Flamingo (Fmi, or Celsr2), which antagonized canonical Wnt signals. Under stress noncanonical Wnt signaling is suppressed, but canonical Wnt signaling is turned on for HSC activation to support their postinjury recovery (Lento et al., 2014; Nemeth, Topol, Anderson, Yang, & Bodine, 2007; Sugimura et al., 2012).

Fibroblast growth factor (FGF) signaling promotes HSC expansion in vitro (de Haan et al., 2003; Zhang & Lodish, 2005). Although genetically deleting *Fgfr1* from HSCs or FGF-2 from the niche did not affect HSC hematopoiesis, those FGF signal-defective mice have severely impaired HSC regeneration and hematopoietic recovery after chemotherapeutic stress. This is due to the dynamic switch of FGF pathway between homeostasis and stress. In response to chemotherapeutic stress, FGF ligands were increased in BM niche and FGF receptors were increased in HSCs to support HSC proliferation through activation of AKT and NF-κB mechanisms. Interestingly, FGF signal was also shown to play a role in mobilizing HSCs to the spleen for EMH through regulating CXCR4–CXCL12 pathway, which further facilitated HSC regeneration under stress (Itkin et al., 2012; Zhao et al., 2014, 2012).

Epidermal growth factor (EGF) signaling was also shown to promote HSC regeneration postinjury, possibly through repression of the proapoptotic protein PUMA (Doan et al., 2013). Deletion or blockage of the adhesion molecule E-selectin was shown to promote HSC survival and regeneration postmyeloablative stress through manipulating the interaction between HSCs and their vascular niche (Winkler et al., 2012). However, the effects from EGF and E-selectin are not limited to stressed conditions, as they also contribute to HSC maintenance during homeostasis.

5. HSC NICHES BEYOND BM

Adult hematopoiesis occurs primarily in the BM of mammals. However, a wide range of hematopoietic stresses including myelofibrosis (Abdel-Wahab & Levine, 2009), anemia (Bennett, Pinkerton, Cudkowicz, & Bannerman, 1968; Cheshier, Prohaska, & Weissman, 2007), pregnancy (Fowler & Nash, 1968; Nakada et al., 2014), infection (Baldridge, King, Boles, Weksberg, & Goodell, 2010; Burberry et al., 2014), myeloablation (Morrison, Wright, & Weissman, 1997), and myocardial infarction (Dutta et al., 2012) can induce EMH in which a subset of HSCs are mobilized to other organs, such as spleen and liver, to expand hematopoiesis. EMH may occur whenever the hematopoietic demand exceeds what can be produced by the BM. Within spleen, the major EMH organ, the splenic red pulp is a prominent site of EMH

(Freedman & Saunders, 1981; Johns & Christopher, 2012; Lowell, Niwa, Soriano, & Varmus, 1996; Tavassoli & Weiss, 1973) and the splenic white pulp is responsible for lymphocyte activation in both mice and humans (Mueller & Germain, 2009). Consistent with this, two HSC niche factors, SCF and SDF-1, were found to be expressed by stromal cells in the splenic red pulp (Inra et al., 2015). Conditional deletion of *Scf* from spleen endothelial cells and *Scf* or *Sdf-1* from $Tcf21^+$ perivascular stromal cells severely impaired spleen EMH and reduced blood cell counts in response to chemotherapy, pregnancy, or acute blood loss (Inra et al., 2015). Thus, similar to BM, endothelial cells and perivascular stromal cells are critical components of the HSC niche in the spleen. Notably, SCF or SDF-1 from endothelial cells and/or perivascular stromal cells is dispensable for normal hematopoiesis in the spleen (Inra et al., 2015). Consistent with this, HSCs residing in spleen do not express EPCR protein, a marker that label all functional HSCs in the BM (Gur-Cohen et al., 2015). Thus, spleen likely functions as an acute HSC niche rather than a place for long-term HSC maintenance.

SCF and SDF-1 are also highly expressed in the sinusoidal endothelial cells in liver (Inra et al., 2015) which suggests a potential role of the liver perivascular niche in regulating EMH, although more direct experiments are required to test this possibility.

6. MATURE HEMATOPOIETIC CELLS AS HSC NICHE

Differentiated cells can constitute the niche for their mother stem cells was first discovered in small intestine system in which mature paneth cells form a niche for $Lgr5^+$ intestinal stem cell (Sato et al., 2011). Consistent with this, transit-amplifying cells were reported to form a dynamic niche for hair follicle stem cell regulation (Hsu, Li, & Fuchs, 2014). These studies indicated that an appropriate feedback mechanism from mature blood cells for HSC regulation is critical to control the output of hematopoiesis (Day & Link, 2014). MKs were found physically associated with HSCs in a nonrandom fashion in the BM. Functional studies showed that MKs directly regulate HSC quiescence through multiple HSC-regulating factors including TGFβ, CXCL4, and THPO (Bruns et al., 2014; Heazlewood et al., 2013; Nakamura-Ishizu, Takubo, Fujioka, & Suda, 2014; Zhao et al., 2014). More importantly, MKs can balance TGFβ and FGF output, thereby regulating hematopoietic homeostasis and regeneration following chemotherapeutic stress, respectively, in vivo (Blank & Karlsson, 2015; Zhao et al., 2014). This suggests that MKs constitute an even more specialized HSC niche, which can control HSC cell cycle activity based on demands, serving as a sensor to guide HSC switching between quiescence and regeneration (Schepers, Campbell, & Passegue, 2015).

More interestingly, MK niche also contributes to the architectural maintenance of stromal niches. Studies suggest that MKs express multiple vascular and osteoblast regulatory factors that influence the number and function of both osteoblastic and vascular niches in the BM. Mature MKs are primarily located adjacent to blood vessels, where they extend transendothelial pseudopods or migrate through the endothelium to produce platelets (Avecilla et al., 2004; Hamada et al., 1998; Wang, Liu, & Groopman, 1998). MKs are a major source of both proangiogenetic and antiangiogenic factors, which makes them a regulator for angiogenesis and maintenance of blood vessel integrity (Kopp et al., 2006; Mohle, Green, Moore, Nachman, & Rafii, 1997; Varner, 2006). MKs have also been proposed for HSC niche reconstitution when both vascular niche and endosteal niche were damaged by myelosuppression (Hooper et al., 2009; Jiang, Bonig, Ulyanova, Chang, & Papayannopoulou, 2009; Kunisaki et al., 2013). MKs are tightly associated with BM blood vessels postchemotherapeutic stress (Avecilla et al., 2004; Zhao et al., 2014) and contribute to blood vessel recovery through thrombospondin 1, an endogenous angiogenesis inhibitor (Kopp et al., 2006). Furthermore, post whole-body irradiation, surviving MKs relocated to the endosteal surface of trabecular bone where they promoted osteoblast proliferation but inhibited osteoclast formation through producing growth factors, including PDGF-β and FGF-2, which facilities stem cell engraftment postirradiation (Beeton, Bord, Ireland, & Compston, 2006; Ciovacco, Cheng, Horowitz, & Kacena, 2010; Dominici et al., 2009; Olson et al., 2013).

7. PERSPECTIVE

Recent experimentation has validated the niche concept and built up the structure of the molecular and cellular nature of the HSC niches in the BM. In the future, manipulating these stem cell niches will provide promising possibilities to benefit hematopoietic maintenance and regeneration. Continued progress in heightened understanding of the key players of HSC niches will open exciting new avenues to improve regenerative medicine and to treat hematopoietic disorders. This knowledge will also be applied to engineer ex vivo niches for HSC expansion that facilitates stem cell therapy in clinics.

Despite the great achievements that we have made in our understanding of HSC niche over the past 13 years, several fundamental questions remain to be addressed in the future. For example, the BM niche has to sense the demand of hematopoiesis during homeostasis and under postinjury stress to guild HSCs switching between quiescent and proliferating state. How HSC states are dynamically regulated by their niches is still unknown. Second, although the EMH niche in the spleen has been identified, it remains completely unknown how the splenic niche is activated upon a certain type of stress and what is the functional difference between BM and splenic niche in regulating hematopoiesis. Finally, our ultimate goal of understanding the extrinsic mechanisms that regulate HSC self-renewal is to recreate the environment in vitro that allows HSC expansion. However, the niche cells and niche factors that have been identified in vivo so far appear to be essential yet insufficient to support HSC self-renewal in vitro. More efforts should be made to integrate our theoretical achievements into preclinical practice.

REFERENCES

Abdel-Wahab, O. I., & Levine, R. L. (2009). Primary myelofibrosis: Update on definition, pathogenesis, and treatment. *Annual Review of Medicine*, *60*, 233–245.

Acar, M., Kocherlakota, K. S., Murphy, M. M., Peyer, J. G., Oguro, H., Inra, C. N., et al. (2015). Deep imaging of bone marrow shows non-dividing stem cells are mainly perisinusoidal. *Nature*, *526*(7571), 126–130.

Arai, F., Hirao, A., Ohmura, M., Sato, H., Matsuoka, S., Takubo, K., et al. (2004). Tie2/angiopoietin-1 signaling regulates hematopoietic stem cell quiescence in the bone marrow niche. *Cell*, *118*(2), 149–161.

Avecilla, S. T., Hattori, K., Heissig, B., Tejada, R., Liao, F., Shido, K., et al. (2004). Chemokine-mediated interaction of hematopoietic progenitors with the bone marrow vascular niche is required for thrombopoiesis. *Nature Medicine*, *10*, 64–71.

Baldridge, M. T., King, K. Y., Boles, N. C., Weksberg, D. C., & Goodell, M. A. (2010). Quiescent haematopoietic stem cells are activated by IFN-gamma in response to chronic infection. *Nature*, *465*(7299), 793–797.

Beeton, C. A., Bord, S., Ireland, D., & Compston, J. E. (2006). Osteoclast formation and bone resorption are inhibited by megakaryocytes. *Bone*, *39*(5), 985–990.

Bennett, M., Pinkerton, P. H., Cudkowicz, G., & Bannerman, R. M. (1968). Hemopoietic progenitor cells in marrow and spleen of mice with hereditary iron deficiency anemia. *Blood*, *32*(6), 908–921.

Blank, U., & Karlsson, S. (2015). TGF-beta signaling in the control of hematopoietic stem cells. *Blood*, *125*(23), 3542–3550.

Bowers, M., Zhang, B., Ho, Y., Agarwal, P., Chen, C. C., & Bhatia, R. (2015). Osteoblast ablation reduces normal long-term hematopoietic stem cell self-renewal but accelerates leukemia development. *Blood*, *125*(17), 2678–2688.

Broudy, V. C. (1997). Stem cell factor and hematopoiesis. *Blood*, *90*(4), 1345–1364.

Bruns, I., Lucas, D., Pinho, S., Ahmed, J., Lambert, M. P., Kunisaki, Y., et al. (2014). Megakaryocytes regulate hematopoietic stem cell quiescence through CXCL4 secretion. *Nature Medicine*, *20*(11), 1315–1320.

Burberry, A., Zeng, M. Y., Ding, L., Wicks, I., Inohara, N., Morrison, S. J., et al. (2014). Infection mobilizes hematopoietic stem cells through cooperative NOD-like receptor and Toll-like receptor signaling. *Cell Host & Microbe*, *15*(6), 779–791.

Calvi, L. M., Adams, G. B., Weibrecht, K. W., Weber, J. M., Olson, D. P., Knight, M. C., et al. (2003). Osteoblastic cells regulate the haematopoietic stem cell niche. *Nature*, *425*(6960), 841–846.

Chen, J. Y., Miyanishi, M., Wang, S. K., Yamazaki, S., Sinha, R., Kao, K. S., et al. (2016). Hoxb5 marks long-term haematopoietic stem cells and reveals a homogenous perivascular niche. *Nature*, *530*(7589), 223–227.

Cheng, T., Rodrigues, N., Shen, H., Yang, Y., Dombkowski, D., Sykes, M., et al. (2000). Hematopoietic stem cell quiescence maintained by p21cip1/waf1. *Science*, *287*(5459), 1804–1808.

Cheshier, S. H., Prohaska, S. S., & Weissman, I. L. (2007). The effect of bleeding on hematopoietic stem cell cycling and self-renewal. *Stem Cells and Development*, *16*(5), 707–717.

Ciovacco, W. A., Cheng, Y. H., Horowitz, M. C., & Kacena, M. A. (2010). Immature and mature megakaryocytes enhance osteoblast proliferation and inhibit osteoclast formation. *Journal of Cellular Biochemistry*, *109*(4), 774–781.

Cobas, M., Wilson, A., Ernst, B., Mancini, S. J., MacDonald, H. R., Kemler, R., et al. (2004). Beta-catenin is dispensable for hematopoiesis and lymphopoiesis. *The Journal of Experimental Medicine*, *199*(2), 221–229.

Day, R. B., & Link, D. C. (2014). Megakaryocytes in the hematopoietic stem cell niche. *Nature Medicine*, *20*(11), 1233–1234.

de Haan, G., Weersing, E., Dontje, B., van Os, R., Bystrykh, L. V., Vellenga, E., et al. (2003). In vitro generation of long-term repopulating hematopoietic stem cells by fibroblast growth factor-1. *Developmental Cell*, *4*, 241–251.

Dexter, T. M., Allen, T. D., & Lajtha, L. G. (1977). Conditions controlling the proliferation of haemopoietic stem cells in vitro. *Journal of Cellular Physiology*, *91*(3), 335–344.

Ding, L., & Morrison, S. J. (2013). Haematopoietic stem cells and early lymphoid progenitors occupy distinct bone marrow niches. *Nature*, *495*(7440), 231–235.

Ding, L., Saunders, T. L., Enikolopov, G., & Morrison, S. J. (2012). Endothelial and perivascular cells maintain haematopoietic stem cells. *Nature*, *481*(7382), 457–462.

Doan, P. L., Himburg, H. A., Helms, K., Russell, J. L., Fixsen, E., Quarmyne, M., et al. (2013). Epidermal growth factor regulates hematopoietic regeneration after radiation injury. *Nature Medicine*, *19*(3), 295–304.

Dominici, M., Rasini, V., Bussolari, R., Chen, X., Hofmann, T. J., Spano, C., et al. (2009). Restoration and reversible expansion of the osteoblastic hematopoietic stem cell niche after marrow radioablation. *Blood*, *114*(11), 2333–2343.

Dutta, P., Courties, G., Wei, Y., Leuschner, F., Gorbatov, R., Robbins, C. S., et al. (2012). Myocardial infarction accelerates atherosclerosis. *Nature*, *487*(7407), 325–329.

Foudi, A., Hochedlinger, K., Van Buren, D., Schindler, J. W., Jaenisch, R., Carey, V., et al. (2009). Analysis of histone 2B-GFP retention reveals slowly cycling hematopoietic stem cells. *Nature Biotechnology*, *27*, 84–90.

Fowler, J. H., & Nash, D. J. (1968). Erythropoiesis in the spleen and bone marrow of the pregnant mouse. *Developmental Biology*, *18*(4), 331–353.

Freedman, M. H., & Saunders, E. F. (1981). Hematopoiesis in the human spleen. *American Journal of Hematology*, *11*(3), 271–275.

Gazit, R., Mandal, P. K., Ebina, W., Ben-Zvi, A., Nombela-Arrieta, C., Silberstein, L. E., et al. (2014). Fgd5 identifies hematopoietic stem cells in the murine bone marrow. *The Journal of Experimental Medicine*, *211*(7), 1315–1331.

Godin, I., & Cumano, A. (2002). The hare and the tortoise: An embryonic haematopoietic race. *Nature Reviews. Immunology*, *2*(8), 593–604.

Greenbaum, A., Hsu, Y. M., Day, R. B., Schuettpelz, L. G., Christopher, M. J., Borgerding, J. N., et al. (2013). CXCL12 in early mesenchymal progenitors is required for haematopoietic stem-cell maintenance. *Nature, 495*(7440), 227–230.

Guezguez, B., Campbell, C. J., Boyd, A. L., Karanu, F., Casado, F. L., Di Cresce, C., et al. (2013). Regional localization within the bone marrow influences the functional capacity of human HSCs. *Cell Stem Cell, 13*(2), 175–189.

Gur-Cohen, S., Itkin, T., Chakrabarty, S., Graf, C., Kollet, O., Ludin, A., et al. (2015). PAR1 signaling regulates the retention and recruitment of EPCR-expressing bone marrow hematopoietic stem cells. *Nature Medicine, 21*(11), 1307–1317.

Hamada, T., Mohle, R., Hesselgesser, J., Hoxie, J., Nachman, R. L., Moore, M. A., et al. (1998). Transendothelial migration of megakaryocytes in response to stromal cell-derived factor 1 (SDF-1) enhances platelet formation. *The Journal of Experimental Medicine, 188*(3), 539–548.

Heazlewood, S. Y., Neaves, R. J., Williams, B., Haylock, D. N., Adams, T. E., & Nilsson, S. K. (2013). Megakaryocytes co-localise with hemopoietic stem cells and release cytokines that up-regulate stem cell proliferation. *Stem Cell Research, 11*(2), 782–792.

Hooper, A. T., Butler, J. M., Nolan, D. J., Kranz, A., Iida, K., Kobayashi, M., et al. (2009). Engraftment and reconstitution of hematopoiesis is dependent on VEGFR2-mediated regeneration of sinusoidal endothelial cells. *Cell Stem Cell, 4*(3), 263–274.

Hsu, Y. C., Li, L., & Fuchs, E. (2014). Transit-amplifying cells orchestrate stem cell activity and tissue regeneration. *Cell, 157*(4), 935–949.

Inra, C. N., Zhou, B. O., Acar, M., Murphy, M. M., Richardson, J., Zhao, Z., et al. (2015). A perisinusoidal niche for extramedullary haematopoiesis in the spleen. *Nature, 527*(7579), 466–471.

Itkin, T., Gur-Cohen, S., Spencer, J. A., Schajnovitz, A., Ramasamy, S. K., Kusumbe, A. P., et al. (2016). Distinct bone marrow blood vessels differentially regulate haematopoiesis. *Nature, 532*(7599), 323–328.

Itkin, T., Ludin, A., Gradus, B., Gur-Cohen, S., Kalinkovich, A., Schajnovitz, A., et al. (2012). FGF-2 expands murine hematopoietic stem and progenitor cells via proliferation of stromal cells, c-Kit activation, and CXCL12 down-regulation. *Blood, 120*(9), 1843–1855.

Jeannet, G., Scheller, M., Scarpellino, L., Duboux, S., Gardiol, N., Back, J., et al. (2008). Long-term, multilineage hematopoiesis occurs in the combined absence of {beta}-catenin and {gamma}-catenin. *Blood, 111*, 142–149.

Jiang, Y., Bonig, H., Ulyanova, T., Chang, K., & Papayannopoulou, T. (2009). On the adaptation of endosteal stem cell niche function in response to stress. *Blood, 114*(18), 3773–3782.

Johns, J. L., & Christopher, M. M. (2012). Extramedullary hematopoiesis: A new look at the underlying stem cell niche, theories of development, and occurrence in animals. *Veterinary Pathology, 49*(3), 508–523.

Kent, D., Copley, M., Benz, C., Dykstra, B., Bowie, M., & Eaves, C. (2008). Regulation of hematopoietic stem cells by the steel factor/KIT signaling pathway. *Clinical Cancer Research, 14*(7), 1926–1930.

Kiel, M. J., Yilmaz, O. H., Iwashita, T., Terhorst, C., & Morrison, S. J. (2005). SLAM family receptors distinguish hematopoietic stem and progenitor cells and reveal endothelial niches for stem cells. *Cell, 121*(7), 1109–1121.

Koch, U., Wilson, A., Cobas, M., Kemler, R., Macdonald, H. R., & Radtke, F. (2008). Simultaneous loss of beta- and gamma-catenin does not perturb hematopoiesis or lymphopoiesis. *Blood, 111*(1), 160–164.

Kollet, O., Dar, A., Shivtiel, S., Kalinkovich, A., Lapid, K., Sztainberg, Y., et al. (2006). Osteoclasts degrade endosteal components and promote mobilization of hematopoietic progenitor cells. *Nature Medicine, 12*, 657–664.

Kopp, H. G., Hooper, A. T., Broekman, M. J., Avecilla, S. T., Petit, I., Luo, M., et al. (2006). Thrombospondins deployed by thrombopoietic cells determine angiogenic switch and extent of revascularization. *The Journal of Clinical Investigation, 116*(12), 3277–3291.

Kunisaki, Y., Bruns, I., Scheiermann, C., Ahmed, J., Pinho, S., Zhang, D., et al. (2013). Arteriolar niches maintain haematopoietic stem cell quiescence. *Nature, 502*, 637–643.

Larsson, J., Blank, U., Helgadottir, H., Bjornsson, J. M., Ehinger, M., Goumans, M. J., et al. (2003). TGF-beta signaling-deficient hematopoietic stem cells have normal self-renewal and regenerative ability in vivo despite increased proliferative capacity in vitro. *Blood, 102*, 3129–3135.

Lento, W., Ito, T., Zhao, C., Harris, J. R., Huang, W., Jiang, C., et al. (2014). Loss of beta-catenin triggers oxidative stress and impairs hematopoietic regeneration. *Genes and Development, 28*(9), 995–1004.

Lessard, J., Faubert, A., & Sauvageau, G. (2004). Genetic programs regulating HSC specification, maintenance and expansion. *Oncogene, 23*(43), 7199–7209.

Levesque, J. P., Hendy, J., Takamatsu, Y., Simmons, P. J., & Bendall, L. J. (2003). Disruption of the CXCR4/CXCL12 chemotactic interaction during hematopoietic stem cell mobilization induced by GCSF or cyclophosphamide. *The Journal of Clinical Investigation, 111*(2), 187–196.

Lo Celso, C., Fleming, H. E., Wu, J. W., Zhao, C. X., Miake-Lye, S., Fujisaki, J., et al. (2009). Live-animal tracking of individual haematopoietic stem/progenitor cells in their niche. *Nature, 457*(7225), 92–96.

Lowell, C. A., Niwa, M., Soriano, P., & Varmus, H. E. (1996). Deficiency of the Hck and Src tyrosine kinases results in extreme levels of extramedullary hematopoiesis. *Blood, 87*(5), 1780–1792.

Ma, Q., Jones, D., Borghesani, P. R., Segal, R. A., Nagasawa, T., Kishimoto, T., et al. (1998). Impaired B-lymphopoiesis, myelopoiesis, and derailed cerebellar neuron migration in CXCR4- and SDF-1-deficient mice. *Proceedings of the National Academy of Sciences of the United States of America, 95*(16), 9448–9453.

Massague, J. (2012). TGFbeta signalling in context. *Nature Reviews. Molecular Cell Biology, 13*(10), 616–630.

Medvinsky, A., & Dzierzak, E. (1996). Definitive hematopoiesis is autonomously initiated by the AGM region. *Cell, 86*(6), 897–906.

Mendez-Ferrer, S., Michurina, T. V., Ferraro, F., Mazloom, A. R., Macarthur, B. D., Lira, S. A., et al. (2010). Mesenchymal and haematopoietic stem cells form a unique bone marrow niche. *Nature, 466*(7308), 829–834.

Mendez-Ferrer, S., Scadden, D. T., & Sanchez-Aguilera, A. (2015). Bone marrow stem cells: Current and emerging concepts. *The Annals of the New York Academy of Sciences, 1335*, 32–44.

Mikkola, H. K., & Orkin, S. H. (2006). The journey of developing hematopoietic stem cells. *Development, 133*(19), 3733–3744.

Mohle, R., Green, D., Moore, M. A., Nachman, R. L., & Rafii, S. (1997). Constitutive production and thrombin-induced release of vascular endothelial growth factor by human megakaryocytes and platelets. *Proceedings of the National Academy of Sciences of the United States of America, 94*(2), 663–668.

Morrison, S. J., & Scadden, D. T. (2014). The bone marrow niche for haematopoietic stem cells. *Nature, 505*(7483), 327–334.

Morrison, S. J., Wright, D. E., & Weissman, I. L. (1997). Cyclophosphamide/granulocyte colony-stimulating factor induces hematopoietic stem cells to proliferate prior to mobilization. *Proceedings of the National Academy of Sciences of the United States of America, 94*(5), 1908–1913.

Mueller, S. N., & Germain, R. N. (2009). Stromal cell contributions to the homeostasis and functionality of the immune system. *Nature Reviews. Immunology, 9*(9), 618–629.

Nakada, D., Oguro, H., Levi, B. P., Ryan, N., Kitano, A., Saitoh, Y., et al. (2014). Oestrogen increases haematopoietic stem-cell self-renewal in females and during pregnancy. *Nature*, *505*(7484), 555–558.

Nakamura-Ishizu, A., Takubo, K., Fujioka, M., & Suda, T. (2014). Megakaryocytes are essential for HSC quiescence through the production of thrombopoietin. *Biochemical and Biophysical Research Communications*, *454*(2), 353–357.

Nemeth, M. J., Topol, L., Anderson, S. M., Yang, Y., & Bodine, D. M. (2007). Wnt5a inhibits canonical Wnt signaling in hematopoietic stem cells and enhances repopulation. *Proceedings of the National Academy of Sciences of the United States of America*, *104*(39), 15436–15441.

Nombela-Arrieta, C., Pivarnik, G., Winkel, B., Canty, K. J., Harley, B., Mahoney, J. E., et al. (2013). Quantitative imaging of haematopoietic stem and progenitor cell localization and hypoxic status in the bone marrow microenvironment. *Nature Cell Biology*, *15*(5), 533–543.

Ogawa, M. (1993). Differentiation and proliferation of hematopoietic stem cells. *Blood*, *81*(11), 2844–2853.

Oguro, H., Ding, L., & Morrison, S. J. (2013). SLAM family markers resolve functionally distinct subpopulations of hematopoietic stem cells and multipotent progenitors. *Cell Stem Cell*, *13*(1), 102–116.

Olson, T. S., Caselli, A., Otsuru, S., Hofmann, T. J., Williams, R., Paolucci, P., et al. (2013). Megakaryocytes promote murine osteoblastic HSC niche expansion and stem cell engraftment after radioablative conditioning. *Blood*, *121*, 5238–5249.

Omatsu, Y., Seike, M., Sugiyama, T., Kume, T., & Nagasawa, T. (2014). Foxc1 is a critical regulator of haematopoietic stem/progenitor cell niche formation. *Nature*, *508*(7497), 536–540.

Perry, J. M., He, X. C., Sugimura, R., Grindley, J. C., Haug, J. S., Ding, S., et al. (2011). Cooperation between both Wnt/{beta}-catenin and PTEN/PI3K/Akt signaling promotes primitive hematopoietic stem cell self-renewal and expansion. *Genes and Development*, *25*(18), 1928–1942.

Raaijmakers, M. H., Mukherjee, S., Guo, S., Zhang, S., Kobayashi, T., Schoonmaker, J. A., et al. (2010). Bone progenitor dysfunction induces myelodysplasia and secondary leukaemia. *Nature*, *464*(7290), 852–857.

Reya, T., Duncan, A. W., Ailles, L., Domen, J., Scherer, D. C., Willert, K., et al. (2003). A role for Wnt signalling in self-renewal of haematopoietic stem cells. *Nature*, *423*(6938), 409–414.

Sato, T., van Es, J. H., Snippert, H. J., Stange, D. E., Vries, R. G., van den Born, M., et al. (2011). Paneth cells constitute the niche for Lgr5 stem cells in intestinal crypts. *Nature*, *469*(7330), 415–418.

Scadden, D. T. (2006). The stem-cell niche as an entity of action. *Nature*, *441*(7097), 1075–1079.

Schepers, K., Campbell, T. B., & Passegue, E. (2015). Normal and leukemic stem cell niches: Insights and therapeutic opportunities. *Cell Stem Cell*, *16*(3), 254–267.

Schofield, R. (1978). The relationship between the spleen colony-forming cell and the haemopoietic stem cell. *Blood Cells*, *4*, 7–25.

Siminovitch, L., McCulloch, E. A., & Till, J. E. (1963). The distribution of colony-forming cells among spleen colonies. *Journal of Cellular Physiology*, *62*, 327–336.

Sprent, J., & Tough, D. F. (1994). Lymphocyte life-span and memory. *Science*, *265*(5177), 1395–1400.

Sugimura, R., He, X. C., Venkatraman, A., Arai, F., Box, A., Semerad, C., et al. (2012). Noncanonical Wnt signaling maintains hematopoietic stem cells in the niche. *Cell*, *150*(2), 351–365.

Sugiyama, T., Kohara, H., Noda, M., & Nagasawa, T. (2006). Maintenance of the hematopoietic stem cell pool by CXCL12-CXCR4 chemokine signaling in bone marrow stromal cell niches. *Immunity*, *25*(6), 977–988.

Taichman, R. S., & Emerson, S. G. (1994). Human osteoblasts support hematopoiesis through the production of granulocyte colony-stimulating factor. *The Journal of Experimental Medicine*, *179*(5), 1677–1682.

Tavassoli, M., & Weiss, L. (1973). An electron microscopic study of spleen in myelofibrosis with myeloid metaplasia. *Blood*, *42*(2), 267–279.

Till, J. E., McCulloch, E. A., & Siminovitch, L. (1964). A stochastic model of stem cell proliferation, based on the growth of spleen colony-forming cells. *Proceedings of the National Academy of Sciences of the United States of America*, *51*, 29–36.

Varner, J. A. (2006). The sticky truth about angiogenesis and thrombospondins. *The Journal of Clinical Investigation*, *116*(12), 3111–3113.

Wang, J. F., Liu, Z. Y., & Groopman, J. E. (1998). The alpha-chemokine receptor CXCR4 is expressed on the megakaryocytic lineage from progenitor to platelets and modulates migration and adhesion. *Blood*, *92*(3), 756–764.

Willert, K., Brown, J. D., Danenberg, E., Duncan, A. W., Weissman, I. L., Reya, T., et al. (2003). Wnt proteins are lipid-modified and can act as stem cell growth factors. *Nature*, *423*, 448–452.

Wilson, A., Laurenti, E., Oser, G., van der Wath, R. C., Blanco-Bose, W., Jaworski, M., et al. (2008). Hematopoietic stem cells reversibly switch from dormancy to self-renewal during homeostasis and repair. *Cell*, *135*(6), 1118–1129.

Wilson, A., & Trumpp, A. (2006). Bone-marrow haematopoietic-stem-cell niches. *Nature Reviews. Immunology*, *6*(2), 93–106.

Winkler, I. G., Barbier, V., Nowlan, B., Jacobsen, R. N., Forristal, C. E., Patton, J. T., et al. (2012). Vascular niche E-selectin regulates hematopoietic stem cell dormancy, self renewal and chemoresistance. *Nature Medicine*, *18*(11), 1651–1657.

Xie, Y., Yin, T., Wiegraebe, W., He, X. C., Miller, D., Stark, D., et al. (2009). Detection of functional haematopoietic stem cell niche using real-time imaging. *Nature*, *457*(7225), 97–101.

Yamazaki, S., Ema, H., Karlsson, G., Yamaguchi, T., Miyoshi, H., Shioda, S., et al. (2011). Nonmyelinating Schwann cells maintain hematopoietic stem cell hibernation in the bone marrow niche. *Cell*, *147*(5), 1146–1158.

Yamazaki, S., Iwama, A., Takayanagi, S., Eto, K., Ema, H., & Nakauchi, H. (2009). TGF-beta as a candidate bone marrow niche signal to induce hematopoietic stem cell hibernation. *Blood*, *113*(6), 1250–1256.

Yoshihara, H., Arai, F., Hosokawa, K., Hagiwara, T., Takubo, K., Nakamura, Y., et al. (2007). Thrombopoietin/MPL signaling regulates hematopoietic stem cell quiescence and interaction with the osteoblastic niche. *Cell Stem Cell*, *1*(6), 685–697.

Zhang, C. C., & Lodish, H. F. (2005). Murine hematopoietic stem cells change their surface phenotype during ex vivo expansion. *Blood*, *105*(11), 4314–4320.

Zhang, J., Niu, C., Ye, L., Huang, H., He, X., Tong, W. G., et al. (2003). Identification of the haematopoietic stem cell niche and control of the niche size. *Nature*, *425*(6960), 836–841.

Zhao, M., & Li, L. (2015). Osteoblast ablation burns out functional stem cells. *Blood*, *125*(17), 2590–2591.

Zhao, M., Perry, J. M., Marshall, H., Venkatraman, A., Qian, P., He, X. C., et al. (2014). Megakaryocytes maintain homeostatic quiescence and promote post-injury regeneration of hematopoietic stem cells. *Nature Medicine*, *20*(11), 1321–1326.

Zhao, M., Ross, J. T., Itkin, T., Perry, J. M., Venkatraman, A., Haug, J. S., et al. (2012). FGF signaling facilitates postinjury recovery of mouse hematopoietic system. *Blood*, *120*(9), 1831–1842.

Zhou, B. O., Ding, L., & Morrison, S. J. (2015). Hematopoietic stem and progenitor cells regulate the regeneration of their niche by secreting Angiopoietin-1. *eLife*, *4*, e05521.
Zhou, F., Li, X., Wang, W., Zhu, P., Zhou, J., He, W., et al. (2016). Tracing haematopoietic stem cell formation at single-cell resolution. *Nature*, *533*(7604), 487–492.
Zhou, B. O., Yue, R., Murphy, M. M., Peyer, J. G., & Morrison, S. J. (2014). Leptin-receptor-expressing mesenchymal stromal cells represent the main source of bone formed by adult bone marrow. *Cell Stem Cell*, *15*(2), 154–168.

CHAPTER THREE

The Role of the CNS in the Regulation of HSCs

A. García-García*,†,‡, S. Méndez-Ferrer*,†,‡,1
*Wellcome Trust-Medical Research Council Cambridge Stem Cell Institute, University of Cambridge, Cambridge, United Kingdom
†National Health Service Blood and Transplant, Cambridge, United Kingdom
‡Centro Nacional de Investigaciones Cardiovasculares Carlos III (CNIC), Madrid, Spain
[1]Corresponding author: e-mail address: sm2116@medschl.cam.ac.uk

Contents

1. THE AUTONOMIC NERVOUS SYSTEM

The peripheral nervous system connects the CNS (brain and spinal cord) with innervated organs, serving as a bidirectional communication platform between CNS and peripheral organs. The peripheral nervous system is classically divided into the somatic nervous system and the autonomic nervous system. The somatic nervous system is responsible for voluntary movements of the body and it mainly comprises motoneurons that innervate the skeletal muscle to activate muscle contraction/relaxation and sensory fibers that innervate tissues and transmit sensory signals to the CNS. The autonomic nervous system is instead responsible for modulating those involuntary actions that ensure the proper functioning of our body, contributing to

Advances in Stem Cells and their Niches, Volume 1
ISSN 2468-5097
http://dx.doi.org/10.1016/bs.asn.2016.12.002

homeostasis. Therefore, the autonomic nervous system innervates many internal organs, including lymphoid organs, like the spleen or the BM (Calvo, 1968).

The autonomic nervous system comprises two branches: the sympathetic nervous system (SNS) and the parasympathetic nervous system (PNS). Although both branches frequently exhibit opposing effects in target organs, this is not always the case. Sympathetic signaling is generally associated with rapid activating responses, whereas the PNS usually triggers slow, dampening responses. For instance, in the context of the immune response, the SNS triggers an initial inflammatory response to injury, while the PNS activates the antiinflammatory pathways at the end of the process, to resolve the inflammation. In both cases the organization of the ANS implies the existence of a "two-step efferent pathway." A preganglionic neuron synapses onto a postganglionic neuron (usually in a peripheral ganglion), which releases a postsynaptic neurotransmitter in the peripheral organ. The sympathetic branch, which has its preganglionic origin in the thoracolumbar region of the spinal cord, is also often called "noradrenergic." In contrast, the parasympathetic branch, which has its origin in cranio/sacral areas, has been referred to as "cholinergic." This alternative nomenclature is based on the postsynaptic neurotransmitter, which is normally adrenaline/noradrenaline (SNS), which belong to the family of catecholamines, and acetylcholine (PNS). However, this nomenclature can sometimes lead to confusion since acetylcholine is the presynaptic neurotransmitter for both branches and some sympathetic fibers use acetylcholine (instead of noradrenaline) as postsynaptic neurotransmitter.

2. NEURAL CONTRIBUTIONS TO HEMATOPOIESIS: MULTIPLE AND FROM DEVELOPMENT

Neural signals have been found to contribute to hematopoiesis from the very emergence of HSCs. Sympathetic signals were found to contribute to HSC emergence in the aorta-gonad-mesonephros (AGM) region (Fitch et al., 2012). Nerve fibers colonize limb buds during late stages of mouse embryo development, associated with blood vessels (Calvo & Haas, 1969; Miller & McCuskey, 1973). Staining against pan-neuronal markers like PGP 9.5 and GAP-43 have shown the presence of nerve fibers in the perichondrium and periosteum of the rodent limb buds at E19. Blood vessels serve as a gateway for the innervation entering the BM.

Adult HSCs are originated in embryonic arteries of mammals, being more abundant in the AGM area of the embryo around E10. Subsequently, HSCs progressively colonize different organs (mainly the fetal liver, where HSCs expand) prior to their migration to the BM starting at E15 (Dzierzak & Speck, 2008). HSC migration to BM persists during the first postnatal weeks (Wolber et al., 2002), a period in which HSCs also switch from the proliferative state that these cells exhibit during the fetal liver stage to the quiescent state found in the adult BM. Dissecting how potential microenvironmental cues induce HSC quiescence during this period might facilitate preserving long-term HSC self-renewal.

It is interesting to note that BM mesenchymal stem/progenitor cells (BMSCs) are heterogeneous in markers and function. We found that BMSCs from long bones are heterogeneous in origin, with some of them having a neural crest origin in the perinatal life of mice. Compared with mesoderm-derived BMSCs involved in fetal skeletogenesis, quiescent neural crest-derived BMSCs secreted higher level of Cxcl12 and had as main function to attract HSCs and establish the BM niche (Isern et al., 2014). Impaired developmental migration of these neural crest derivatives along the nerves in Erbb3 KO mice (Riethmacher et al., 1997) was associated with impaired BM colonization by hematopoietic stem/progenitor cells (HSPCs) (Isern et al., 2014). Thus, the neural crest, as a source of peripheral neurons, Schwann cells and BMSCs, might contribute to establish BM HSC niches (Fig. 1).

Nonmyelinating Schwann cells, which unsheathe sympathetic nerve fibers, were found to contribute to maintain HSC quiescence. These Schwann cells were found to be an important source of BM activated transforming growth factor beta (TGF-β), which binds to TGF-β type II receptor in HSCs triggering TGF-β/Smad-mediated quiescence. As a consequence, sympathetic denervation led to a quick reduction of BM HSCs (Yamazaki et al., 2011).

HSCs might exhibit neurotrophic dependence from development, since they express at different developmental stages the receptor RET, a coreceptor for neurotrophic factors, and RET seems to promote HSC survival and proliferation (Fonseca-Pereira et al., 2014).

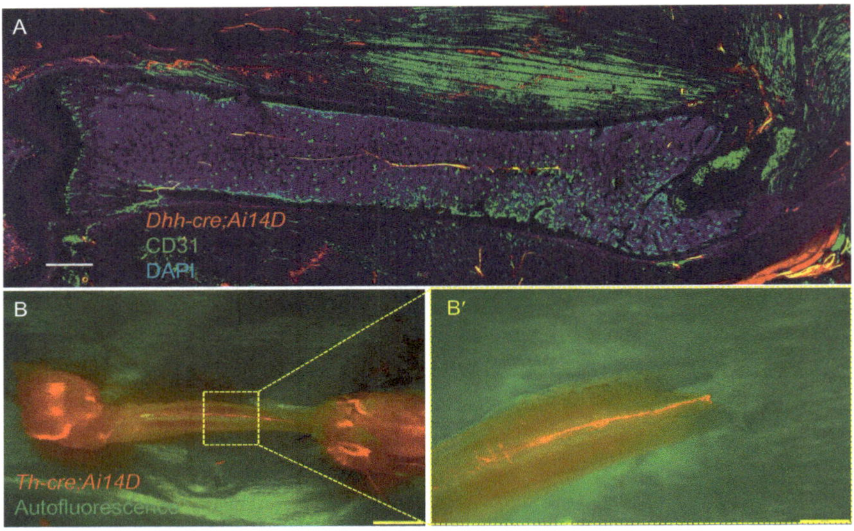

Fig. 1 BM innervation. (A) Immunofluorescence of $CD31^+$ blood vessels (*green*) in thick femoral BM section from 1-week-old *Dhh-cre;Ai14D* mouse with genetically labeled Schwann cells (*red*). Nuclei were counterstained with DAPI (*blue*). Scale, 500 μm. (B) Sympathetic noradrenergic fibers (*red*) genetically labeled in the femur from an adult *Th-cre;Ai14D* mouse. (B′) Higher magnification of the sympathetic noradrenergic fiber (*red*) boxed in B. The green autofluorescence of background is shown for better contrast. Scale, 2 mm.

3. PARASYMPATHETIC INNERVATION OF LYMPHOID ORGANS

The presence of parasympathetic innervation in the spleen has remained debated for many years, probably influenced by interspecies differences. In the beginning of 20th century Dale HH and Dudley HW isolated histamine and acetylcholine from the spleen of the ox and the horse, suggesting that the PNS might innervate the spleen. Subsequent studies showed that adrenergic and cholinergic drugs are able to induce changes in the spleen of the cod (Nilsson & Grove, 1974). Retrograde labeling experiments performed during the late 1990s suggested the existence of parasympathetic innervation in the spleen of the cat (Chen, Itoh, Sun, Miki, & Takeuchi, 1996). In rodents, one group has reported the presence of parasympathetic nerves entering the spleen based on retrograde tracing experiments (Branen et al., 2004; Buijs, van der Vliet, Garidou, Huitinga, & Escobar, 2008), but most groups working in this field doubt about the existence of parasympathetic fibers in the spleen

(Bellinger, Lorton, Hamill, Felten, & Felten, 1993; Cano, Sved, Rinaman, Rabin, & Card, 2001; Nance & Sanders, 2007), although some cholinergic fibers of spinal origin have been found around arterioles and in the white pulp of the spleen (Gautron et al., 2013). However, there is strong evidence for a cholinergic antiinflammatory pathway in the spleen initiated in the PNS (Huston et al., 2006; Rosas-Ballina et al., 2008; Vida, Pena, Deitch, & Ulloa, 2011). This pathway involves an indirect regulation by the vagus nerve over the celiac plexus, from where the splenic nerve emerges and innervates the spleen. While it was clear that the PNS regulates the immune response in the spleen, the splenic nerve is fundamentally catecholaminergic and most researchers believe that the spleen lacks significant PNS innervation. So what was the source of acetylcholine in the spleen? This enigma was solved a few years later, when an acetylcholine-producing T cell population was found in the spleen, which could produce the neurotransmitter required for the immune response in the absence of cholinergic nerve terminals (Rosas-Ballina et al., 2011). Along this line, the group finding evidence of the existence of PNS fibers in the spleen also reported recently an antiinflammatory role for this innervation (Kooijman, Meurs, et al., 2015). Nevertheless, this has generated significant controversy in the field between supporters and detractors of parasympathetic murine spleen innervation, regarding the stringency of the methods and the validity of the conclusions (Anderson, McKinley, Martelli, & McAllen, 2015; Kooijman, de Jonge, & Rensen, 2015).

Whereas the parasympathetic innervation of the spleen remains debated, up to now there has been almost no neuroanatomical evidence of a parasympathetic supply to the BM (Nance & Sanders, 2007). Among the scarce evidence suggesting a possible parasympathetic innervation in the BM is immunohistochemistry for choline acetyltransferase (ChAT) in the rat BM (Artico et al., 2002). Another study used the cholinergic marker vesicular acetylcholine transporter (VAChT), and retrograde tracing with pseudorabies virus and suggested the presence of PNS terminals transmitting bone-anabolic signals from the brain and derived from the sacral intermediolateral cell column and the central autonomic nucleus (Bajayo et al., 2012). However, the possibility of an indirect vagal input (as it occurs in the spleen) through the sympathetic postganglionary terminals innervating the BM cannot be excluded. Along this line, another group has also reported a bone-anabolic effect of PNS but in this case by regulating bone mass accrual indirectly, through the inhibition of brain sympathetic tone (Shi et al., 2010). These sympathetic fibers that regulate bone remodeling also innervate the BM, suggesting the possibility of hematopoietic regulation by the PNS at the central level.

Vasoactive intestinal peptide (VIP) is a neurotransmitter mostly associated with parasympathetic signaling. VIP has been identified in nerve fibers supplying the spleen and the BM. Only residual VIP-immunoreactive fibers have been found along splenic vasculature, where they might modulate lymphocyte function in the inflammatory response (Chevendra & Weaver, 1992). On the other hand, VIP+ fibers have been found in the BM and the thin tissue layer surrounding the bone (periosteum) in rats (Ahmed, Bjurholm, Kreicbergs, & Schultzberg, 1993). These VIP+ fibers appeared more abundant in the growth plate and the intervertebral discs. VIP receptor type 1 (VPAC1) has been detected in human megakaryocytes (Park, Olson, Ercal, Summers, & O'Dorisio, 1996), where an autocrine effect of VIP has been suggested to inhibit the proliferation and inducing the differentiation of megakaryocytes (Nam, Case, Hostager, & O'Dorisio, 2009). VPAC1 is also expressed in human HSPCs; however, VIP seems to have opposite effects on the proliferation and clonogenic capacity of HSPCs isolated from human BM or human umbilical cord blood (Kawakami et al., 2004; Rameshwar et al., 2002). It is also important to note that this neuropeptide is not solely present in parasympathetic terminals. A net of sympathetic VIP-immunoreactive nerve fibers innervate the periosteum and bone in several species of mammals (Hohmann, Elde, Rysavy, Einzig, & Gebhard, 1986). In contrast, VIP+ fibers in the spleen were not affected by sympathectomy (Bellinger et al., 1997), suggesting that these fibers are parasympathetic.

4. SYMPATHETIC INNERVATION OF LYMPHOID ORGANS

Sympathetic postganglionic terminals mainly use catecholamines as neurotransmitters. Catecholamines are synthesized from tyrosine, which is first transformed to dihydroxyphenylalanine by the rate-limiting enzyme, tyrosine hydroxylase (TH), and then to dopamine. Dopamine is loaded into vesicles inside sympathetic terminals and converted to noradrenaline by the enzyme β-hydroxylase (DBH). Although noradrenaline is the main neurotransmitter released by sympathetic fibers in the hematopoietic organs, noradrenaline can also be converted into adrenaline. Adrenaline has an important role as sympathetic effector in the humoral pathway (described later). The two main enzymes involved in noradrenaline synthesis, TH and DBH, have been classically used to label sympathetic fibers in the BM and the spleen.

As described earlier, the splenic nerve is the only widely accepted source of innervation in the spleen. The splenic nerve consists almost exclusively of postganglionic sympathetic axons in the different species studied (De Potter, Kurzawa, Miserez, & Coen, 1995; Felten, Ackerman, Wiegand, & Felten, 1987; Fried et al., 1986; Kinney, Cohen, & Felten, 1994; Saito, 1990). Noradrenergic terminals of the splenic nerve emerge from the celiac/superior mesenteric plexus ganglion (Rosas-Ballina et al., 2008) and branch out inside the spleen with the vasculature (Bellinger, Felten, Lorton, & Felten, 1989).

Similarly, sympathetic nerve fibers travel into the BM in association with blood vessels. The nutrient vein and artery penetrate the BM where they branch to form the central vein and artery. Radial arteries arising from the central artery connect with the sinuses broadly distributed in the marrow through smaller arterioles (Travlos, 2006). Some radial arteries branch into specific arterioles located in the endosteum, recently described as "transition zone vessels" (Kusumbe, Ramasamy, & Adams, 2014), because they connect arteriolar with sinusoidal vessels and exhibit mixed properties. Periosteal arteries can eventually cross the bone and connect with the transition zone vessels as well. Immunostaining for TH has shown that noradrenergic fibers mainly accompany arteriolar vessels entering the BM but once inside they sprout to innervate the sinusoids (Mach et al., 2002; Tabarowski, Gibson-Berry, & Felten, 1996). Therefore, endothelial cells, nerve terminals, and perivascular stromal cells are in intimate contact and abundant gap junctions have been detected among them. These observations led to the hypothesis that a "neuroreticular complex" might function as a synctitium-like unit in the BM (Yamazaki & Allen, 1990; Fig. 2).

Contrasting the broad evidence of neuronal-derived noradrenaline in lymphoid organs, the contribution of neuropeptide Y (NPY), another sympathetic neurotransmitter, has remained debated. Chemical sympathectomy experiments argue for a nonneuronal NPY source in the spleen (Ericsson et al., 1987; Romano, Felten, Felten, & Olschowka, 1991). In BM, several groups have reported the presence of some NPY+ nerve fibers (Mach et al., 2002; Tabarowski et al., 1996; Yamazaki & Allen, 1990).

Regardless of the source of NPY, recent studies have suggested a functional role for NPY in the BM. Previous studies on the effect of NPY in bone physiology reported opposite roles for NPY in bone formation depending on the type of receptor. Mice lacking Y2 receptor have increased

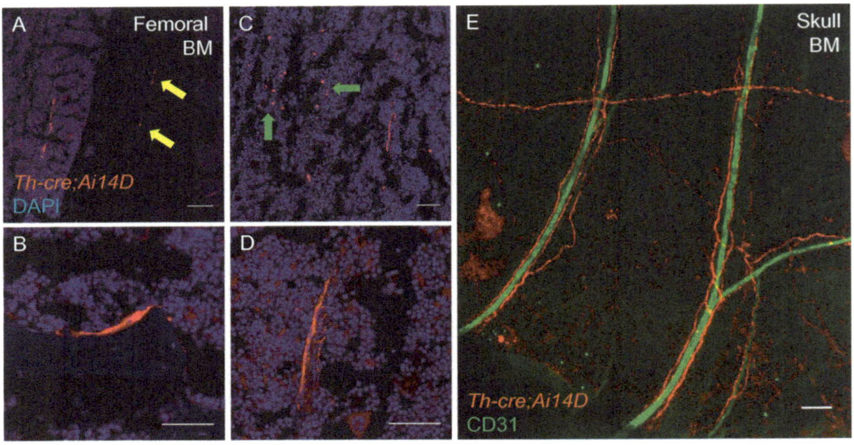

Fig. 2 Adrenergic sympathetic fibers are found in endosteal, nonendosteal, and periosteal locations. (A–D) Sympathetic noradrenergic fibers (*red*) genetically labeled in the femur from an adult *Th-cre;Ai14D* mouse (*red*) in periosteal (A, *yellow arrows*), endosteal (B), and nonendosteal (C and D) locations. Note the presence of some TH+ nonneuronal cells in the BM (C, *green arrows*). Nuclei are counterstained with DAPI (*blue*). (E) Immunofluorescence of $CD31^+$ blood vessels (*green*) in the skull BM of an adult *Th-cre;Ai14D* mouse with labeled sympathetic noradrenergic fibers (*red*). (A, C, and E) Scale, 100 μm. (B and D) Scale, 50 μm.

bone mass osteoprogenitor cells, suggesting a negative role for NPY on bone formation (Lee et al., 2015; Lundberg et al., 2007; Teixeira et al., 2009). On the other hand, NPY inhibits Y1 receptor expression and favors osteoblastic differentiation (Teixeira et al., 2009). Thus, NPY contribution in bone homeostasis would require be validated in further studies. NPY/Y1 signaling in macrophages confers neuroprotection and helps to restore BM dysfunction after chemotherapy-induced neural damage (Park et al., 2015). In addition, it is able to induce HSPC mobilization through Y1 receptor expression in osteoblasts and matrix metalloproteinase-9 (MMP-9) activity (Park et al., 2016).

5. TOWARD FUNCTIONALLY DEFINING THE ADRENERGIC REGULATION OF HEMATOPOIESIS

Although the studies described earlier provided neuroanatomical evidence of sympathetic BM innervation, the functional relevance of these fibers in hematopoiesis remained unknown until the 1990s. Maestroni et al. studied the role of pineal gland-derived hormones in the regulation of immunity. One such hormone with a key function in regulating circadian rhythm in humans is melatonin (Maestroni, 1993). Melatonin synthesis is under the regulation of the SNS, so these authors examined the effect of adrenergic drugs on hematopoiesis. They found that chemical sympathectomy and adrenergic drugs affected the hematopoietic reconstitution (measured as peripheral blood leukocytes) following myeloablation (Maestroni, Conti, & Pedrinis, 1992). Moreover, noradrenaline had protective effects against the chemotherapeutic agent carboplatin (Maestroni, Togni, & Covacci, 1997). These effects were mostly explained by direct effects dependent on alpha1-adrenergic receptors expressed in the hematopoitic cells (Maestroni & Conti, 1994a). Adrenergic antagonists (α1-adrenergic and β-adrenergic blockers) were able to enhance myelopoiesis in the BM and the spleen, whereas they dampened lymphopoiesis. Further studies demonstrated that this effect is mediated by α1-adrenergic receptor in hematopoietic cells, specifically in pre-B cells, and involves TGF-β signaling (Maestroni & Conti, 1994b). Although most of these experiments were performed in vitro or under myeloablative conditions, the same group also measured sympathetic neurotransmitters (noradrenaline and dopamine) in the BM of mice over 24 h. These authors found both neural and nonneural contributions to BM catechomalines and hypothesized that circadian oscillations in the levels of these neurotransmitters might have implications in hematopoiesis. Therefore, these pioneering studies established the first possible functional links between the SNS and hematopoiesis.

6. DAY–NIGHT OSCILLATIONS OF HEMATOPOIESIS

Hematopoietic cells, including HSPC, are in constant circulation between the BM and the periphery. This hematopoietic traffic does not only contribute to homogenize the hematopoietic process in different compartments, but it also helps to regenerate the HSC pool contributing to the maintenance of hematopoiesis throughout life (Abkowitz, Robinson, Kale, Long, & Chen, 2003; Smith, Weissman, & Heimfeld, 1991). Importantly, some studies at the beginning of this century reported that this "normal traffic" could be altered using certain cytokines and factors. This raised the possibility of use this approach to "artificially move" HSPCs into the bloodstream (a process known as mobilization, in contrast of the normal HSPC egress) and facilitate HSPC collection by apheresis (Chen et al., 2006).

Among all the mobilizing agents there is one that has been studied and used much more, to finally become the gold-standard method to mobilize HSPCs for transplantation procedures: the granulocyte-colony stimulating factor (G-CSF) (Molineux, Pojda, Hampson, Lord, & Dexter, 1990). However, subsequent experiments using chimeras have demonstrated that G-CSF does not act directly on HSPCs, but multiple indirect effects are required for efficient G-CSF-induced HSC mobilization (Liu, Poursine-Laurent, & Link, 2000). G-CSF increases the levels of BM proteases, like MMP-9, cathepsin G, and other serine proteases. These proteases can degrade key chemoattractant molecules involved in HSC maintenance, like the stromal cell-derived factor 1 (SDF-1/CXCL12) and its receptor CXCR4 (Kollet et al., 2006; Levesque, Hendy, Takamatsu, Simmons, & Bendall, 2003). However, protease activity disruption is not enough to abrogate G-CSF-mediated CXCL12 decrease and HSPC mobilization (Levesque et al., 2004), suggesting the participation of other signals in G-CSF-induced HSC mobilization.

Some of these signals were unexpectedly found to be derived from the nervous system. Galactocerebrosides are glycolipids present in the myelin sheaths of Schwann cells in myelinating fibers of the ANS. The enzyme UDP-galactose:ceramide galactosyl-transferase (Ctg) is responsible for the transfer of UDP-galactose to ceramide forming galactosylceramide, which is de main galactocerebroside found in myelin. Katayama et al. reported that G-CSF-induced mobilization was impaired in $Ctg^{-/-}$ mouse model, despite normal BM protease. The HSC mobilization defect was correlated with suppressed bone-lining osteoblasts and decreased Cxcl12 expression due to impaired sympathetic transmission to the BM (Katayama et al., 2006). This work expanded the previously described adrenergic regulation of osteoblasts during bone remodeling (Fu, Patel, Bradley, Wagner, & Karsenty, 2005) to the relevance of this regulation for HPSC mobilization. Noradrenergic signals do not only act on osteoblaststic cells, but have also direct on the proliferation, migration and engraftment of human HSPC (CD34+ cells) through β2-adrenergic receptor expressed in these cells (Spiegel et al., 2007).

The rate of DNA synthesis and progenitor activity (colony-forming units in culture) in the murine and human BM exhibits circadian oscillations (Aardal & Laerum, 1983; Smaaland, Sothern, Laerum, & Abrahamsen, 2002) and also rely on β-adrenergic signaling (Byron, 1972). However, the potential mechanisms underlying these circadian fluctuations were unknown. Earlier studies had shown that circulating catecholamine levels undergo circadian fluctuations, suggesting that sympathetic activity follows circadian rhythms (Sauerbier & von Mayersbach, 1977). Indeed, sympathetic metabolites (noradrenaline and dopamine) were later found to follow a daily rhythmicity with maximum values reached during the night (Maestroni et al., 1998). Since the SNS transmits signals from the suprachiasmatic nucleus—the central pacemaker of circadian rhythms at organism level—to peripheral organs as the liver (Terazono et al., 2003), we studied the possible physiological regulation of HSCs by the CNS through the SNS.

We showed that HSC egress to the bloodstream follows circadian oscillations dependent on sympathetic signaling (Mendez-Ferrer, Lucas, Battista, & Frenette, 2008). Postganglionic sympathetic terminals release NE in the BM, activating the β3-adrenergic receptor expressed on particular stromal cells that we later identified as BMSCs (Mendez-Ferrer et al., 2010). This activation caused Cxcl12 downregulation in BM stromal cells. This pathway allows to explain how light signals transmitted from the retina to the suprachiasmatic nucleus can be passed on to the BM microenvironment. Circulating HSPCs fluctuate in antiphase with CXCL2 expression and both oscillations were dramatically altered when light periods and/or sympathetic signaling were disrupted (Mendez-Ferrer et al., 2008).

These circadian oscillations are also present in HSPC human traffic but inverted in time, since mice are noctural and humans are diurnal (Lucas, Battista, Shi, Isola, & Frenette, 2008). Therefore, HSCs are preferentially released into the bloodstream during the resting period in both species. Although the functional role of circulating HSCs remains to be established, HSC traffic allows for homogeneous hematopoiesis in different compartments, and it might also help to regenerate HSC niches (Abkowitz et al., 2003; Smith et al., 1991). Another potential advantage of circulating HSCs might be their "on demand" hematopoietic differentiation on peripheral tissues which they infiltrate (Massberg et al., 2007).

7. SYMPATHETIC SIGNALING IN HEMATOLOGICAL MALIGNANCIES

The chemotherapy used to treat cancer can damage normal cells, including those forming the microenvironment. Especially sensitive were found to be sympathetic nerve fibers. Damage to these fibers impaired BM regeneration and its rescue by neurotrophic factors or deletion of Trp53 in sympathetic fibers improved hematopoietic regeneration (Lucas et al., 2013).

Neuropathy has also been associated with impaired hematopoiesis in other scenarios. A significant reduction of BM innervation has been found in immune diseases, such as diabetes mellitus (Ozeran, Wagner, Reimer, & Hill, 1970). This sympathetic neuropathy has been associated with reduced circulating endothelial progenitor cells (EPCs) and inflammatory alterations related to the generation and release of monocytes in murine and human models (Busik et al., 2009; Hu et al., 2013). Not only EPCs but also HSCs are poorly mobilized in response to G-CSF in diabetic mice and patients. Abnormal sympathetic signaling underlies this HSPC mobilization defect in diabetes (Albiero et al., 2014; Ferraro et al., 2011).

We have provided evidence that chronic BM inflammation found in myeloproliferative neoplasms causes a local neuropathy that stimulates disease progression (Arranz et al., 2014). Using a mouse model expressing the human Janus Kinase 2 mutation (JAK2 V617F) in HSCs and human MPN samples, we found a quick reduction in sympathetic fibers, their associated Schwann cells, and nestin+ BMSCs (and thus HSPC maintenance factors as Cxcl12). This stromal damage was caused by an inflammatory storm initiated by mutant HSCs through interleukin-1β signaling. Rescue experiments with β3-adrenergic agonists managed to block MPN progression in the mouse model correlated with niche rescue and restoration of adrenergic signaling on nestin+ BMSCs. This study suggests that neural BM components might be therapeutic targets in myeloproliferative neoplasms.

Sympathetic neuropathy has also been reported in the pathogenesis of another myeloproliferative disorder: acute myeloid leukemia (AML) (Hanoun et al., 2014). In this case, the β2-adrenergic receptor seemed to underlie some of the microenvironmental alterations, including increased number of osteoblastic biased-nestin+ BMSCs (yet osteoblast numbers were decreased). However, neural interventions did not significantly affect AML evolution in this study.

Recently, Chen et al. have measured the expression levels and clinical significances of nerve-related molecules and T helper-associated molecules (IL-17 and Foxp3) in the BM of AML patients (Chen et al., 2016). Intriguingly, they showed some interesting correlations in line with that proposed previously. Expression levels of nestin, TH, GFAP, and IL-17 were reduced, while Foxp3 and Foxp3/IL-17 ratios were increased. They also associate higher survival of AML patients with high expression of nestin in the BM. This study does not only confirm previous publications but it involves T helper immunology via Foxp3/IL-17 signaling in neuropathy associated to AML.

8. SENSORY INNERVATION OF LYMPHOID ORGANS

Not only parasympathetic and sympathetic fibers have been reported to innervate lymphoid organs, but also sensory afferent terminals have been described in these organs. Most sensory neurons have their bodies in the dorsal root ganglion of the spinal nerve, although some sensory neurons are part of the vagus nerve. Sensory fibers are more complex at the molecular level, what hampers their classification (Lallemend & Ernfors, 2012; Marmigere & Ernfors, 2007; Usoskin et al., 2015). Those found in the BM mainly contain peptidergic neurotransmitters, as calcitonin gene-related peptide (CGRP) and Substance P. Classically, it had been thought that BM sensory fibers basically transmitted nociceptive information from the periphery to the CNS (Bjurholm, Kreicbergs, Brodin, & Schultzberg, 1988). However, it is now recognized the role of sensory neuropeptides as modulators of hematopoiesis. Sensory innervation is also present in the spleen, where it is involved in multiple process related to the immune response. These aspects, which are outside of the scope of this work, have been recently reviewed (Jung, Levesque, & Ruitenberg, 2017).

Substance P is the most notable member of neuropeptides family called Tachykinin and binds to the three tachykinin receptors: NK-1R, NK-2R, and NK-3R. Initial in vitro evidences in late 1980s indicated the influence of Substance P on myelopoiesis in the inflammatory context (Matsuda, Kawakita, Kiso, Nakano, & Kitamura, 1989; Moore, Osmand, Dunn, Joshi, & Rouse, 1988). Substance P has stimulatory effect on hematopoietic progenitors (erythroid and granulocytic progenitors) in short-term methylcellulose BM cultures. This effect might be mediated through the release of hematopoietic growth factors like IL-1, IL-3, IL-6, and granulocyte-macrophage colony-stimulating factor (GM-CSF), and via adherent BM cells (Rameshwar, Ganea, & Gascon, 1993, 1994). In addition, Substance P might act directly on BM stromal cells by altering the production of IL-7 and stem cell factor (SCF) (Manske, Sullivan, & Andersen, 1995). However, both Substance P and neurokinin-A (another member of tachykinins) can be released by stromal cells, suggesting a possible nonneural source. Neurokinin-A, in addition, inhibits myelopoiesis through MIP-1α and TGF-β signaling (Rameshwar & Gascon, 1996). Lately, a work from the same group has shown that increased levels of Cxcl12 can lead to Substance P production in BMSCs (Corcoran, Patel, & Rameshwar, 2007). This means another direct link between a different neurotransmitter and the key HSC maintenance chemokine in the BM.

Some recent studies have suggested a protective role of Substance P in BM recovery after irradiation and BMT. Substance P might reduce radiation-derived damage in mouse and human BMSCs by targeting an apoptotic pathway (An et al., 2011). Another recent study has reported that the genetic deletion of NK-1R impairs lymphoid reconstitution after BMT, suggesting that sensory fibers could regulate hematopoiesis (Berger et al., 2013). Finally, a recent study has linked Substance P with impaired HSC mobilization in diabetes. Nociceptive neuropathy was accompanied by reduced mobilization of NK-1R-expressing HSPCs upon G-CSF treatment (Dang et al., 2015).

Substance P might be involved also in malignant hematopoiesis. Several studies suggest a role for Substance P in childhood acute lymphoblastic leukemia (Nowicki & Miskowiak, 2003; Nowicki, Ostalska-Nowicka, Konwerska, & Miskowiak, 2006).

Instead, CGRP might inhibit early B cell development through both direct effects and indirect regulation by BMSCs (Fernandez, Knopf, & McGillis, 2000; McGillis, Rangnekar, & Ciallella, 1995). In vivo experiments validated these in vitro results and provided evidence of a regulatory role for CGRP in early B cell development in the intact animal (Schlomer, Storey, Ciornei, & McGillis, 2007). Several studies implicated CGRP in myelopoiesis as well. Capsaicin-induced sensory denervation resulted in a considerable reduction in CFU-GM activity in the murine BM (Broome, Whetton, & Miyan, 2000). In addition, human HSPCs (CD34+ cells) were able to respond directly to CGRP by undergoing myeloid differentiation (Harzenetter et al., 2002).

9. CONCLUSIONS AND FUTURES PERSPECTIVES

Accumulating data over the last three decades have contributed to decipher some connections between the CNS and several hematopoietic organs that allows for tissue responses integrated with systemic needs. The CNS exchanges information with lymphoid organs through peripheral innervation and secreted neuroendocrine factors.

In this chapter, we have reviewed how the CNS can modulate the functionality of the BM and the spleen, two of the main lymphoid organs. Although classical studies from the last century described the patterns of innervation in the lymphoid organs, it was not until the beginning of the

21st century that functional assays exposed the real neural regulation of hematopoiesis. However, most studies have been focused on the sympathetic regulation of BM HSC niches. We have summarized how sympathetic efferent terminals can regulate different aspects of HSC function, like development, traffic, proliferation, or malignant behavior. However, the possible relevance of other types of innervation (i.e., parasympathetic and sensory) remains much less explored. Similarly, the possible contribution of nerve-associated Schwann cells in other pathological scenarios remains unknown.

Although we have focused on the neural regulation, it is broadly accepted that neurotransmitters can be nonsynaptically released in the BM microenvironment (Vizi & Elenkov, 2002). To understand how different cell populations can integrate these signals in a spatiotemporal context is something still elusive. On the other hand, it is known that neural fibers consist of multiple axons, each of which can innervate a different target and thus, induce different effects in several populations at the same time. Emerging technologies, like optogenetics, might facilitate the specific neuromodulation of innervated organs.

Finally, based on the published literature, we have discussed how the CNS modulates hematopoietic function, but we cannot ignore the bidirectional nature of this connection. For instance, Klein and Rubin reviewed the parallel roles of CXCL12-CXCR4 axis in the immune and nervous system (Klein & Rubin, 2004). Like in the hematopoiesis system, this pathway regulates cellular movement, proliferation, plasticity and survival of neurons. Indeed, CXCL2-CXCR4 is now considered as a hub point to integrate homeostatic neuroimmune functions. On the other hand, many studies underscore the role of the brain as a receiver and integrator element of peripheral inflammatory stimuli. Can afferent terminals similarly gather information regarding hematopoietic function? Can CNS responses be adjusted to these stimuli? Future studies along these lines will increase our understanding of neural-hematopoietic connection in health and disease.

ACKNOWLEDGMENTS

A.G.G. is the recipient of fellowships from the Ramón Areces and La Caixa Foundations. This work was supported by core support grants from the Wellcome Trust and MRC to the Cambridge Stem Cell Institute, the Spanish Ministry of Economy and Competitiveness (SAF-2011-30308), Pro-CNIC Foundation, Severo Ochoa Center of Excellence award SEV-2015-0505 to CNIC, TerCel (Spanish Cell Therapy Network), Ramón y Cajal Program Grant RYC-2009-04703, Marie Curie Career Integration Grants FP7-PEOPLE-2011-RG-294096 and H2020-MSCA-IF-2015-708411, ConSEPOC-Comunidad de Madrid

S2010/BMD-2542, National Health Institute Blood and Transplant (United Kingdom), and Horizon2020 ERC-2014-CoG-64765 Grant to S.M.-F., who was also supported in part by an International Early Career Scientist grant of the Howard Hughes Medical Institute.

The authors truly apologize for the omission of relevant literature due to space constrictions.

Authorships: A.G.G. prepared the figures and wrote the manuscript. S.M.-F. wrote the manuscript. The authors have no conflict of interest to declare.

REFERENCES

Aardal, N. P., & Laerum, O. D. (1983). Circadian variations in mouse bone marrow. *Experimental Hematology, 11*(9), 792–801.

Abkowitz, J. L., Robinson, A. E., Kale, S., Long, M. W., & Chen, J. (2003). Mobilization of hematopoietic stem cells during homeostasis and after cytokine exposure. *Blood, 102*(4), 1249–1253. http://dx.doi.org/10.1182/blood-2003-01-0318.

Ahmed, M., Bjurholm, A., Kreicbergs, A., & Schultzberg, M. (1993). Neuropeptide Y, tyrosine hydroxylase and vasoactive intestinal polypeptide-immunoreactive nerve fibers in the vertebral bodies, discs, dura mater, and spinal ligaments of the rat lumbar spine. *Spine (Phila Pa 1976), 18*(2), 268–273.

Albiero, M., Poncina, N., Tjwa, M., Ciciliot, S., Menegazzo, L., Ceolotto, G., … Fadini, G. P. (2014). Diabetes causes bone marrow autonomic neuropathy and impairs stem cell mobilization via dysregulated p66Shc and Sirt1. *Diabetes, 63*(4), 1353–1365. http://dx.doi.org/10.2337/db13-0894.

An, Y. S., Lee, E., Kang, M. H., Hong, H. S., Kim, M. R., Jang, W. S., … Yi, J. Y. (2011). Substance P stimulates the recovery of bone marrow after the irradiation. *Journal of Cellular Physiology, 226*(5), 1204–1213. http://dx.doi.org/10.1002/jcp.22447.

Anderson, C., McKinley, M., Martelli, D., & McAllen, R. (2015). Letter to the editor: Parasympathetic innervation of the rodent spleen? *American Journal of Physiology. Heart and Circulatory Physiology, 309*(12), H2158. http://dx.doi.org/10.1152/ajpheart.00766.2015.

Arranz, L., Sanchez-Aguilera, A., Martin-Perez, D., Isern, J., Langa, X., Tzankov, A., … Mendez-Ferrer, S. (2014). Neuropathy of haematopoietic stem cell niche is essential for myeloproliferative neoplasms. *Nature, 512*(7512), 78–81. http://dx.doi.org/10.1038/nature13383.

Artico, M., Bosco, S., Cavallotti, C., Agostinelli, E., Giuliani-Piccari, G., Sciorio, S., … Vitale, M. (2002). Noradrenergic and cholinergic innervation of the bone marrow. *International Journal of Molecular Medicine, 10*(1), 77–80.

Bajayo, A., Bar, A., Denes, A., Bachar, M., Kram, V., Attar-Namdar, M., … Bab, I. (2012). Skeletal parasympathetic innervation communicates central IL-1 signals regulating bone mass accrual. *Proceedings of the National Academy of Sciences of the United States of America, 109*(38), 15455–15460. http://dx.doi.org/10.1073/pnas.1206061109.

Bellinger, D. L., Felten, S. Y., Lorton, D., & Felten, D. L. (1989). Origin of noradrenergic innervation of the spleen in rats. *Brain, Behavior, and Immunity, 3*(4), 291–311.

Bellinger, D. L., Lorton, D., Hamill, R. W., Felten, S. Y., & Felten, D. L. (1993). Acetylcholinesterase staining and choline acetyltransferase activity in the young adult rat spleen: Lack of evidence for cholinergic innervation. *Brain, Behavior, and Immunity, 7*(3), 191–204. http://dx.doi.org/10.1006/brbi.1993.1021.

Bellinger, D. L., Lorton, D., Horn, L., Brouxhon, S., Felten, S. Y., & Felten, D. L. (1997). Vasoactive intestinal polypeptide (VIP) innervation of rat spleen, thymus, and lymph nodes. *Peptides, 18*(8), 1139–1149.

Berger, A., Frelin, C., Shah, D. K., Benveniste, P., Herrington, R., Gerard, N. P., ... Paige, C. J. (2013). Neurokinin-1 receptor signalling impacts bone marrow repopulation efficiency. *PloS One*, *8*(3). e58787. http://dx.doi.org/10.1371/journal.pone.0058787

Bjurholm, A., Kreicbergs, A., Brodin, E., & Schultzberg, M. (1988). Substance P- and CGRP-immunoreactive nerves in bone. *Peptides*, *9*(1), 165–171.

Branen, L., Hovgaard, L., Nitulescu, M., Bengtsson, E., Nilsson, J., & Jovinge, S. (2004). Inhibition of tumor necrosis factor-alpha reduces atherosclerosis in apolipoprotein E knockout mice. *Arteriosclerosis, Thrombosis, and Vascular Biology*, *24*(11), 2137–2142. http://dx.doi.org/10.1161/01.ATV.0000143933.20616.1b.

Broome, C. S., Whetton, A. D., & Miyan, J. A. (2000). Neuropeptide control of bone marrow neutrophil production is mediated by both direct and indirect effects on CFU-GM. *British Journal of Haematology*, *108*(1), 140–150.

Buijs, R. M., van der Vliet, J., Garidou, M. L., Huitinga, I., & Escobar, C. (2008). Spleen vagal denervation inhibits the production of antibodies to circulating antigens. *PloS One*, *3*(9). e3152. http://dx.doi.org/10.1371/journal.pone.0003152.

Busik, J. V., Tikhonenko, M., Bhatwadekar, A., Opreanu, M., Yakubova, N., Caballero, S., ... Grant, M. B. (2009). Diabetic retinopathy is associated with bone marrow neuropathy and a depressed peripheral clock. *The Journal of Experimental Medicine*, *206*(13), 2897–2906. http://dx.doi.org/10.1084/jem.20090889.

Byron, J. W. (1972). Evidence for α-adrenergic receptor initiating DNA synthesis in haemopoietic stem cells. *Experimental Cell Research*, *71*(1), 228–232.

Calvo, W. (1968). The innervation of the bone marrow in laboratory animals. *The American Journal of Anatomy*, *123*(2), 315–328. http://dx.doi.org/10.1002/aja.1001230206.

Calvo, W., & Haas, R. J. (1969). On the histogenesis of the bone marrow in the rat. Innervation, stroma and their relations to hemopoiesis. *Zeitschrift für Zellforschung und Mikroskopische Anatomie*, *95*(3), 377–395.

Cano, G., Sved, A. F., Rinaman, L., Rabin, B. S., & Card, J. P. (2001). Characterization of the central nervous system innervation of the rat spleen using viral transneuronal tracing. *The Journal of Comparative Neurology*, *439*(1), 1–18. http://dx.doi.org/10.1002/cne.1331.

Chen, X. H., Itoh, M., Sun, W., Miki, T., & Takeuchi, Y. (1996). Localization of sympathetic and parasympathetic neurons innervating pancreas and spleen in the cat. *Journal of the Autonomic Nervous System*, *59*(1–2), 12–16.

Chen, J., Larochelle, A., Fricker, S., Bridger, G., Dunbar, C. E., & Abkowitz, J. L. (2006). Mobilization as a preparative regimen for hematopoietic stem cell transplantation. *Blood*, *107*(9), 3764–3771. http://dx.doi.org/10.1182/blood-2005-09-3593.

Chen, C., Liu, Y., Hua, M., Li, X., Ji, C., & Ma, D. (2016). Neuropathy correlated with imbalanced Foxp3/IL-17 in bone marrow microenvironment of patients with acute myeloid leukemia. *Oncotarget*, 7(17), 24455–24465. http://dx.doi.org/10.18632/oncotarget.8227.

Chevendra, V., & Weaver, L. C. (1992). Distributions of neuropeptide Y, vasoactive intestinal peptide and somatostatin in populations of postganglionic neurons innervating the rat kidney, spleen and intestine. *Neuroscience*, *50*(3), 727–743.

Corcoran, K. E., Patel, N., & Rameshwar, P. (2007). Stromal derived growth factor-1alpha: Another mediator in neural-emerging immune system through Tac1 expression in bone marrow stromal cells. *Journal of Immunology*, *178*(4), 2075–2082.

Dang, Z., Maselli, D., Spinetti, G., Sangalli, E., Carnelli, F., Rosa, F., ... Madeddu, P. (2015). Sensory neuropathy hampers nociception-mediated bone marrow stem cell release in mice and patients with diabetes. *Diabetologia*, *58*(11), 2653–2662. http://dx.doi.org/10.1007/s00125-015-3735-0.

De Potter, W. P., Kurzawa, R., Miserez, B., & Coen, E. P. (1995). Evidence against differential release of noradrenaline, neuropeptide Y, and dopamine-beta-hydroxylase from adrenergic nerves in the isolated perfused sheep spleen. *Synapse*, *19*(2), 67–76. http://dx.doi.org/10.1002/syn.890190202.

Dzierzak, E., & Speck, N. A. (2008). Of lineage and legacy: The development of mammalian hematopoietic stem cells. *Nature Immunology*, *9*(2), 129–136. http://dx.doi.org/10.1038/ni1560.
Ericsson, A., Schalling, M., McIntyre, K. R., Lundberg, J. M., Larhammar, D., Seroogy, K., ... Persson, H. (1987). Detection of neuropeptide Y and its mRNA in megakaryocytes: Enhanced levels in certain autoimmune mice. *Proceedings of the National Academy of Sciences of the United States of America*, *84*(16), 5585–5589.
Felten, D. L., Ackerman, K. D., Wiegand, S. J., & Felten, S. Y. (1987). Noradrenergic sympathetic innervation of the spleen: I. Nerve fibers associate with lymphocytes and macrophages in specific compartments of the splenic white pulp. *Journal of Neuroscience Research*, *18*(1), 28–36. http://dx.doi.org/10.1002/jnr.490180107 118–121.
Fernandez, S., Knopf, M. A., & McGillis, J. P. (2000). Calcitonin-gene related peptide (CGRP) inhibits interleukin-7-induced pre-B cell colony formation. *Journal of Leukocyte Biology*, *67*(5), 669–676.
Ferraro, F., Lymperi, S., Mendez-Ferrer, S., Saez, B., Spencer, J. A., Yeap, B. Y., ... Scadden, D. T. (2011). Diabetes impairs hematopoietic stem cell mobilization by altering niche function. *Science Translational Medicine*, *3*(104), 104ra101. http://dx.doi.org/10.1126/scitranslmed.3002191. 3/104/104ra101 [pii].
Fitch, S. R., Kimber, G. M., Wilson, N. K., Parker, A., Mirshekar-Syahkal, B., Gottgens, B., ... Ottersbach, K. (2012). Signaling from the sympathetic nervous system regulates hematopoietic stem cell emergence during embryogenesis. *Cell Stem Cell*, *11*(4), 554–566. http://dx.doi.org/10.1016/j.stem.2012.07.002.
Fonseca-Pereira, D., Arroz-Madeira, S., Rodrigues-Campos, M., Barbosa, I. A., Domingues, R. G., Bento, T., ... Veiga-Fernandes, H. (2014). The neurotrophic factor receptor RET drives haematopoietic stem cell survival and function. *Nature*, *514*(7520), 98–101. http://dx.doi.org/10.1038/nature13498.
Fried, G., Terenius, L., Brodin, E., Efendic, S., Dockray, G., Fahrenkrug, J., ... Hokfelt, T. (1986). Neuropeptide Y, enkephalin and noradrenaline coexist in sympathetic neurons innervating the bovine spleen. Biochemical and immunohistochemical evidence. *Cell and Tissue Research*, *243*(3), 495–508.
Fu, L., Patel, M. S., Bradley, A., Wagner, E. F., & Karsenty, G. (2005). The molecular clock mediates leptin-regulated bone formation. *Cell*, *122*(5), 803–815.
Gautron, L., Rutkowski, J. M., Burton, M. D., Wei, W., Wan, Y., & Elmquist, J. K. (2013). Neuronal and nonneuronal cholinergic structures in the mouse gastrointestinal tract and spleen. *The Journal of Comparative Neurology*, *521*(16), 3741–3767. http://dx.doi.org/10.1002/cne.23376.
Hanoun, M., Zhang, D., Mizoguchi, T., Pinho, S., Pierce, H., Kunisaki, Y., ... Frenette, P. S. (2014). Acute myelogenous leukemia-induced sympathetic neuropathy promotes malignancy in an altered hematopoietic stem cell niche. *Cell Stem Cell*, *15*(3), 365–375. http://dx.doi.org/10.1016/j.stem.2014.06.020.
Harzenetter, M. D., Keller, U., Beer, S., Riedl, C., Peschel, C., & Holzmann, B. (2002). Regulation and function of the CGRP receptor complex in human granulopoiesis. *Experimental Hematology*, *30*(4), 306–312.
Hohmann, E. L., Elde, R. P., Rysavy, J. A., Einzig, S., & Gebhard, R. L. (1986). Innervation of periosteum and bone by sympathetic vasoactive intestinal peptide-containing nerve fibers. *Science*, *232*(4752), 868–871.
Hu, P., Thinschmidt, J. S., Yan, Y., Hazra, S., Bhatwadekar, A., Caballero, S., ... Grant, M. B. (2013). CNS inflammation and bone marrow neuropathy in type 1 diabetes. *The American Journal of Pathology*, *183*(5), 1608–1620. http://dx.doi.org/10.1016/j.ajpath.2013.07.009.
Huston, J. M., Ochani, M., Rosas-Ballina, M., Liao, H., Ochani, K., Pavlov, V. A., ... Ulloa, L. (2006). Splenectomy inactivates the cholinergic antiinflammatory pathway

during lethal endotoxemia and polymicrobial sepsis. *The Journal of Experimental Medicine, 203*(7), 1623–1628. http://dx.doi.org/10.1084/jem.20052362.

Isern, J., Garcia-Garcia, A., Martin, A. M., Arranz, L., Martin-Perez, D., Torroja, C., … Mendez-Ferrer, S. (2014). The neural crest is a source of mesenchymal stem cells with specialized hematopoietic stem cell niche function. *eLife3*. e03696. http://dx.doi.org/10.7554/eLife.03696

Jung, W. C., Levesque, J. P., & Ruitenberg, M. J. (2017). It takes nerve to fight back: The significance of neural innervation of the bone marrow and spleen for immune function. *Seminars in Cell & Developmental Biology, 61*, 60–70. http://dx.doi.org/10.1016/j.semcdb.2016.08.010.

Katayama, Y., Battista, M., Kao, W. M., Hidalgo, A., Peired, A. J., Thomas, S. A., & Frenette, P. S. (2006). Signals from the sympathetic nervous system regulate hematopoietic stem cell egress from bone marrow. *Cell, 124*(2), 407–421. http://dx.doi.org/10.1016/j.cell.2005.10.041.

Kawakami, M., Kimura, T., Kishimoto, Y., Tatekawa, T., Baba, Y., Nishizaki, T., … Soma, T. (2004). Preferential expression of the vasoactive intestinal peptide (VIP) receptor VPAC1 in human cord blood-derived CD34+CD38- cells: Possible role of VIP as a growth-promoting factor for hematopoietic stem/progenitor cells. *Leukemia, 18*(5), 912–921. http://dx.doi.org/10.1038/sj.leu.2403330.

Kinney, K. S., Cohen, N., & Felten, S. Y. (1994). Noradrenergic and peptidergic innervation of the amphibian spleen: Comparative studies. *Developmental and Comparative Immunology, 18*(6), 511–521.

Klein, R. S., & Rubin, J. B. (2004). Immune and nervous system CXCL12 and CXCR4: Parallel roles in patterning and plasticity. *Trends in Immunology, 25*(6), 306–314. http://dx.doi.org/10.1016/j.it.2004.04.002.

Kollet, O., Dar, A., Shivtiel, S., Kalinkovich, A., Lapid, K., Sztainberg, Y., … Lapidot, T. (2006). Osteoclasts degrade endosteal components and promote mobilization of hematopoietic progenitor cells. *Nature Medicine, 12*(6), 657–664.

Kooijman, S., de Jonge, W. J., & Rensen, P. C. (2015). Reply to "Letter to the editor: Parasympathetic innervation of the rodent spleen?" *American Journal of Physiology Heart and Circulatory Physiology, 309*(12), H2159. http://dx.doi.org/10.1152/ajpheart.00805.2015.

Kooijman, S., Meurs, I., van Beek, L., Khedoe, P. P., Giezekamp, A., Pike-Overzet, K., … Rensen, P. C. (2015). Splenic autonomic denervation increases inflammatory status but does not aggravate atherosclerotic lesion development. *American Journal of Physiology Heart and Circulatory Physiology, 309*(4), H646–H654. http://dx.doi.org/10.1152/ajpheart.00787.2014.

Kusumbe, A. P., Ramasamy, S. K., & Adams, R. H. (2014). Coupling of angiogenesis and osteogenesis by a specific vessel subtype in bone. *Nature, 507*(7492), 323–328. http://dx.doi.org/10.1038/nature13145.

Lallemend, F., & Ernfors, P. (2012). Molecular interactions underlying the specification of sensory neurons. *Trends in Neurosciences, 35*(6), 373–381. http://dx.doi.org/10.1016/j.tins.2012.03.006.

Lee, N. J., Nguyen, A. D., Enriquez, R. F., Luzuriaga, J., Bensellam, M., Laybutt, R., … Herzog, H. (2015). NPY signalling in early osteoblasts controls glucose homeostasis. *Molecular Metabolism, 4*(3), 164–174. http://dx.doi.org/10.1016/j.molmet.2014.12.010.

Levesque, J. P., Hendy, J., Takamatsu, Y., Simmons, P. J., & Bendall, L. J. (2003). Disruption of the CXCR4/CXCL12 chemotactic interaction during hematopoietic stem cell mobilization induced by GCSF or cyclophosphamide. *The Journal of Clinical Investigation, 111*(2), 187–196.

Levesque, J. P., Liu, F., Simmons, P. J., Betsuyaku, T., Senior, R. M., Pham, C., & Link, D. C. (2004). Characterization of hematopoietic progenitor mobilization in protease-deficient mice. *Blood, 104*(1), 65–72.

Liu, F., Poursine-Laurent, J., & Link, D. C. (2000). Expression of the G-CSF receptor on hematopoietic progenitor cells is not required for their mobilization by G-CSF. *Blood*, *95*(10), 3025–3031.

Lucas, D., Battista, M., Shi, P. A., Isola, L., & Frenette, P. S. (2008). Mobilized hematopoietic stem cell yield depends on species-specific circadian timing. *Cell Stem Cell*, *3*(4), 364–366. http://dx.doi.org/10.1016/j.stem.2008.09.004.

Lucas, D., Scheiermann, C., Chow, A., Kunisaki, Y., Bruns, I., Barrick, C., ... Frenette, P. S. (2013). Chemotherapy-induced bone marrow nerve injury impairs hematopoietic regeneration. *Nature Medicine*, *19*(6), 695–703. http://dx.doi.org/10.1038/nm.3155.

Lundberg, P., Allison, S. J., Lee, N. J., Baldock, P. A., Brouard, N., Rost, S., ... Herzog, H. (2007). Greater bone formation of Y2 knockout mice is associated with increased osteoprogenitor numbers and altered Y1 receptor expression. *The Journal of Biological Chemistry*, *282*(26), 19082–19091. http://dx.doi.org/10.1074/jbc.M609629200.

Mach, D. B., Rogers, S. D., Sabino, M. C., Luger, N. M., Schwei, M. J., Pomonis, J. D., ... Mantyh, P. W. (2002). Origins of skeletal pain: Sensory and sympathetic innervation of the mouse femur. *Neuroscience*, *113*(1), 155–166.

Maestroni, G. J. (1993). The immunoneuroendocrine role of melatonin. *Journal of Pineal Research*, *14*(1), 1–10.

Maestroni, G. J., & Conti, A. (1994a). Modulation of hematopoiesis via alpha 1-adrenergic receptors on bone marrow cells. *Experimental Hematology*, *22*(3), 313–320.

Maestroni, G. J., & Conti, A. (1994b). Noradrenergic modulation of lymphohematopoiesis. *International Journal of Immunopharmacology*, *16*(2), 117–122.

Maestroni, G. J., Conti, A., & Pedrinis, E. (1992). Effect of adrenergic agents on hematopoiesis after syngeneic bone marrow transplantation in mice. *Blood*, *80*(5), 1178–1182.

Maestroni, G. J., Cosentino, M., Marino, F., Togni, M., Conti, A., Lecchini, S., & Frigo, G. (1998). Neural and endogenous catecholamines in the bone marrow. Circadian association of norepinephrine with hematopoiesis? *Experimental Hematology*, *26*(12), 1172–1177.

Maestroni, G. J., Togni, M., & Covacci, V. (1997). Norepinephrine protects mice from acute lethal doses of carboplatin. *Experimental Hematology*, *25*(6), 491–494.

Manske, J. M., Sullivan, E. L., & Andersen, S. M. (1995). Substance P mediated stimulation of cytokine levels in cultured murine bone marrow stromal cells. *Advances in Experimental Medicine and Biology*, *383*, 53–64.

Marmigere, F., & Ernfors, P. (2007). Specification and connectivity of neuronal subtypes in the sensory lineage. *Nature Reviews. Neuroscience*, *8*(2), 114–127. http://dx.doi.org/10.1038/nrn2057.

Massberg, S., Schaerli, P., Knezevic-Maramica, I., Kollnberger, M., Tubo, N., Moseman, E. A., ... von Andrian, U. H. (2007). Immunosurveillance by hematopoietic progenitor cells trafficking through blood, lymph, and peripheral tissues. *Cell*, *131*(5), 994–1008. http://dx.doi.org/10.1016/j.cell.2007.09.047. S0092-8674(07)01356-6 [pii].

Matsuda, H., Kawakita, K., Kiso, Y., Nakano, T., & Kitamura, Y. (1989). Substance P induces granulocyte infiltration through degranulation of mast cells. *Journal of Immunology*, *142*(3), 927–931.

McGillis, J. P., Rangnekar, V., & Ciallella, J. R. (1995). A role for calcitonin gene related peptide (CGRP) in the regulation of early B lymphocyte differentiation. *Canadian Journal of Physiology and Pharmacology*, *73*(7), 1057–1064.

Mendez-Ferrer, S., Lucas, D., Battista, M., & Frenette, P. S. (2008). Haematopoietic stem cell release is regulated by circadian oscillations. *Nature*, *452*(7186), 442–447. http://dx.doi.org/10.1038/nature06685.

Mendez-Ferrer, S., Michurina, T. V., Ferraro, F., Mazloom, A. R., Macarthur, B. D., Lira, S. A., ... Frenette, P. S. (2010). Mesenchymal and haematopoietic stem cells form a unique bone marrow niche. *Nature*, *466*(7308), 829–834. http://dx.doi.org/10.1038/nature09262.

Miller, M. L., & McCuskey, R. S. (1973). Innervation of bone marrow in the rabbit. *Scandinavian Journal of Haematology*, *10*(1), 17–23.
Molineux, G., Pojda, Z., Hampson, I. N., Lord, B. I., & Dexter, T. M. (1990). Transplantation potential of peripheral blood stem cells induced by granulocyte colony-stimulating factor. *Blood*, *76*(10), 2153–2158.
Moore, R. N., Osmand, A. P., Dunn, J. A., Joshi, J. G., & Rouse, B. T. (1988). Substance P augmentation of CSF-1-stimulated in vitro myelopoiesis. A two-signal progenitor restricted, tuftsin-like effect. *Journal of Immunology*, *141*(8), 2699–2703.
Nam, C., Case, A. J., Hostager, B. S., & O'Dorisio, M. S. (2009). The role of vasoactive intestinal peptide (VIP) in megakaryocyte proliferation. *Journal of Molecular Neuroscience*, *37*(2), 160–167. http://dx.doi.org/10.1007/s12031-008-9119-x.
Nance, D. M., & Sanders, V. M. (2007). Autonomic innervation and regulation of the immune system (1987–2007). *Brain, Behavior, and Immunity*, *21*(6), 736–745. http://dx.doi.org/10.1016/j.bbi.2007.03.008.
Nilsson, S., & Grove, D. J. (1974). Adrenergic and cholinergic innervation of the spleen of the cod: Gadus morhua. *European Journal of Pharmacology*, *28*(1), 135–143.
Nowicki, M., & Miskowiak, B. (2003). Substance P—A potent risk factor in childhood lymphoblastic leukaemia. *Leukemia*, *17*(6), 1096–1099. http://dx.doi.org/10.1038/sj.leu.2402920.
Nowicki, M., Ostalska-Nowicka, D., Konwerska, A., & Miskowiak, B. (2006). The predicting role of substance P in the neoplastic transformation of the hypoplastic bone marrow. *Journal of Clinical Pathology*, *59*(9), 935–941. http://dx.doi.org/10.1136/jcp.2005.034629.
Ozeran, R. S., Wagner, G. R., Reimer, T. R., & Hill, R. A. (1970). Neuropathy of the sympathetic nervous system associated with diabetes mellitus. *Surgery*, *68*(6), 953–958.
Park, M. H., Jin, H. K., Min, W. K., Lee, W. W., Lee, J. E., Akiyama, H., … Bae, J. S. (2015). Neuropeptide Y regulates the hematopoietic stem cell microenvironment and prevents nerve injury in the bone marrow. *The EMBO Journal*, *34*(12), 1648–1660. http://dx.doi.org/10.15252/embj.201490174.
Park, M. H., Lee, J. K., Kim, N., Min, W. K., Lee, J. E., Kim, K. T., … Bae, J. S. (2016). Neuropeptide Y induces hematopoietic stem/progenitor cell mobilization by regulating matrix metalloproteinase-9 activity through Y1 receptor in osteoblasts. *Stem Cells*, *34*(8), 2145–2156. http://dx.doi.org/10.1002/stem.2383.
Park, S. K., Olson, T. A., Ercal, N., Summers, M., & O'Dorisio, M. S. (1996). Characterization of vasoactive intestinal peptide receptors on human megakaryocytes and platelets. *Blood*, *87*(11), 4629–4635.
Rameshwar, P., Ganea, D., & Gascon, P. (1993). In vitro stimulatory effect of substance P on hematopoiesis. *Blood*, *81*(2), 391–398.
Rameshwar, P., Ganea, D., & Gascon, P. (1994). Induction of IL-3 and granulocyte-macrophage colony-stimulating factor by substance P in bone marrow cells is partially mediated through the release of IL-1 and IL-6. *Journal of Immunology*, *152*(8), 4044–4054.
Rameshwar, P., & Gascon, P. (1996). Induction of negative hematopoietic regulators by neurokinin-A in bone marrow stroma. *Blood*, *88*(1), 98–106.
Rameshwar, P., Gascon, P., Oh, H. S., Denny, T. N., Zhu, G., & Ganea, D. (2002). Vasoactive intestinal peptide (VIP) inhibits the proliferation of bone marrow progenitors through the VPAC1 receptor. *Experimental Hematology*, *30*(9), 1001–1009.
Riethmacher, D., Sonnenberg-Riethmacher, E., Brinkmann, V., Yamaai, T., Lewin, G. R., & Birchmeier, C. (1997). Severe neuropathies in mice with targeted mutations in the ErbB3 receptor. *Nature*, *389*(6652), 725–730. http://dx.doi.org/10.1038/39593.

Romano, T. A., Felten, S. Y., Felten, D. L., & Olschowka, J. A. (1991). Neuropeptide-Y innervation of the rat spleen: Another potential immunomodulatory neuropeptide. *Brain, Behavior, and Immunity*, *5*(1), 116–131.

Rosas-Ballina, M., Ochani, M., Parrish, W. R., Ochani, K., Harris, Y. T., Huston, J. M., … Tracey, K. J. (2008). Splenic nerve is required for cholinergic antiinflammatory pathway control of TNF in endotoxemia. *Proceedings of the National Academy of Sciences of the United States of America*, *105*(31), 11008–11013. http://dx.doi.org/10.1073/pnas.0803237105.

Rosas-Ballina, M., Olofsson, P. S., Ochani, M., Valdes-Ferrer, S. I., Levine, Y. A., Reardon, C., … Tracey, K. J. (2011). Acetylcholine-synthesizing T cells relay neural signals in a vagus nerve circuit. *Science*, *334*(6052), 98–101. http://dx.doi.org/10.1126/science.1209985.

Saito, H. (1990). Innervation of the guinea pig spleen studied by electron microscopy. *The American Journal of Anatomy*, *189*(3), 213–235. http://dx.doi.org/10.1002/aja.1001890305.

Sauerbier, I., & von Mayersbach, H. (1977). Circadian variation of catecholamines in human blood. *Hormone and Metabolic Research*, *9*(6), 529–530. http://dx.doi.org/10.1055/s-0028-1095589.

Schlomer, J. J., Storey, B. B., Ciornei, R. T., & McGillis, J. P. (2007). Calcitonin gene-related peptide inhibits early B cell development in vivo. *Journal of Leukocyte Biology*, *81*(3), 802–808. http://dx.doi.org/10.1189/jlb.0306229.

Shi, Y., Oury, F., Yadav, V. K., Wess, J., Liu, X. S., Guo, X. E., … Karsenty, G. (2010). Signaling through the M(3) muscarinic receptor favors bone mass accrual by decreasing sympathetic activity. *Cell Metabolism*, *11*(3), 231–238.

Smaaland, R., Sothern, R. B., Laerum, O. D., & Abrahamsen, J. F. (2002). Rhythms in human bone marrow and blood cells. *Chronobiology International*, *19*(1), 101–127.

Smith, L. G., Weissman, I. L., & Heimfeld, S. (1991). Clonal analysis of hematopoietic stem-cell differentiation in vivo. *Proceedings of the National Academy of Sciences of the United States of America*, *88*(7), 2788–2792.

Spiegel, A., Shivtiel, S., Kalinkovich, A., Ludin, A., Netzer, N., Goichberg, P., … Lapidot, T. (2007). Catecholaminergic neurotransmitters regulate migration and repopulation of immature human CD34+ cells through Wnt signaling. *Nature Immunology*, *8*(10), 1123–1131. http://dx.doi.org/10.1038/ni1509.

Tabarowski, Z., Gibson-Berry, K., & Felten, S. Y. (1996). Noradrenergic and peptidergic innervation of the mouse femur bone marrow. *Acta Histochemica*, *98*(4), 453–457. http://dx.doi.org/10.1016/S0065-1281(96)80013-4.

Teixeira, L., Sousa, D. M., Nunes, A. F., Sousa, M. M., Herzog, H., & Lamghari, M. (2009). NPY revealed as a critical modulator of osteoblast function in vitro: New insights into the role of Y1 and Y2 receptors. *Journal of Cellular Biochemistry*, *107*(5), 908–916. http://dx.doi.org/10.1002/jcb.22194.

Terazono, H., Mutoh, T., Yamaguchi, S., Kobayashi, M., Akiyama, M., Udo, R., … Shibata, S. (2003). Adrenergic regulation of clock gene expression in mouse liver. *Proceedings of the National Academy of Sciences of the United States of America*, *100*(11), 6795–6800.

Travlos, G. S. (2006). Normal structure, function, and histology of the bone marrow. *Toxicologic Pathology*, *34*(5), 548–565. http://dx.doi.org/10.1080/01926230600939856.

Usoskin, D., Furlan, A., Islam, S., Abdo, H., Lonnerberg, P., Lou, D., … Ernfors, P. (2015). Unbiased classification of sensory neuron types by large-scale single-cell RNA sequencing. *Nature Neuroscience*, *18*(1), 145–153. http://dx.doi.org/10.1038/nn.3881.

Vida, G., Pena, G., Deitch, E. A., & Ulloa, L. (2011). alpha7-Cholinergic receptor mediates vagal induction of splenic norepinephrine. *Journal of Immunology*, *186*(7), 4340–4346. http://dx.doi.org/10.4049/jimmunol.1003722.

Vizi, E. S., & Elenkov, I. J. (2002). Nonsynaptic noradrenaline release in neuro-immune responses. *Acta Biologica Hungarica, 53*(1–2), 229–244. http://dx.doi.org/10.1556/ABiol.53.2002.1-2.21.
Wolber, F. M., Leonard, E., Michael, S., Orschell-Traycoff, C. M., Yoder, M. C., & Srour, E. F. (2002). Roles of spleen and liver in development of the murine hematopoietic system. *Experimental Hematology, 30*(9), 1010–1019.
Yamazaki, K., & Allen, T. D. (1990). Ultrastructural morphometric study of efferent nerve terminals on murine bone marrow stromal cells, and the recognition of a novel anatomical unit: The "neuro-reticular complex" *The American Journal of Anatomy, 187*(3), 261–276.
Yamazaki, S., Ema, H., Karlsson, G., Yamaguchi, T., Miyoshi, H., Shioda, S., … Nakauchi, H. (2011). Nonmyelinating Schwann cells maintain hematopoietic stem cell hibernation in the bone marrow niche. *Cell, 147*(5), 1146–1158. http://dx.doi.org/10.1016/j.cell.2011.09.053.

CHAPTER FOUR

Imaging the Hematopoietic Stem Cell Niche

D. Duarte*,†, C. Lo Celso*,†,1
*Imperial College London, London, United Kingdom
†The Francis Crick Institute, London, United Kingdom
[1]Corresponding author: e-mail address: c.lo-celso@imperial.ac.uk

Contents

1. INTRODUCTION

Adult hematopoiesis occurs primarily in the bone marrow (BM), where hematopoietic stem cells (HSCs) reside. Through asymmetric division, HSCs can self-renew, thereby maintaining a long-lived pool of stem cells, as well as differentiate into progenitors, thereby sustaining hematopoietic development. Progenitor cells give rise to lineage-restricted blood cells (Becker, Mc, & Till, 1963; McCulloch & Till, 1960; Till & Mc, 1961) that play significant physiological roles in immune surveillance, oxygen transport, tissue repair, and clotting. Schofield (1978) proposed that cellular microenvironments in the BM are associated with and critical to maintain tissue-fixed HSCs. According to this hypothesis, after division HSCs are

Advances in Stem Cells and their Niches, Volume 1
ISSN 2468-5097
http://dx.doi.org/10.1016/bs.asn.2017.01.002

retained in the niche, while progenitors are spatially dislodged and differentiate. The stem cell niche hypothesis was since experimentally explored, accepted, and applied to other tissues, such as the epidermis and intestinal epithelium, and to pathological conditions, such as cancer. The BM niche is reviewed in depth in several recent reviews (Boulais & Frenette, 2015; Hoggatt, Kfoury, & Scadden, 2016; Morrison & Scadden, 2014).

A logical consequence of these studies was that the spatial organization of HSCs and their interactions with neighboring cells needed to be better understood. This obviated the importance of microscopy to directly visualize the niche.

2. EXPERIMENTAL MODELS ALLOWING VISUALIZATION OF HSCs

The classic definition of HSCs relies on functional assays (i.e., the long-term capacity of regenerating the whole hematopoietic tissue, assessed by serial transplantation), and the true stem cell can only be identified retrospectively, at which time the HSC does no longer exist to be assayed and visualized. This obstacle is partially circumvented by using phenotypic markers that can easily identify hematopoietic cell populations that are more or less enriched in HSCs. The definition of such markers using flow cytometry was pioneered by the Weissman group using the mouse as a model and later refined by others (Bryder, Rossi, & Weissman, 2006). For example, in mice, hematopoietic progenitors are identified based on their lack of markers associated with blood cell lineage commitment (lineage negative, Lin^-). This population is further enriched for progenitors based on the expression of stem cell antigen (Sca-1) and stem cell growth factor receptor (cKit). The Lin^-, $cKit^+$, $Sca\text{-}1^+$ (LKS) fraction can be additionally split according to the expression of the members of the signaling lymphocyte activation molecule family, CD48 and CD150. This top-down strategy is commonly applied to identify a LKS $CD48^-$/low $CD150^+$ population that is highly enriched in HSCs (Kiel et al., 2005).

HSCs can be directly labeled with antibodies and visualized by immunofluorescence (Kiel et al., 2005). A big advantage of this approach is that cells are labeled in situ, which allows the search for resident HSCs and bypasses the need of transplantation. Nonetheless, the number of markers that can be used in microscopy with confidence is limited compared with flow cytometry. This is for a number of practical reasons, including the presence of autofluorescence and the overlap between excitation and emission spectra of available fluorophores, which causes fluorescence spillover that is harder to compensate for, compared to flow cytometry. Fortunately, emerging techniques such as spectral unmixing (Cutrale et al., 2017) offer a partial solution to this problem. An additional and fundamental problem with this approach is the use of a fixed sample that prevents the dynamic study of the niche. This is discussed in detail later.

Another option is to cell sort genetically labeled HSCs and transplant them into recipient mice. This can be achieved by using donor mice that express a bright fluorophore, such as actin-green fluorescent protein (Actin-GFP) mouse (Koechlein et al., 2016). This alternative allows good cell definition and prevents extensive manipulation of the sorted HSCs. However, the fluorophores available are limited to existing strains. Alternatively, cells can be labeled ex vivo with intracellular dyes that allow in vivo cell tracking, such as 5-(and-6)-carboxyfluorescein diacetate succinimidyl ester (Nilsson, Johnston, & Coverdale, 2001). There are also useful lipophilic cyanine dyes that have a large range of excitation and emission spectra. These include the Vybrant 1,1′-dioctadecil-3,3,3′-tetramethylindodicarbocyanine perchlorate (DiD) or 1,1′-dilinoleyl-3,3,3′,3′-tetramethylindocarbocyanine perchlorate (DiI) dyes (Fig. 1A) (Lo Celso et al., 2009). Altogether, these strategies limit the study of HSCs to the transplantation setting, which can give great insight into the process of cell homing and hematopoietic reconstitution, but does not allow the study of resident HSCs and their steady-state

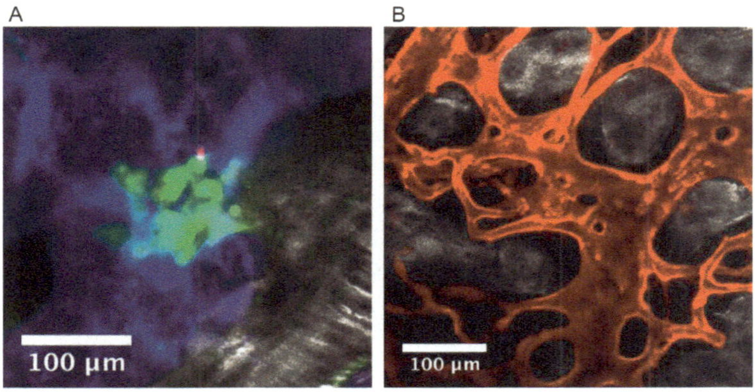

Fig. 1 Combined confocal and multiphoton intravital imaging of the calvarium bone marrow. (A) A DiD-labeled LKS cell observed in a Col2.3-GFP recipient. Transplanted the day before, it homed to a highly vascularized endosteal area and sits next to an osteoblast. (B) Labeling of bone marrow endothelial cells with antibodies against CD31. (A) *Green*, GFP^+ osteoblasts; *purple*, DiD^+ cell; *blue*, vasculature; (B) *Red*, endothelial cells; *gray*, SHG bone collagen.

behavior. Furthermore, recipients need to be lethally irradiated to decrease the number of BM cells and enable the engraftment of transplanted HSCs. As a consequence, these animals have depleted numbers of hematopoietic cells and damaged stromal cells, especially endothelial cells (Hooper et al., 2009) and osteoblasts (Dominici et al., 2009). An alternative to myeloablation is the use of double heterozygous Kit^{W}/Kit^{W-v} mice, as they do not require irradiaton for engraftment to occur. This strain carries point mutations that lead to significant cKit deficiency and make hematopoietic cells less efficient in competing for space (Migliaccio, Carta, & Migliaccio, 1999). Proficient HSCs can therefore easily occupy niche areas in the BM. Yet Kit^{W}/Kit^{W-v} mice are infertile and hard to breed, demanding complex and time-consuming crossings. A substitute approach is the use of less deleterious cKit mutants, such as the Kit^{W41}/Kit^{W41}, combined with sublethal irradiation (Miller & Eaves, 1997).

More recently, reporter mice were developed in which fluorophore expression is driven by genes that are highly enriched in HSCs. Two particular genes recently identified through gene expression screenings of HSCs were α-catulin (Acar et al., 2015) and Hoxb5 (Chen et al., 2016). According to these studies, limiting dilution assays showed a frequency of HSCs with multilineage capacity of 1 in 6.7 α-catulin-GFP^{+} cells and 1 in 2.1 Hoxb5-tri-$mCherry^{hi}$ cells. Such mice are very promising to prospectively identify resident adult HSCs that can be easily imaged. Other reporters marking candidate HSCs include Fgd5 (Gazit et al., 2014), Hoxb4 (Hills et al., 2011), and Tie2 (Ito et al., 2016). In most cases, however, either other cells express the same marker, making the identification of comparatively very rare HSCs more complex, or not all HSCs express the marker, raising the question whether there might be functional (and anatomical) differences between labeled and unlabeled HSCs,

3. EXPERIMENTAL MODELS ALLOWING VISUALIZATION OF BM NICHES

Several chemical signals (e.g., CXC chemokine ligand (CXCL)12, stem cell factor (SCF)) produced by hematopoietic and stromal cells are involved in the maintenance of HSCs. Megakaryocytes, macrophages, and regulatory T cells have been described to be associated with HSCs. Several mesenchymal cells have been shown one by one to be able to affect HSC numbers and function: bone-forming osteoblasts, their progenitors, and terminally differentiated osteocytes, endothelial cells, CXCL12 abundant reticular cells, leptin receptor-expressing stromal cells, and nestin-expressing mesenchymal stem cells (MSCs) (Morrison & Scadden, 2014). Through the use of genetically modified mouse models, it has been possible to visualize and manipulate these cells and the niche factors produced by them. For example, the Cre-Lox system allows deletion of target genes in specific cells, either constitutively or following induction at specific times. This strategy and several transgenic lines used in HSC research are reviewed in Joseph et al. (2013). Using such approaches it was possible to show, for example, that expansion and deletion of osteoblasts were associated with higher and lower numbers of HSCs, respectively (Calvi et al., 2003; Visnjic et al., 2004). Alternatively, HSCs have been shown to be in close contact with sinusoidal and arteriolar vessels (Kiel et al., 2005; Kunisaki et al., 2013). It should be noted that there is redundancy among stromal cells in the production of niche factors, such as SCF. Furthermore, the bone-lining endosteum and bone-rich metaphysis are highly vascularized, so that the so-called endosteal niches are also perivascular (Ellis et al., 2011). It is widely accepted that understanding the spatial distribution of HSCs in relation to their surroundings is fundamental to study the relevance of certain niche cells. In this regard, reporter mice allow for the identification and prospective isolation of such cells. Because the loxP cassettes are normally inserted in the Rosa26 locus, it is possible to cross them with other Rosa26 strains expressing fluorescent proteins, such as GFP, and in this way image a large number of reporter mice. There are, however, strains that are particularly popular in HSC niche studies. To image mature osteoblasts, the Col2.3-GFP mouse is often used (Fig. 1A). These mice express enhanced GFP under the control of the 2.3 kb rat procollagen type 1 alpha 1 (Col1a1) promoter (Kalajzic et al., 2002). Alternatively, osteoblast progenitors can be visualized using Osx1-CreGFP mice, where the transgene is expressed under the regulation of the osterix promoter (Rodda & McMahon, 2006). Nestin-expressing perivascular MSCs can be identified using the Nestin-GFP mouse, where the expression of the intermediate filament nestin drives GFP (Mignone, Kukekov, Chiang, Steindler, & Enikolopov, 2004), and CXCL12-expressing cells, primarily perivascular, can be visualized using the CXCL12-DsRed knock-in mice (Ding & Morrison, 2013). Endothelial cells can be studied through the use of endothelial-specific promoters, such as Flk1 and VE-Cadherin reporter mice. Yet, most of the studies label these cells using antibodies specific for vascular markers such as CD31, VE-Cadherin, laminin, and endomucin (Kunisaki et al., 2013; Kusumbe, Ramasamy, & Adams, 2014; Nombela-

Arrieta et al., 2013) (Fig. 1B). Such antibodies can be either directly applied to sections or intravenously injected in the live animal to perform intravital microscopy (IVM) or ex vivo whole mounts analyses. Alternatives that provide excellent discrimination of endothelial cells include Dil-labeled acetylated low-density lipoprotein (Dil-Ac-LDL) and fluorescently conjugated *Griffonia simplicifolia* isolectin B4 (Kunisaki et al., 2013; Lassailly, Foster, Lopez-Onieva, Currie, & Bonnet, 2013). In IVM studies, blood vessels are often identified through the use of intravascular dyes, such as fluorescently tagged bovine serum albumin (BSA) or dextrans (Hawkins et al., 2016) (Fig. 1A). To increase the number of colors analyzed, far red and near-infrared Quantom dots (Qdots), such as Qdot 655 and Qdot 800, can also be used (Lo Celso et al., 2009). Importantly, immune labeling is not restricted to endothelial cells and can be applied to other putative niche cells, such as angiopoietin-1-expressing stromal cells (Zhou, Ding, & Morrison, 2015) and to many of the classical niches, such as osteoblasts marked by antiosteopontin antibodies (Calvi et al., 2003). Finally, nonlinear optical microscopy, specifically second harmonic generation (SHG), allows to image collagen 1 fibers that compose the bone itself (Fig. 1) (Zipfel, Williams, & Webb, 2003).

Although these strategies are often used in isolation, the combination of reporter mice, antibodies, and SHG signals makes possible the simultaneous imaging of different BM microenvironmental components either in sections or in live imaging (Fig. 1).

4. OPTICAL MICROSCOPY APPROACHES

The BM cellular environment is visualized through microscopy. The microscope, which literally permits "to see" (-scope) what is "small" (micro), increases both resolution and contrast to resolve images invisible to the naked eye. Good examples of landmark discoveries in biology made possible by the microscope include the observation of bacteria and other microorganisms by Leeuwenhoek, the observation of phagocytosis by Metchikoff, and the description of structures of the central nervous system by Ramon y Cajal.

Different microscopy approaches are available and result from theoretical and technical advances combined with the development of dyes, fluorochromes, model organisms, and sample preparation methods. While wide-field microscopy is appropriate for thin specimens (e.g., two-dimensional (2D) cell cultures and very thin tissue sections), as it lacks depth discrimination. It illuminates the sample in its entire width and depth, and is therefore not suitable to image thick or highly scattering samples, such as bone. Alternatively, laser scanning confocal microscopy (LSCM) is able to obtain good quality three-dimensional (3D) information. LSCM acquires a series of optical sections, by rejecting the light from out-of-focus *z* planes and is therefore ideal for thick specimens. It has been often used for imaging tissue sections and for IVM. Nevertheless, LSCM is limited to the tissue surface, and for depths over 100 μm, there is increased light scattering that limits the contrast and signal strength. Imaging of deeper planes is better achieved by multiphoton (MP) microscopy (Denk, Strickler, & Webb, 1990). MP or two-photon microscopy can image the tissue at depths ranging from 150 μm, such as in the calvarium (Lo Celso et al., 2009) to 500 μm deep in soft tissues like the lymph nodes (Halin, Mora, Sumen, & von Andrian, 2005). Because MP microscopy uses near-infrared light (700–1000 nm), it allows not only deeper fluorescence excitation but also less phototoxicity and photobleaching, in comparison to LSCM (Helmchen & Denk, 2005). Two-photon excitation also allows better 3D resolution, photolytic release of caged molecules, SHG to observe the bone, and the excitation of near-infrared Qdots, which increases the range of signals generated. Due to their ability to excite fluorescence in thick tissues, LSCM and particularly MP microscopy have been used either alone or in combination to perform IVM and capture biological processes in four dimensions (4D). The use of IVM has contributed to significant discoveries in stem cell biology in several fields of study (Brown & Greco, 2014). For example, noninvasive IVM was used to show how the stem cell niche in the hair follicles of the skin is structurally and functionally compartmentalized (Rompolas et al., 2012). Also, MP live imaging combined with a chronically implanted abdominal imaging window was used to demonstrate how intestinal stem cells compete and are organized in relation to niche cells (Ritsma et al., 2014). IVM studies have also revolutionized the fields of neuroscience, cancer, and immunology (Ellenbroek & van Rheenen, 2014; Helmchen & Denk, 2005; Sumen, Mempel, Mazo, & von Andrian, 2004).

5. IMAGING HSC NICHES: FROM 2D TO 3D SAMPLES

Histological studies are very frequently performed to analyze the position of HSCs in the BM. Such studies use bone sections that have typically between 5 and 20 μm of thickness and offer 2D information alone (Fig. 2). Serial sectioning to generate 3D reconstructions of the whole tissue is technically difficult and unpractical. Because the bone is an extremely hard tissue,

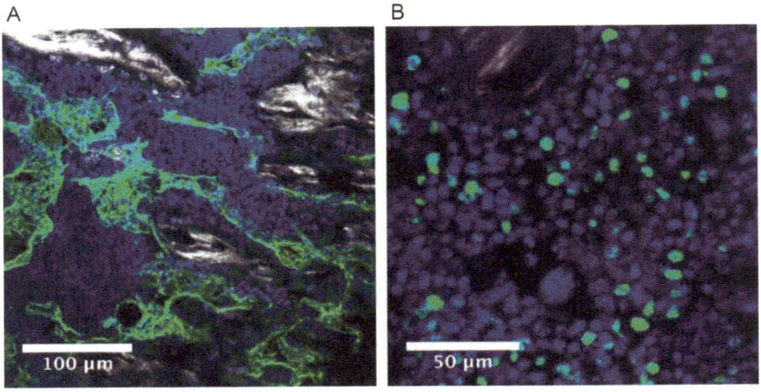

Fig. 2 Immunofluorescence of long bone sections. (A) Antiendomucin antibody (*green*) identifies blood vessels (*green*) surrounded by bone marrow cells marked by DAPI (*blue*) in a trabecular area. (B) Cell proliferation is assessed by labeling with anti-Ki-67 antibodies (*green*). *Green*, vessels (A) and Ki-67+ proliferating cells (B); *blue*, DAPI/nuclei; *gray*, SHG bone collagen.

decalcification is a common step of sample preparation, particularly for obtaining thick sections. Apart from being a long process that can last several weeks, decalcification leads to loss of intrinsic fluorescence, normal tissue morphology, and protein antigenicity. Fortunately, recent protocols have shortened and optimized this process and are able to generate images with excellent quality (Kusumbe, Ramasamy, Starsichova, & Adams, 2015). Another commonly used strategy relies on the tape-mediated transfer of bone cryosections, which can be achieved using either the CryoJane system (Fig. 2) (Hawkins et al., 2016; Kiel et al., 2005) or the Kawamoto method (Kawamoto, 2003). Nevertheless, the main disadvantage of histological analysis is that it generates static images that lack temporal dimension necessary to detect microenvironmental interactions. Single time point analysis can generate misleading data that either suggests false or excludes true associations between HSCs and niche cells. Another problem is their quantification. A cell-to-cell interaction is usually classified as such for cells that appear to be physically in contact with each other or that are interspaced by a subjective distance (e.g., 1–12 cell diameters), where there is a hypothetical gradient of secreted niche factors. These two problems can be partially resolved by using randomly generated virtual cells (Acar et al., 2015; Chen et al., 2016; Hawkins et al., 2016; Kunisaki et al., 2013) and nonbiased automatic quantification of the shortest distance between two cells (Khorshed et al., 2015) in custom-made platforms or commercially available software (e.g., Imaris, Definiens). Importantly, these quantification methods can also be applied to 3D (Hawkins et al., 2016).

The main advantage of histological analysis is that it avoids niche perturbation or transplantation and allows the study of HSCs in their steady state. Another advantage is that multiple antibodies can be used to label not only HSCs but also multiple stromal components simultaneously. Once a permeabilization step is included in the specimen preparation, it is also possible to stain for intracellular proteins that inform on cellular processes, such as proliferation, apoptosis, and hypoxia. Cell proliferation can be detected through the labeling of nuclear Ki-67 (Fig. 2B) and proliferation cell nuclear antigen (Kunisaki et al., 2013). Cell dormancy can also be evaluated by pulse-chase experiments with thymidine analogues that are incorporated in newly synthesized DNA. The detection of the analogues 5-bromo-2′-deoxyuridine and 5-ethynyl-29-deoxyuridine is achieved through anti-BrdU staining and through a Click-IT®-based chemical reaction, respectively (Kusumbe et al., 2015). Cells undergoing apoptosis can be visualized by labeling proteins that are activated during this process, such as caspase-3, or through terminal deoxynucleotidyl transferase dUTP nick end labeling, which detects fragmented DNA (Hawkins et al., 2016). Hypoxia has been traditionally studied through the in vivo incorporation of pimonidazole in cells in low-oxygen conditions followed by immunolabeling (Nombela-Arrieta et al., 2013). Importantly, immunofluorescence studies can be approached with the same type of analysis used for flow cytometry, but to obtain refined spatial information. A recently developed technique, laser scanning cytometry, takes this concept further and merges LSCM with an automated flow cytometry-like acquisition (Harnett, 2007). While traditional immunofluorescence imaging focuses on specific fields of view, laser scanning cytometry expands the image acquisition to much larger areas, thereby decreasing biases inherent to subjective area selection. Using this technology, Nombela-Arrieta and colleagues were able to show that phenotypic hematopoietic stem and progenitor cells (HSPCs) preferentially localized next to endosteal blood vessels, although hypoxic HSPCs were found throughout the BM cavity (Nombela-Arrieta et al., 2013). Along this line, Kiel et al. found $Lin^-CD41^-CD48^-CD150^+$ HSCs localizing adjacently to sinusoids (Kiel et al., 2005). In contrast, other studies have used histological sections to show that quiescent HSCs are directly in contact with osteoblasts (Arai et al., 2004) and enriched by the endosteum (Wilson et al., 2004; Zhang et al., 2003).

In recent years, several groups have perfected imaging of thick BM sections (over 300 μm thickness). Typically, the bone of interest is either cryosectioned from both sides or cut in half and stained (Acar et al., 2015; Kunisaki et al., 2013). Alternatively, BM plugs are obtained through flushing with a syringe (Acar et al., 2015; Chen et al., 2016). In this way, it is possible to optically section the sample and 3D reconstruct what is often called a BM whole-mount. Bones used for whole-mount preparation include the sternum (Kunisaki et al., 2013), femur (Nombela-Arrieta et al., 2013), and tibia (Acar et al., 2015; Chen et al., 2016). This technique can be accompanied by tissue clearing, initially developed for and popularized in neuroscience studies. The clearing process makes deeper areas optically accessible but can significantly alter tissue morphology and intrinsic fluorescence and requires long troubleshooting for the selection of the best method. Successful clearing strategies used for the bone to visualize HSCs include modified Murray's clear (Acar et al., 2015; Becker, Jahrling, Saghafi, & Dodt, 2013), 3DISCO (Acar et al., 2015; Erturk et al., 2012), and CUBIC (Chen et al., 2016; Susaki et al., 2014). Kunisaki and colleagues were able to show that in sternum whole mounts, dormant $Lin^{-}CD41^{-}CD48^{-}CD150^{+}$ HSCs located near arterioles, which are surrounded by other important stromal cells, such as smooth muscle cells, $Nestin^{+}$ MSCs, and sympathetic nerve fibers (Kunisaki et al., 2013). This observation was recently opposed by a study from Acar et al. (2015). In this report, the authors used optically cleared tibias to perform deep imaging of the BM and show that nondividing α-catulin-GFP^{+} HSCs did not have a specific spatial patterning in relation to arterioles. Interestingly, Chen et al., using Hoxb5 as an HSC reporter and cleared BM plugs of the tibia, propose that in fact long-term HSCs are virtually all next to VE-$Cadherin^{+}$ blood vessels (Chen et al., 2016). Several reasons can explain the differences between the reports, including the different bones analyzed (tibia vs sternum) and the different methods used to identify HSCs (phenotypic $Lin^{-}CD41^{-}CD48^{-}CD150^{+}$ HSCs vs α-$catulin^{+}$ HSCs vs $Hoxb5^{+}$ HSCs).

6. INTRAVITAL MICROSCOPY

While histological studies and in vitro studies of live cells allow detailed manipulation of experimental conditions and high imaging resolution, they often fail to detail the complex dynamics of biological processes in a meaningful way. This is particularly true in situations where the environmental context is pivotal. One such example is the immunological synapse formation between cytotoxic T cells (CTLs) and target cells. In this regard, the active killing of tumor cells by CTLs in vitro takes about 5 min (Stinchcombe, Bossi, Booth, & Griffiths, 2001). In striking contrast, in vivo imaging showed CTL-mediated killing of peptide-pulsed B cells in the lymph node and of tumor cells in the tumor microenvironment lasting up to 25 min and 6 h, respectively (Breart, Lemaitre, Celli, & Bousso, 2008; Mempel et al., 2006). Another example that illustrates the importance of intravital imaging is the migration of tumor cells, as originally reviewed by Condeelis and Segall (2003). They describe how the in vivo speed of migrating breast adenocarcinoma cells is greater than 3 μm/min, which is 10 times superior compared to recordings of these cells migrating in 2D plates and 30 times higher compared to migration in 3D matrices. Altogether, these observations highlight how only IVM can correctly capture dynamic events that are important in the study of HSC–niche interactions. These include: the influence of the tissue microenvironment (i.e., the study of HSCs in the BM), the timescale of biological processes (for example, how long does it take for an HSC to divide and differentiate), and the cell behavior (for example, is the cell immotile or migratory, how is it moving and at what speed).

Early attempts at performing live imaging of the BM were made by McClugage, McCuskey, and Meineke (1971) and McCuskey, McClugage, and Younker (1971). In these studies, a metallic chamber was transversely inserted in the tibias of rabbits and coupled with a conventional wide-field microscope. Notably, chronic imaging was possible and the same areas could be revisited. The images obtained had low resolution and low contrast, and the BM cells and structures were identified solely based on their morphology. Nevertheless, the authors were able to draw important conclusions from functional manipulation of the hematopoietic system. For example, they observed that blood vessels could support bone regeneration through a vascular network that was primarily venular in nature. They also concluded that arterioles and capillaries supported hematopoiesis stimulated by bleeding and erythropoietin administration. It is remarkable that in more recent years, much more highly refined studies have confirmed the central role of BM blood vessels in bone formation (Kusumbe et al., 2014) and hematopoiesis (Hooper et al., 2009; Kobayashi et al., 2010).

In 2009, several imaging studies aimed to visualize the dynamics of phenotypically purified HSCs in their natural BM environment (Kohler et al., 2009; Lo Celso et al., 2009; Xie et al., 2009). In one of this studies, Xie et al. developed a method for ex vivo imaging of stem cells, consisting in imaging of sagitally sectioned tibias and femurs of mice, previously transplanted with phenotypic HSCs (Xie et al., 2009). While using the long bones, the imaging was done in a severely damaged tissue. Nevertheless, this study suggested that the endosteal area is converted into a stimulatory environment that promotes HSC expansion after irradiation. At the same time, IVM of single engrafting HSCs was achieved by focusing on the BM contained in the frontal bones of the mouse skull (calvarium) and generated 3D measurements indicating that engrafting HSCs selectively localize in proximity of osteoblastic cells and endosteal vessels, while their progeny are more distal (Lo Celso et al., 2009). Multiple groups have developed methods that allow for the in vivo imaging of the long bones. Kohler et al. thinned the tibeal compact bone to reveal differential positioning and activity of young and aged HSCs (Kohler et al., 2009). Lewandowski et al. used a miniaturized endoscope to capture what remains the only live imaging of central marrow in the femur, albeit achieved with an invasive imaging technique (Lewandowski et al., 2010). Very recently, Pitt and colleagues used two-photon live imaging of shaved tibias to enquire about the interaction between T-cell acute lymphoblastic leukemia (T-ALL) cells and CXCL12-producing stromal cells in the BM (Pitt et al., 2015). Chronic imaging of long bones has been achieved through the development of imaging windows analogous to others used in the brain and abdomen. These imaging windows have been applied to both the thinned femur (Kim, Lin, Brown, Hosaka, & Scott, 2017) and the thinned tibia (Kim et al., 2017). Although real-time imaging of the femur and tibia has the advantage of studying stem and progenitor cell behavior in bones that sustain the majority of the adult mouse steady-state hematopoiesis, it also presents significant problems that limit their present utility. One of these is the skeletal shaving required to make the marrow space underlying the bone optically accessible. A bone thinning of around 200 μm is done with a micro drill and can easily lead to tissue disruption and bleeding. Another pitfall of the method is that it maintains a bias towards the recording of events occurring near the endosteum. Finally, the image resolution obtained with IVM of long bones is often lower in comparison to calvarium imaging.

Calvarium BM imaging was pioneered by the von Andrian group, using epifluorescence microscopy (Mazo et al., 1998). One of the main advantages of this method is that because the bone plate is very thin, shaving is not required. Therefore, the BM can be imaged noninvasively and without altering the tissue architecture. Recent refinements have shown that further higher quality in the imaging can be achieved after laser-mediated osteotomy of the calvarium (Turcotte, Alt, Mortensen, & Lin, 2014). Here, the laser is expected to cause less damage than the conventional drills used on the long bones; however, this approach would not be easily implemented to shave those much thicker areas. One of the concerns often raised with the use of the calvarium for the study of HSCs and their niches is how its BM compares with well-described compartments in long bones. This has been disputed by Lassailly et al. who demonstrated that while the calvarium is a flat bone resulting from intramembranous ossification, it is as representative as the femur and the tibia in the study of hematopoiesis (Lassailly et al., 2013).

Building on the original work by Mazo and coworkers, IVM of the calvarium was also used to image platelet formation from megakaryocytes (Junt et al., 2007), T cells (Cavanagh et al., 2005), B cells (Cariappa et al., 2005), and dendritic cells (Sapoznikov et al., 2008).

This technique was later expanded by us and by the groups of David Scadden and Charles Lin to obtain images of the BM with higher resolution (Lo Celso et al., 2009; Sipkins et al., 2005). Using LSCM alone of the skull's frontal bones, Sipkins et al. showed preferential localization of hematopoietic progenitor, stem cells and leukemia cells to specific BM endothelial microdomains expressing high levels of E-selectin and CXCL-12 (or stromal cell-derived factor (SDF)-1), which facilitates homing by binding to CXCR4 (Sipkins et al., 2005). Through the combination of LSCM with MP microscopy, the number of components analyzed simultaneously was increased, and the first in vivo measurement of transplanted HSCs localization in relation to the osteoblastic and vascular niche compartments reported (Lo Celso et al., 2009).

Recently, we have developed a protocol that enables hours long time-lapsing and repeated cycles of imaging and recovery using a calvarial window and high-precision mouse holder (Scott, Akinduro, & Lo, 2014). Using this approach, it was possible to uncover the in vivo migratory pattern of some HSCs (Rashidi et al., 2014). In contract to steady state, transplanted HSCs, which are stationary, infection-exposed HSCs, are motile and engage with larger surrounding BM niches within the first 24 h from injection (Rashidi et al., 2014). Using this technical platform, we have also recently described the niche-agnostic behavior of developing and chemoresistant T-ALL cells (Hawkins et al., 2016). Importantly, the study combined high temporal and spatial resolution recording of cells with large 3D tilescans of the entirety of the bone cavity. This type of approach will contribute to future studies aiming to frame single cells' behavior within the dynamics of the tissue they reside in.

7. EMERGING APPROACHES IN IVM

The evidence referenced above shows how the complex anatomy of the mammalian BM and the interactions of HSCs with their niches are being progressively better understood. Although the combination of intravital imaging with antibodies, dyes, transgenic mouse models, drugs, and irradiation allows a comprehensive study of many dynamic processes in the BM, we are still lacking fundamental information about the nature of HSC–niche interactions. This includes in vivo data on gene expression, activation/repression of certain intracellular signaling pathways, and protein secretion, among others.

Live imaging of the calvarium has recently contributed to the functional assessment of BM areas and blood vessels. Spencer and colleagues used two-photon phosphorescence lifetime microscopy to directly measure the local oxygen tension in the BM of live mice (Spencer et al., 2014). Unexpectedly, they found that the perisinusoidal zone was more hypoxic when compared to the endosteal area, preferentially irrigated by arterioles covered with Nestin^{+} cells. Another study by Itkin et al. also used IVM of the calvarium to question how different types of BM blood vessels regulate hematopoiesis (Itkin et al., 2016). According to their study, sinusoids are more permeable to reactive oxygen species (ROS) that in turn induce HSPCs to migrate and differentiate. In contrast arterial vessels are less permeable and therefore maintain HSCs in a less activated state. New questions will be likely answered by studies such as that completed by Ito et al., combining in vivo and in vitro analyses to reveal that HSC self-renewal is linked to mitochondrial clearance (Ito et al., 2016). Interestingly, in some experiments the authors were able to further develop calvarium osteotomy to generate a microscopic channel through which they could deposit one or more HSCs directly in the BM cavity. This approach will be useful in the future to investigate how incoming HSCs may select specific niches, whether they may travel long distances before settling, or whether they will tend to remain in the proximity of the microchannel where they are deposited.

Emerging technologies that in our opinion will be useful in the in vivo study of the HSC niche include: fluorescence lifetime imaging of fluorescence resonance energy transfer biosensors (Hirata et al., 2015), such as the caspase 3 sensor used to image apoptotic cells (Breart et al., 2008); laser ablation of HSCs or niche cells similar to dermal papilla ablation in hair follicles (Rompolas et al., 2012); lineage tracing using Confetti (Ritsma et al., 2014) or other systems; fluorescence recovery after photobleaching to study protein dynamics (Erami et al., 2016); photoswitching using constructs such as Dendra2 (Kedrin et al., 2008), to temporally and spatially follow specific cells or cell populations.

In depth analysis of microscopy-generated data will be critical to maximize imaging-based discoveries. An example that nicely illustrates this is a method developed by the group of Philip Bousso called dynamic in situ cytometry (DISC). DISC combines the single-cell phenotypic analysis of flow cytometry with dynamic intravital imaging data, such as cell speed (Moreau & Bousso, 2012). In this way, DISC allows an intuitive display and multiparametric analysis that has been used to study in vivo antigen recognition by T cells (Moreau et al., 2012). Similar approaches in the HSC niche field will undoubtedly be productive and allow further investigating cell position and fate.

8. IMAGING HUMAN HSCs

In contrast to the mouse, the BM niche for human HSCs remains less well studied. This is in part justified by the relative rarity of human samples and by our poor understanding of the human hematopoietic hierarchy. However, the main restraints for the exploration of the human HSC microenvironment are the experimental limitations associated with the manipulation of human HSCs and human BM tissue. In this regard, John Dick' group pioneered the study of human hematopoiesis by xenotransplanting human HSCs into immunodeficient mice (Doulatov, Notta, Laurenti, & Dick, 2012). Human HSCs with self-renewal and multilineage capacity exist at an estimated frequency of 1 in 10^6 BM cells. Similarly to the mouse system, it is possible to enrich for stem cell activity by selecting a population using flow cytometry cell surface markers. The most widely accepted phenotype of human HSCs is $CD34^+CD38^-Thy1^+CD45RA^-$. Recently, Notta et al.

subdivided this population in a CD49f$^+$ HSC fraction with long-term engraftment potential and in a CD49f$^-$ fraction of multipotent progenitor cells (Notta et al., 2011). Phenotypic HSCs can be therefore isolated from human cord blood or adult BM and transplanted into mice that lack certain immune cells or immune cell functions and therefore do not reject human cells. Among the many strains of immunodeficient mice available today, one of the most used is the NOD-SCID-Il2r$\gamma^{-/-}$ (NSG) line (Shultz et al., 2005). These mice have a defective V(D)J recombination and accompanying deficiency of mature B and T cells due to the severe combined immunodeficiency (SCID) mutation, a deficiency of natural killer cells due to the lack of Il-2Rγ and a phagocytic tolerance due to the presence of cross-reactive SIRPα of the nonobese diabetic background (NOD). Transplantation of immunodeficient mice coupled with the analysis of human biopsies and in vitro assays allows for the study of the human HSC microenvironment.

One of the first niche cells shown to be associated with human HSCs in histological analysis was the human perivascular CD146$^+$ MSC (Sacchetti et al., 2007). This study also showed that CD146$^+$ MSCs express the important niche factor angiopoietin-1. This observation anticipated the later descriptions of Nestin$^+$ cells as HSC regulators by Mendez-Ferrer and Paul Frenette (Mendez-Ferrer et al., 2010). More recently, it was shown that human fetal PDGFRα^+ CD51$^+$ cells expressing nestin represent a subset of CD146$^+$ MSCs that are enriched for HSC-supporting activity (Pinho et al., 2013). In a recent study, Guezguez, Bhatia, and coworkers performed elegant immunofluorescence imaging of bone sections from human BM biopsies and from immunodeficient mice transplanted with human CD34$^+$ cells and showed that human HSCs are preferentially located in the trabecular bone (Guezguez et al., 2013). The authors nicely compared the whole-tissue distribution of stem vs progenitor cells and concluded that HSCs were more concentrated in endosteal areas.

As discussed earlier, imaging of the BM space in sections does not provide information on cell behavior over time. To address this point, Dominique Bonnet and colleagues have recently adapted the IVM setup used to image the mouse calvarium (Lo Celso et al., 2009) to visualize human HSCs transplanted into NSG mice. Foster et al. showed that while hematopoietic progenitor cells remain motile after homing to the BM, HSCs are only transiently migratory and eventually settle into hypothetical niche areas (Foster et al., 2015). This suggests that, as observed for transplanted mouse HSCs (Rashidi et al., 2014), long-lasting engagement with the niche is important for hematopoietic reconstitution (Lapidot, Dar, & Kollet,

2005). Along this line, Rak et al. used live imaging of the calvarium and observed that the cell adhesion molecule cytohesin-1 is fundamental for the stable endosteal localization of human HSCs (Rak et al., 2016).

Future studies focusing on the imaging of human hematopoietic niches will benefit from recently developed mouse strains. Exciting examples are the NSGW41 mice, capable of sustaining human erythro-megakaryopoiesis (Rahmig et al., 2016) and the next-generation humanized mice containing knock-in alleles that express human genes, such as thrombopoietin (TPO) (Theocharides, Rongvaux, Fritsch, Flavell, & Manz, 2016). The latter are important because they mimic the human microenvironment in the recipient mouse BM and are able to support benign and malignant hematopoiesis. One example of such strain is the MIS$^{(KI)}$TRG containing knock-ins for five human genes: macrophage colony-stimulating factor, IL-3, granulocyte-macrophage colony-stimulating factor (GM-CSF), TPO, and SIRPα (Das et al., 2016). Furthermore, increased technical improvements in IVM, such as in vivo labeling of human cells and increased temporal resolution of time-lapse acquisition (Hawkins et al., 2016) will help the dynamic study of human HSCs.

9. IMAGING HSC NICHES IN OTHER MODEL ORGANISMS

The study of the hematopoietic niche is not restricted to the mouse and can also be done using alternative model systems, such as *Drosophila melanogaster* and zebrafish.

In comparison to vertebrates, the hematopoietic system of *Drosophila* is more rudimentary and misses many cell types and key niche components, such as a vascular network. Nevertheless, important conserved hematopoietic regulators are shared, such as RUNX, GATA, Notch, and Wnt (Crozatier & Vincent, 2011). Analogously to vertebrates, the fly has waves of hematopoiesis during development that give rise to mature blood cells, termed hemocytes. The second of these waves occurs in the larval lymph gland, which is the most studied hematopoietic organ in *Drosophila* (Jung, Evans, Uemura, & Banerjee, 2005). It is possible to investigate the hematopoietic microenvironment in *Drosophila* since the differentiation of hemocytes is spatially and temporarily regulated in the lymph gland (Jung et al., 2005). Imaging of this niche using different fly strains and antibodies combined with refined genetic and molecular tools allows for the analysis of involved signaling pathways with a level of genetic manipulation that is not possible in mice. Important contributions from the fly field include, but are not restricted to, the role of ROS in the regulation of hematopoietic progenitor differentiation (Owusu-Ansah & Banerjee, 2009), the importance of differentiated cell-derived signals in the maintenance of progenitors (Mondal et al., 2011), the role of Insulin/TOR (Benmimoun, Polesello, Waltzer, & Haenlin, 2012) and Slit/Robo (Morin-Poulard et al., 2016) signaling pathways, and the role of the extracellular matrix (Grigorian, Liu, Banerjee, & Hartenstein, 2013) in the maintenance of hematopoietic progenitors.

A model organism that has been extensively used for the study of the HSC niche is the zebrafish (*Danio rerio*). In zebrafish, the definitive wave of hematopoiesis is characterized by HSCs emerging from the ventral wall of the dorsal aorta and migrating first to the caudal hematopoietic tissue and then to the thymus and the kidney marrow, which is the fish equivalent of the mammalian adult BM (Paik & Zon, 2010). Apart from the conservation of several important hematopoietic programs, zebrafish has also a high level of microenvironmental complexity. It has blood circulation, with endothelial cells, mesenchymal perivascular cells, and macrophages known to contribute to HSC regulation. A big advantage of using zebrafish is its particular suitability for live imaging using wide-field and confocal microscopy. The embryo is naturally optically accessible and transparent adult zebrafish, such as the *Casper* (White et al., 2008) and *Tra/Nac* (Krauss, Astrinidis, Frohnhofer, Walderich, & Nusslein-Volhard, 2013) fish lines are also available. Furthermore, there are several fish strains that report on stromal populations and that mark HSCs with different levels of specificity, such as the CD41 line (Ma, Zhang, Lin, Italiano, & Handin, 2011) and the recently introduced Runx line (Tamplin et al., 2015). The combination of these tools with the easy genetic manipulation of zebrafish enables the in vivo study of resident and transplanted HSCs. The field of developmental hematopoiesis has greatly benefited from live imaging of the zebrafish. A good example is the high-resolution recording of HSC emergence from the hemogenic endothelium (Kissa & Herbomel, 2010). Among the several reports using zebrafish to investigate the interactions between HSCs and niche cells, we highlight the recent study by Tamplin et al. (2015). In this study, confocal live imaging and correlative light and electron microscopy were performed to visualize previously unseen dynamics of the perivascular niche. The authors show endothelial cells wrapping around single HSCs and mesenchymal stromal cells engaging with stem and progenitors cells and directing their divisions. It is expected that the identification of better refined cell surface markers and development of antibodies will further facilitate the use of zebrafish in the study of the HSC niche.

10. CONCLUSION

Our understanding of how HSCs are regulated by specific microenvironments has been greatly illuminated by their direct observation. We discussed advantages and limitations of imaging fixed samples and IVM. Importantly, the combination of different methods provides the most complete picture and is therefore the most desirable. It is expected that the development of new live imaging tools that assess cell function and the further analysis of human HSCs will support future studies. The use of alternative model organisms to the mouse, such as the zebrafish, is also expected to provide new insights about the HSC niche. Finally, it should be emphasized that the development of BM imaging can also contribute to the study of bone physiology, immunity and of several pathological conditions, most notably hematological malignancies and metastasis.

ACKNOWLEDGMENTS

D.D. is a recipient of a GABBA FCT fellowship (SFRH/BD/52195/2013), and C.L.C. is supported by ERC (ERC-2013-Stg 337066), BBSRC (BB/L023776/1), and Bloodwise (15040 and 15031).

REFERENCES

Acar, M., Kocherlakota, K. S., Murphy, M. M., Peyer, J. G., Oguro, H., Inra, C. N., et al. (2015). Deep imaging of bone marrow shows non-dividing stem cells are mainly perisinusoidal. *Nature*, *526*(7571), 126–130.

Arai, F., Hirao, A., Ohmura, M., Sato, H., Matsuoka, S., Takubo, K., et al. (2004). Tie2/angiopoietin-1 signaling regulates hematopoietic stem cell quiescence in the bone marrow niche. *Cell*, *118*(2), 149–161.

Becker, K., Jahrling, N., Saghafi, S., & Dodt, H. U. (2013). Immunostaining, dehydration, and clearing of mouse embryos for ultramicroscopy. *Cold Spring Harbor Protocols*, *8*, 743–744.

Becker, A. J., Mc, C. E., & Till, J. E. (1963). Cytological demonstration of the clonal nature of spleen colonies derived from transplanted mouse marrow cells. *Nature*, *197*, 452–454.

Benmimoun, B., Polesello, C., Waltzer, L., & Haenlin, M. (2012). Dual role for Insulin/TOR signaling in the control of hematopoietic progenitor maintenance in Drosophila. *Development*, *139*(10), 1713–1717.

Boulais, P. E., & Frenette, P. S. (2015). Making sense of hematopoietic stem cell niches. *Blood*, *125*(17), 2621–2629.

Breart, B., Lemaitre, F., Celli, S., & Bousso, P. (2008). Two-photon imaging of intratumoral CD8+ T cell cytotoxic activity during adoptive T cell therapy in mice. *The Journal of Clinical Investigation*, *118*(4), 1390–1397.

Brown, S., & Greco, V. (2014). Stem cells in the wild: Understanding the World of stem cells through intravital imaging. *Cell Stem Cell, 15*(6), 683–686.
Bryder, D., Rossi, D. J., & Weissman, I. L. (2006). Hematopoietic stem cells: The paradigmatic tissue-specific stem cell. *The American Journal of Pathology, 169*(2), 338–346.
Calvi, L. M., Adams, G. B., Weibrecht, K. W., Weber, J. M., Olson, D. P., Knight, M. C., et al. (2003). Osteoblastic cells regulate the haematopoietic stem cell niche. *Nature, 425*(6960), 841–846.
Cariappa, A., Mazo, I. B., Chase, C., Shi, H. N., Liu, H., Li, Q., et al. (2005). Perisinusoidal B cells in the bone marrow participate in T-independent responses to blood-borne microbes. *Immunity, 23*(4), 397–407.
Cavanagh, L. L., Bonasio, R., Mazo, I. B., Halin, C., Cheng, G., van der Velden, A. W., et al. (2005). Activation of bone marrow-resident memory T cells by circulating, antigen-bearing dendritic cells. *Nature Immunology, 6*(10), 1029–1037.
Chen, J. Y., Miyanishi, M., Wang, S. K., Yamazaki, S., Sinha, R., Kao, K. S., et al. (2016). Hoxb5 marks long-term haematopoietic stem cells and reveals a homogenous perivascular niche. *Nature, 530*(7589), 223–227.
Condeelis, J., & Segall, J. E. (2003). Intravital imaging of cell movement in tumours. *Nature Reviews. Cancer, 3*(12), 921–930.
Crozatier, M., & Vincent, A. (2011). Drosophila: A model for studying genetic and molecular aspects of haematopoiesis and associated leukaemias. *Disease Models & Mechanisms, 4*(4), 439–445.
Cutrale, F., Trivedi, V., Trinh, L. A., Chiu, C. L., Choi, J. M., Artiga, M. S., et al. (2017). Hyperspectral phasor analysis enables multiplexed 5D in vivo imaging. *Nature Methods, 4*(2), 149–152.
Das, R., Strowig, T., Verma, R., Koduru, S., Hafemann, A., Hopf, S., et al. (2016). Microenvironment-dependent growth of preneoplastic and malignant plasma cells in humanized mice. *Nature Medicine, 22*(11), 1351–1357.
Denk, W., Strickler, J. H., & Webb, W. W. (1990). Two-photon laser scanning fluorescence microscopy. *Science, 248*(4951), 73–76.
Ding, L., & Morrison, S. J. (2013). Haematopoietic stem cells and early lymphoid progenitors occupy distinct bone marrow niches. *Nature, 495*(7440), 231–235.
Dominici, M., Rasini, V., Bussolari, R., Chen, X., Hofmann, T. J., Spano, C., et al. (2009). Restoration and reversible expansion of the osteoblastic hematopoietic stem cell niche after marrow radioablation. *Blood, 114*(11), 2333–2343.
Doulatov, S., Notta, F., Laurenti, E., & Dick, J. E. (2012). Hematopoiesis: A human perspective. *Cell Stem Cell, 10*(2), 120–136.
Ellenbroek, S. I., & van Rheenen, J. (2014). Imaging hallmarks of cancer in living mice. *Nature Reviews. Cancer, 14*(6), 406–418.
Ellis, S. L., Grassinger, J., Jones, A., Borg, J., Camenisch, T., Haylock, D., et al. (2011). The relationship between bone, hemopoietic stem cells, and vasculature. *Blood, 118*(6), 1516–1524.
Erami, Z., Herrmann, D., Warren, S. C., Nobis, M., McGhee, E. J., Lucas, M. C., et al. (2016). Intravital FRAP imaging using an E-cadherin-GFP mouse reveals disease- and drug-dependent dynamic regulation of cell-cell junctions in live tissue. *Cell Reports, 14*(1), 152–167.
Erturk, A., Becker, K., Jahrling, N., Mauch, C. P., Hojer, C. D., Egen, J. G., et al. (2012). Three-dimensional imaging of solvent-cleared organs using 3DISCO. *Nature Protocols, 7*(11), 1983–1995.
Foster, K., Lassailly, F., Anjos-Afonso, F., Currie, E., Rouault-Pierre, K., & Bonnet, D. (2015). Different motile behaviors of human hematopoietic stem versus progenitor cells at the osteoblastic niche. *Stem Cell Reports, 5*(5), 690–701.

Gazit, R., Mandal, P. K., Ebina, W., Ben-Zvi, A., Nombela-Arrieta, C., Silberstein, L. E., et al. (2014). Fgd5 identifies hematopoietic stem cells in the murine bone marrow. *The Journal of Experimental Medicine*, *211*(7), 1315–1331.

Grigorian, M., Liu, T., Banerjee, U., & Hartenstein, V. (2013). The proteoglycan Trol controls the architecture of the extracellular matrix and balances proliferation and differentiation of blood progenitors in the Drosophila lymph gland. *Developmental Biology*, *384*(2), 301–312.

Guezguez, B., Campbell, C. J., Boyd, A. L., Karanu, F., Casado, F. L., Di Cresce, C., et al. (2013). Regional localization within the bone marrow influences the functional capacity of human HSCs. *Cell Stem Cell*, *13*(2), 175–189.

Halin, C., Mora, J. R., Sumen, C., & von Andrian, U. H. (2005). In vivo imaging of lymphocyte trafficking. *Annual Review of Cell and Developmental Biology*, *21*, 581–603.

Harnett, M. M. (2007). Laser scanning cytometry: Understanding the immune system in situ. *Nature Reviews. Immunology*, 7(11), 897–904.

Hawkins, E. D., Duarte, D., Akinduro, O., Khorshed, R. A., Passaro, D., Nowicka, M., et al. (2016). T-cell acute leukaemia exhibits dynamic interactions with bone marrow microenvironments. *Nature*, *538*(7626), 518–522.

Helmchen, F., & Denk, W. (2005). Deep tissue two-photon microscopy. *Nature Methods*, *2*(12), 932–940.

Hills, D., Gribi, R., Ure, J., Buza-Vidas, N., Luc, S., Jacobsen, S. E., et al. (2011). Hoxb4-YFP reporter mouse model: A novel tool for tracking HSC development and studying the role of Hoxb4 in hematopoiesis. *Blood*, *117*(13), 3521–3528.

Hirata, E., Girotti, M. R., Viros, A., Hooper, S., Spencer-Dene, B., Matsuda, M., et al. (2015). Intravital imaging reveals how BRAF inhibition generates drug-tolerant microenvironments with high integrin beta1/FAK signaling. *Cancer Cell*, *27*(4), 574–588.

Hoggatt, J., Kfoury, Y., & Scadden, D. T. (2016). Hematopoietic stem cell niche in health and disease. *Annual Review of Pathology*, *11*, 555–581.

Hooper, A. T., Butler, J. M., Nolan, D. J., Kranz, A., Iida, K., Kobayashi, M., et al. (2009). Engraftment and reconstitution of hematopoiesis is dependent on VEGFR2-mediated regeneration of sinusoidal endothelial cells. *Cell Stem Cell*, *4*(3), 263–274.

Itkin, T., Gur-Cohen, S., Spencer, J. A., Schajnovitz, A., Ramasamy, S. K., Kusumbe, A. P., et al. (2016). Distinct bone marrow blood vessels differentially regulate haematopoiesis. *Nature*, *532*(7599), 323–328.

Ito, K., Turcotte, R., Cui, J., Zimmerman, S. E., Pinho, S., Mizoguchi, T., et al. (2016). Self-renewal of a purified Tie2+ hematopoietic stem cell population relies on mitochondrial clearance. *Science*, *354*(6316), 1156–1160.

Joseph, C., Quach, J. M., Walkley, C. R., Lane, S. W., Lo Celso, C., & Purton, L. E. (2013). Deciphering hematopoietic stem cells in their niches: A critical appraisal of genetic models, lineage tracing, and imaging strategies. *Cell Stem Cell*, *13*(5), 520–533.

Jung, S. H., Evans, C. J., Uemura, C., & Banerjee, U. (2005). The Drosophila lymph gland as a developmental model of hematopoiesis. *Development*, *132*(11), 2521–2533.

Junt, T., Schulze, H., Chen, Z., Massberg, S., Goerge, T., Krueger, A., et al. (2007). Dynamic visualization of thrombopoiesis within bone marrow. *Science*, *317*(5845), 1767–1770.

Kalajzic, I., Kalajzic, Z., Kaliterna, M., Gronowicz, G., Clark, S. H., Lichtler, A. C., et al. (2002). Use of type I collagen green fluorescent protein transgenes to identify subpopulations of cells at different stages of the osteoblast lineage. *Journal of Bone and Mineral Research*, *17*(1), 15–25.

Kawamoto, T. (2003). Use of a new adhesive film for the preparation of multi-purpose fresh-frozen sections from hard tissues, whole-animals, insects and plants. *Archives of Histology and Cytology*, *66*(2), 123–143.

Kedrin, D., Gligorijevic, B., Wyckoff, J., Verkhusha, V. V., Condeelis, J., Segall, J. E., et al. (2008). Intravital imaging of metastatic behavior through a mammary imaging window. *Nature Methods, 5*(12), 1019–1021.

Khorshed, R. A., Hawkins, E. D., Duarte, D., Scott, M. K., Akinduro, O. A., Rashidi, N. M., et al. (2015). Automated identification and localization of hematopoietic stem cells in 3D intravital microscopy data. *Stem Cell Reports, 5*(1), 139–153.

Kiel, M. J., Yilmaz, O. H., Iwashita, T., Yilmaz, O. H., Terhorst, C., & Morrison, S. J. (2005). SLAM family receptors distinguish hematopoietic stem and progenitor cells and reveal endothelial niches for stem cells. *Cell, 121*(7), 1109–1121.

Kim, S., Lin, L., Brown, G. A., Hosaka, K., & Scott, E. W. (2016). Extended time-lapse in vivo imaging of tibia bone marrow to visualize dynamic hematopoietic stem cell engraftment. *Leukemia*. http://dx.doi.org/10.1038/leu.2016.354 [Epub ahead of print.

Kissa, K., & Herbomel, P. (2010). Blood stem cells emerge from aortic endothelium by a novel type of cell transition. *Nature, 464*(7285), 112–115.

Kobayashi, H., Butler, J. M., O'Donnell, R., Kobayashi, M., Ding, B. S., Bonner, B., et al. (2010). Angiocrine factors from Akt-activated endothelial cells balance self-renewal and differentiation of haematopoietic stem cells. *Nature Cell Biology, 12*(11), 1046–1056.

Koechlein, C. S., Harris, J. R., Lee, T. K., Weeks, J., Fox, R. G., Zimdahl, B., et al. (2016). High-resolution imaging and computational analysis of haematopoietic cell dynamics in vivo. *Nature Communications*, 7, 12169.

Kohler, A., Schmithorst, V., Filippi, M. D., Ryan, M. A., Daria, D., Gunzer, M., et al. (2009). Altered cellular dynamics and endosteal location of aged early hematopoietic progenitor cells revealed by time-lapse intravital imaging in long bones. *Blood, 114*(2), 290–298.

Krauss, J., Astrinidis, P., Frohnhofer, H. G., Walderich, B., & Nusslein-Volhard, C. (2013). Transparent, a gene affecting stripe formation in Zebrafish, encodes the mitochondrial protein Mpv17 that is required for iridophore survival. *Biology Open, 2*(7), 703–710.

Kunisaki, Y., Bruns, I., Scheiermann, C., Ahmed, J., Pinho, S., Zhang, D., et al. (2013). Arteriolar niches maintain haematopoietic stem cell quiescence. *Nature, 502*(7473), 637–643.

Kusumbe, A. P., Ramasamy, S. K., & Adams, R. H. (2014). Coupling of angiogenesis and osteogenesis by a specific vessel subtype in bone. *Nature, 507*(7492), 323–328.

Kusumbe, A. P., Ramasamy, S. K., Starsichova, A., & Adams, R. H. (2015). Sample preparation for high-resolution 3D confocal imaging of mouse skeletal tissue. *Nature Protocols, 10*(12), 1904–1914.

Lapidot, T., Dar, A., & Kollet, O. (2005). How do stem cells find their way home? *Blood, 106*(6), 1901–1910.

Lassailly, F., Foster, K., Lopez-Onieva, L., Currie, E., & Bonnet, D. (2013). Multimodal imaging reveals structural and functional heterogeneity in different bone marrow compartments: Functional implications on hematopoietic stem cells. *Blood, 122*(10), 1730–1740.

Lewandowski, D., Barroca, V., Duconge, F., Bayer, J., Van Nhieu, J. T., Pestourie, C., et al. (2010). In vivo cellular imaging pinpoints the role of reactive oxygen species in the early steps of adult hematopoietic reconstitution. *Blood, 115*(3), 443–452.

Lo Celso, C., Fleming, H. E., Wu, J. W., Zhao, C. X., Miake-Lye, S., Fujisaki, J., et al. (2009). Live-animal tracking of individual haematopoietic stem/progenitor cells in their niche. *Nature, 457*(7225), 92–96.

Ma, D., Zhang, J., Lin, H. F., Italiano, J., & Handin, R. I. (2011). The identification and characterization of zebrafish hematopoietic stem cells. *Blood, 118*(2), 289–297.

Mazo, I. B., Gutierrez-Ramos, J. C., Frenette, P. S., Hynes, R. O., Wagner, D. D., & von Andrian, U. H. (1998). Hematopoietic progenitor cell rolling in bone marrow microvessels: Parallel contributions by endothelial selectins and vascular cell adhesion molecule 1. *The Journal of Experimental Medicine*, *188*(3), 465–474.

McClugage, S. G., Jr., McCuskey, R. S., & Meineke, H. A. (1971). Microscopy of living bone marrow in Situ. II. Influence of the microenvironment on hemopoiesis. *Blood*, *38*(1), 96–107.

McCulloch, E. A., & Till, J. E. (1960). The radiation sensitivity of normal mouse bone marrow cells, determined by quantitative marrow transplantation into irradiated mice. *Radiation Research*, *13*, 115–125.

McCuskey, R. S., McClugage, S. G., & Younker, W. J. (1971). Microscopy of living bone marrow in situ. *Blood*, *38*(1), 87–95.

Mempel, T. R., Pittet, M. J., Khazaie, K., Weninger, W., Weissleder, R., von Boehmer, H., et al. (2006). Regulatory T cells reversibly suppress cytotoxic T cell function independent of effector differentiation. *Immunity*, *25*(1), 129–141.

Mendez-Ferrer, S., Michurina, T. V., Ferraro, F., Mazloom, A. R., Macarthur, B. D., Lira, S. A., et al. (2010). Mesenchymal and haematopoietic stem cells form a unique bone marrow niche. *Nature*, *466*(7308), 829–834.

Migliaccio, A. R., Carta, C., & Migliaccio, G. (1999). In vivo expansion of purified hematopoietic stem cells transplanted in nonablated W/Wv mice. *Experimental Hematology*, *27*(11), 1655–1666.

Mignone, J. L., Kukekov, V., Chiang, A. S., Steindler, D., & Enikolopov, G. (2004). Neural stem and progenitor cells in nestin-GFP transgenic mice. *The Journal of Comparative Neurology*, *469*(3), 311–324.

Miller, C. L., & Eaves, C. J. (1997). Expansion in vitro of adult murine hematopoietic stem cells with transplantable lympho-myeloid reconstituting ability. *Proceedings of the National Academy of Sciences of the United States of America*, *94*(25), 13648–13653.

Mondal, B. C., Mukherjee, T., Mandal, L., Evans, C. J., Sinenko, S. A., Martinez-Agosto, J. A., et al. (2011). Interaction between differentiating cell- and niche-derived signals in hematopoietic progenitor maintenance. *Cell*, *147*(7), 1589–1600.

Moreau, H. D., & Bousso, P. (2012). DISC—A guide to in vivo cytometry. *IntraVital*, *1*(1), 27–31.

Moreau, H. D., Lemaitre, F., Terriac, E., Azar, G., Piel, M., Lennon-Dumenil, A. M., et al. (2012). Dynamic in situ cytometry uncovers T cell receptor signaling during immunological synapses and kinapses in vivo. *Immunity*, *37*(2), 351–363.

Morin-Poulard, I., Sharma, A., Louradour, I., Vanzo, N., Vincent, A., & Crozatier, M. (2016). Vascular control of the Drosophila haematopoietic microenvironment by Slit/Robo signalling. *Nature Communications*, 7, 11634.

Morrison, S. J., & Scadden, D. T. (2014). The bone marrow niche for haematopoietic stem cells. *Nature*, *505*(7483), 327–334.

Nilsson, S. K., Johnston, H. M., & Coverdale, J. A. (2001). Spatial localization of transplanted hemopoietic stem cells: Inferences for the localization of stem cell niches. *Blood*, *97*(8), 2293–2299.

Nombela-Arrieta, C., Pivarnik, G., Winkel, B., Canty, K. J., Harley, B., Mahoney, J. E., et al. (2013). Quantitative imaging of haematopoietic stem and progenitor cell localization and hypoxic status in the bone marrow microenvironment. *Nature Cell Biology*, *15*(5), 533–543.

Notta, F., Doulatov, S., Laurenti, E., Poeppl, A., Jurisica, I., & Dick, J. E. (2011). Isolation of single human hematopoietic stem cells capable of long-term multilineage engraftment. *Science*, *333*(6039), 218–221.

Owusu-Ansah, E., & Banerjee, U. (2009). Reactive oxygen species prime Drosophila haematopoietic progenitors for differentiation. *Nature*, *461*(7263), 537–541.
Paik, E. J., & Zon, L. I. (2010). Hematopoietic development in the zebrafish. *The International Journal of Developmental Biology*, *54*(6–7), 1127–1137.
Pinho, S., Lacombe, J., Hanoun, M., Mizoguchi, T., Bruns, I., Kunisaki, Y., et al. (2013). PDGFRalpha and CD51 mark human nestin + sphere-forming mesenchymal stem cells capable of hematopoietic progenitor cell expansion. *The Journal of Experimental Medicine*, *210*(7), 1351–1367.
Pitt, L. A., Tikhonova, A. N., Hu, H., Trimarchi, T., King, B., Gong, Y., et al. (2015). CXCL12-producing vascular endothelial niches control acute T cell leukemia maintenance. *Cancer Cell*, *27*(6), 755–768.
Rahmig, S., Kronstein-Wiedemann, R., Fohgrub, J., Kronstein, N., Nevmerzhitskaya, A., Bornhauser, M., et al. (2016). Improved human erythropoiesis and platelet formation in humanized NSGW41 mice. *Stem Cell Reports*, 7(4), 591–601.
Rak, J., Foster, K., Potrzebowska, K., Safaee Talkhoncheh, M., Miharada, N., Komorowska, K., et al. (2016). Cytohesin 1 regulates homing and engraftment of human hematopoietic stem- and progenitor cells. *Blood*, pii: blood-2016-06-720649. http://dx.doi.org/10.1182/blood-2016-06-720649 [Epub ahead of print].
Rashidi, N. M., Scott, M. K., Scherf, N., Krinner, A., Kalchschmidt, J. S., Gounaris, K., et al. (2014). In vivo time-lapse imaging shows diverse niche engagement by quiescent and naturally activated hematopoietic stem cells. *Blood*, *124*(1), 79–83.
Ritsma, L., Ellenbroek, S. I., Zomer, A., Snippert, H. J., de Sauvage, F. J., Simons, B. D., et al. (2014). Intestinal crypt homeostasis revealed at single-stem-cell level by in vivo live imaging. *Nature*, *507*(7492), 362–365.
Rodda, S. J., & McMahon, A. P. (2006). Distinct roles for Hedgehog and canonical Wnt signaling in specification, differentiation and maintenance of osteoblast progenitors. *Development*, *133*(16), 3231–3244.
Rompolas, P., Deschene, E. R., Zito, G., Gonzalez, D. G., Saotome, I., Haberman, A. M., et al. (2012). Live imaging of stem cell and progeny behaviour in physiological hair-follicle regeneration. *Nature*, *487*(7408), 496–499.
Sacchetti, B., Funari, A., Michienzi, S., Di Cesare, S., Piersanti, S., Saggio, I., et al. (2007). Self-renewing osteoprogenitors in bone marrow sinusoids can organize a hematopoietic microenvironment. *Cell*, *131*(2), 324–336.
Sapoznikov, A., Pewzner-Jung, Y., Kalchenko, V., Krauthgamer, R., Shachar, I., & Jung, S. (2008). Perivascular clusters of dendritic cells provide critical survival signals to B cells in bone marrow niches. *Nature Immunology*, *9*(4), 388–395.
Schofield, R. (1978). The relationship between the spleen colony-forming cell and the haemopoietic stem cell. *Blood Cells*, *4*(1–2), 7–25.
Scott, M. K., Akinduro, O., & Lo, Celso C. (2014). In vivo 4-dimensional tracking of hematopoietic stem and progenitor cells in adult mouse calvarial bone marrow. *Journal of Visualized Experiments*, *91*, e51683.
Shultz, L. D., Lyons, B. L., Burzenski, L. M., Gott, B., Chen, X., Chaleff, S., et al. (2005). Human lymphoid and myeloid cell development in NOD/LtSz-scid IL2R gamma null mice engrafted with mobilized human hemopoietic stem cells. *The Journal of Immunology*, *174*(10), 6477–6489.
Sipkins, D. A., Wei, X., Wu, J. W., Runnels, J. M., Cote, D., Means, T. K., et al. (2005). In vivo imaging of specialized bone marrow endothelial microdomains for tumour engraftment. *Nature*, *435*(7044), 969–973.
Spencer, J. A., Ferraro, F., Roussakis, E., Klein, A., Wu, J., Runnels, J. M., et al. (2014). Direct measurement of local oxygen concentration in the bone marrow of live animals. *Nature*, *508*(7495), 269–273.

Stinchcombe, J. C., Bossi, G., Booth, S., & Griffiths, G. M. (2001). The immunological synapse of CTL contains a secretory domain and membrane bridges. *Immunity, 15*(5), 751–761.
Sumen, C., Mempel, T. R., Mazo, I. B., & von Andrian, U. H. (2004). Intravital microscopy: Visualizing immunity in context. *Immunity, 21*(3), 315–329.
Susaki, E. A., Tainaka, K., Perrin, D., Kishino, F., Tawara, T., Watanabe, T. M., et al. (2014). Whole-brain imaging with single-cell resolution using chemical cocktails and computational analysis. *Cell, 157*(3), 726–739.
Tamplin, O. J., Durand, E. M., Carr, L. A., Childs, S. J., Hagedorn, E. J., Li, P., et al. (2015). Hematopoietic stem cell arrival triggers dynamic remodeling of the perivascular niche. *Cell, 160*(1–2), 241–252.
Theocharides, A. P., Rongvaux, A., Fritsch, K., Flavell, R. A., & Manz, M. G. (2016). Humanized hemato-lymphoid system mice. *Haematologica, 101*(1), 5–19.
Till, J. E., & Mc, C. E. (1961). A direct measurement of the radiation sensitivity of normal mouse bone marrow cells. *Radiation Research, 14*, 213–222.
Turcotte, R., Alt, C., Mortensen, L. J., & Lin, C. P. (2014). Characterization of multiphoton microscopy in the bone marrow following intravital laser osteotomy. *Biomedical Optics Express, 5*(10), 3578–3588.
Visnjic, D., Kalajzic, Z., Rowe, D. W., Katavic, V., Lorenzo, J., & Aguila, H. L. (2004). Hematopoiesis is severely altered in mice with an induced osteoblast deficiency. *Blood, 103*(9), 3258–3264.
White, R. M., Sessa, A., Burke, C., Bowman, T., LeBlanc, J., Ceol, C., et al. (2008). Transparent adult zebrafish as a tool for in vivo transplantation analysis. *Cell Stem Cell, 2*(2), 183–189.
Wilson, A., Murphy, M. J., Oskarsson, T., Kaloulis, K., Bettess, M. D., Oser, G. M., et al. (2004). c-Myc controls the balance between hematopoietic stem cell self-renewal and differentiation. *Genes and Development, 18*(22), 2747–2763.
Xie, Y., Yin, T., Wiegraebe, W., He, X. C., Miller, D., Stark, D., et al. (2009). Detection of functional haematopoietic stem cell niche using real-time imaging. *Nature, 457*(7225), 97–101.
Zhang, J., Niu, C., Ye, L., Huang, H., He, X., Tong, W. G., et al. (2003). Identification of the haematopoietic stem cell niche and control of the niche size. *Nature, 425*(6960), 836–841.
Zhou, B. O., Ding, L., & Morrison, S. J. (2015). Hematopoietic stem and progenitor cells regulate the regeneration of their niche by secreting Angiopoietin-1. *eLife, 4*. e05521.
Zipfel, W. R., Williams, R. M., & Webb, W. W. (2003). Nonlinear magic: Multiphoton microscopy in the biosciences. *Nature Biotechnology, 21*(11), 1369–1377.

CHAPTER FIVE

Mechanisms of Hematopoietic Stem and Progenitor Cells Bone Marrow Homing and Mobilization

A. Kumari[1], K. Golan, E. Khatib-Massalha, O. Kollet, T. Lapidot[1]
Weizmann Institute of Science, Rehovot, Israel
[1]Corresponding authors: e-mail address: anju.kumari@weizmann.ac.il; Tsvee.Lapidot@weizmann.ac.il

Contents

1. INTRODUCTION

Under steady-state conditions, most of HSPC are retained in the BM, while only low levels of quiescent progenitors are present in the circulation which replenish the daily needs (Laird, von Andrian, & Wagers, 2008; Lapidot, Dar, & Kollet, 2005). HSPC development and functions in the BM are regulated by the factors/cues from the microenvironment/specialized niche, the current concepts and knowledge about which have been

Advances in Stem Cells and their Niches, Volume 1
ISSN 2468-5097
http://dx.doi.org/10.1016/bs.asn.2016.12.003

described comprehensively by Morrison and Scadden (2014). Osteoblasts and mesenchymal stem/stromal cells (MSCs) are important components of the niche which mediate retention of HSPC in the BM. Osteoclasts in endosteal niche mediate bone remodeling (Kollet et al., 2006), while depletion of MSC induces HSPC egress (Chow et al., 2011; Winkler et al., 2010). Despite technological advancements in high-resolution imaging, the exact location of BM long-term repopulating (LTR)-HSCs and the anatomy of their niches is not completely understood. Immature murine $Lin^{-}Sca^{+}c\text{-}kit^{+}$ (LSK) were found attached to CXCL12 (SDF-1)-expressing stromal cells. More enriched, primitive signaling lymphocyte activation molecule (SLAM) cells (Kiel & Morrison, 2008) inhabit adjacent to stromal cell-derived factor (SDF)-1-expressing CXCL12-abundant reticular (CAR) cells (Sugiyama, Kohara, Noda, & Nagasawa, 2006) and LTR-HSCs functionally expressing endothelial protein C receptor (EPCR) reside in unique anticoagulant vascular niches in the murine BM (Gur-Cohen et al., 2015). Adhesion molecules which are directly involved in cell-to-cell contact and attachment to extracellular matrix (ECM) components play a role in BM HSPC maintenance. Chemotactic SDF-1/CXCR4 axis is crucial for HSPC maintenance in the BM, where SDF-1 provides a retention signal and attracts cells-expressing CXC chemokine receptor 4 (CXCR4). The chemokine SDF-1 is expressed by osteoblasts (Ponomaryov et al., 2000) and endothelial cells and stromal cells, whereas its receptor CXCR4 is expressed by HSPC. During host defense responses and in response to stress, the adhesion interactions between HSPC and the microenvironment are disrupted to induce robust HSPC mobilization to the circulation. In the clinical practices, these effects are mimicked by pharmacological approaches which can desensitize chemotactic responses and impair adhesion interactions between HSPC and their niches, thus enforcing HSPC to egress out of the BM. The critical step in stem cell transplantation is collection from different sources including umbilical cord blood (CB), BM, and mobilized peripheral blood. However, stem cells recovered from CB are not sufficient to support adult hematopoietic transplantation, while stem cells aspiration from BM comprises use of invasive techniques. Therefore, harvesting stem cells mobilized to the peripheral blood is more suitable and preferred approach for transplantation owing to less ethical constrains (unlike use of CB) and its noninvasive nature (unlike use of BM cells). On the other hand, for a successful transplantation, HSPC need to home back to the BM of the recipient following the navigation cues and finally lodge in their specialized niches. This involves increased chemotactic signals which facilitate the attraction of the cells toward the BM. Subsequently, HSPC cross the endothelial barrier, a navigation step which is facilitated by selectins and other adhesion molecules and then the homing cells are directed to reach their specialized niches in the BM. In contrast to mobilization, increased activity of cell adhesion molecules and microenvironment interactions are needed for reestablishing homed stem cells in the BM, their lodgment and repopulation. HSPC mobilization and homing share various regulators and overlapping mechanisms including adhesion to the vascular wall and crossing over the endothelial blood–BM barrier. Newer studies in the stem cell biology providing deeper insights of HSPC maintenance in the BM have helped to improve the current stem cell mobilization regimes. Understanding the mechanisms of HSPC mobilization and homing would be crucial for efficacy of HSPC-based therapies.

2. GETTING HSPC IN THE BM NICHE (HSPC HOMING)

The process of migration of transplanted hematopoietic stem cells through the circulation, crossing the endothelial barrier and finally lodging to place of their origin (i.e., bone marrow) is termed as "homing." This active process is orchestrated by coordinated actions of various adhesion molecules, integrins, selectins, cytokines, chemokines, and receptors. The process of homing is very fast and can be measured within hours. Prior to transplantation, the recipients must undergo preconditioning with irradiation or chemotherapy in order to remove the cycling hematopoietic cells and initiate recruiting signals as discussed ahead. Preconditioning of mice through total body irradiation increased SDF-1 and stem cell factor (SCF) secretion in the BM and spleen within 24–48 h (Cottler-Fox et al., 2003; Ponomaryov et al., 2000; Zhao, Zhan, Burke, & Anderson, 2005). Megakaryocytes in the endosteal region were reported to enhance osteoblasts expansion and HSC engraftment (Dominici et al., 2009; Olson et al., 2013) following irradiation. The various events occurring during HSC homing were unraveled by real-time imaging (Xie et al., 2009). Several reports showed that postirradiation HSCs trafficked to trabecular bone areas and in particular, N-cadherin^{+} preosteoblastic cells, however, no such homing was observed without radiation conditioning. However, this preconditioning might not be crucial for the homing itself but importantly creates a conducible environment which facilitates successful repopulation.

Preconditioning induces BM endothelial barrier damage and enhances the secretion of various chemokines, cytokines, and proteolytic enzymes all of which determine the efficiency and success of migration from the site of transplantation back to BM. Homing assays based on fluorescent dyes allow to track the homing of HSCs at the individual cell level at early time points posttransplant under physiological conditions (Ellis et al., 2011; Nilsson, Johnston, & Coverdale, 2001; Williams & Nilsson, 2009). Recent work from our lab using live imaging showed specialized sites for HSPC and leukocyte homing under physiological condition (Itkin et al., 2016). Here in this section of the chapter we will discuss about role of different factors and the suggestive mechanisms involved in homing.

2.1 CXCR4/SDF-1 Axis Regulates HSPC Homing via Modulating Rac-1 and PKCζ Activity

CXCR4 receptor on HSPC and its ligand SDF-1 (CXCR4/SDF-1) interactions are important for maintaining the HSPC in the BM. However, CXCR4/SDF-1 axis is also crucial in mediating navigation of HSPC from circulation to the BM niches (Avigdor et al., 2004; Lapidot & Kollet, 2010; Lévesque, Helwani, & Winkler, 2010; Peled et al., 1999; Ratajczak, 2015). Elevated levels of CXCR4 show enhanced chemotaxis of HSPC toward SDF-1. It was shown that CXCR4 blocking restricted homing and engraftment of human $CD34^{+}CD38^{-}$ immature human progenitors in NOD/SCID mice (Kollet et al., 2001; Peled et al., 1999). In the same line, homing of fetal liver HSPC from $CXCR4^{-/-}$ mice was severely impaired but was rescued by increasing the cell dose (Foudi et al., 2006). CXCR4 influences homing by modulating the downstream signaling pathways including activation of atypical protein kinase C ζ (PKC-ζ) which further induces protein tyrosine kinase 2 beta (Pyk2) and ERK pathways (Bonig, Priestley, Nilsson, Jiang, & Papayannopoulou, 2004; Petit et al., 2005). A broad-range PKC inhibitor, chelerythrine chloride abrogated homing of both human $CD34^{+}$ cells and murine progenitors to the murine BM (Kollet et al., 2001). PKCζ regulates SDF-1-induced chemotaxis, matrix metallopeptidase 9 (MMP-9) secretion, and adhesion of CD34 cells (Petit et al., 2005). Furthermore, in response to 3′,5′-cyclic adenosine monophosphate (cAMP), it elevates functional CXCR4 levels on immature human $CD34^{+}$ cells. Homing of cAMP-treated $CD34^{+}$ cells was enhanced and it was dependent on activity of RHO family GTPase Rac-1 and PKCζ (Goichberg et al., 2005). Therefore, CXCR4 and PKCζ seem to regulate their levels and functions in a feedback manner. Deficiency of Rac1 in $Lin^{-}KIT^{+}$ cells led to failure in engraftment (Gu et al., 2003) and reduced homing efficiency suggesting involvement of Rac-1 in homing and repopulation. Mechanistically, Rac-1 deletion reduced SDF-1-induced migration and inhibited homing of murine HSPC (Cancelas et al., 2005). Furthermore, depletion of both Rac1 and Rac2 caused vast egress of HSPC to peripheral blood, defects in migration, proliferation, and adhesion (Gu et al., 2003). Taken together, CXCR4 activation, SDF-1 chemotaxis, Rac-1 and PKCζ activity regulate homing and retention of HSPC in the BM (Cancelas et al., 2005; Williams, Zheng, & Cancelas, 2008). Furthermore, Rac GTPase effector p21-activated kinase 2 (Pak2) affects HSPC homing and engraftment (Dorrance et al., 2013) and involved in survival, proliferation, and differentiation of HSC (Zeng et al., 2015). Recently, it has been shown that Pak2 activity and its interaction with PAK-interacting exchange factor-β (β-Pix) regulate the HSPC migration, cytoskeleton integrity, and homing via CDC42 activation (Reddy et al., 2016).

Pretreatment of $CD34^+$ cells with SDF-1α and anti-CXCR4 was shown to increase engraftment in NOD/SCID mice at lower dose; however, higher dose (10- to 100-fold) reduced the repopulation of $CD34^+$ cells (Plett, Frankovitz, Wolber, Abonour, & Orschell-Traycoff, 2002). Homing of human mobilized and CB CD34-enriched cells to the BM and spleen was found to be enhanced in nonirradiated $B2m^{null}$ NOD/SCID mice following intravenously injection of SDF-1 (Dar et al., 2005; Dar, Kollet, & Lapidot, 2006). Priming of HSPC with supernatants of leukapheresis products (SLPs) of granulocyte-colony stimulating factor (G-CSF) mobilized or fibrinogen, fibronectin, soluble VCAM-1, ICAM-1, and urokinase plasminogen activator receptor (uPAR) enhance their chemotactic responses to SDF-1. Further, these factors could prime CXCR4 and increase its association with membrane lipid rafts facilitating homing process in cooperation with matrix metalloproteinase (MMP) (Wysoczynski et al., 2005). Altogether, CXCR4/SDF-1 and their downstream mediators influence HSPC homing.

Another receptor CXCR7 was later found to have high affinity for SDF-1 (Burns et al., 2006). It is expressed by activated endothelial cells, monocytes, mature B cells, and blood and lymphatic endothelial cells (Burns et al., 2006; Infantino, Moepps, & Thelen, 2006; Neusser et al., 2010; Thelen & Thelen, 2008). The expression of this receptor is enhanced under hypoxic conditions (Esencay, Sarfraz, & Zagzag, 2013) owing to the presence of two hypoxia-responsive elements in its promoter region (Tarnowski et al., 2010). CXCR7 specifically acts as a scavenger of

SDF-1 and by downregulating CXCR4 levels modulate CXCR4/SDF1 responses (Hartmann et al., 2008; Naumann et al., 2010). This receptor CXCR7 is required for LFA-1 and VLA-4 integrin's activation under shear flow (Hartmann et al., 2008), and its expression in B cells was inversely correlated with the activation of CXCR4 (Levoye, Balabanian, Baleux, Bachelerie, & Lagane, 2009). CXCR4 and CXCR7 can form dimers (heterodimer and homodimers) and thus suggested to regulate SDF-1-mediated signaling including homing of HSCs (Levoye et al., 2009). Thus, CXCR7 can modulate responses of CXCR4/SDF-1 axis and integrin expression; however, its direct involvement in migration of HSPC and homing is not reported (Hartmann et al., 2008; Naumann et al., 2010; Tarnowski et al., 2010; Uto-Konomi et al., 2013).

2.2 Integrins and Selectins Mediate HPSC Rolling and Adhesion at Transendothelial Barrier

Integrins and selectins comprise the major adhesion molecules involved in mediating rolling and adhesion of homing cells at the endothelial barrier. HSPC infused through intravenous administration must migrate across endothelium (a process called extravasation) during their navigation toward the BM. Circulating HSPC show reduced affinity to endothelium, but their rolling and tethering to the blood vessel wall under blood flow shear stress is assisted by P-selectin and E-selectin expressed on the surface of endothelial cells. E-selectin colocalizes with SDF-1 into specific microdomains in BM and leading HSPC homing to SDF-1-positive vascular microdomains (Sipkins et al., 2005). Furthermore, E-selectin expedites the process of homing by elevating the expression of the adhesion molecules ICAM-1 and VCAM-1 on the endothelial cells, whereas P-selectin provides anchorage sites for the HSCs. Further firm adhesion of HSPC is governed by integrins including VCAM-1 and ICAM-1. We have shown previously that repopulating human stem cells express integrins including LFA-1, VLA-4, and VLA-5 whose expression is induced by SDF-1. Activated integrins (VLA-4 or VLA-5) which subsequently facilitates transition of $CD34^{+}$ cells from rolling state into stable arrest on the endothelium (Peled et al., 2000) and the subsequent progenitors transmigration across the endothelial lining. Neutralization of VLA-4 and VLA-5 by antibodies precluded engraftment of $CD34^{+}$ cells whereas that of LFA-1 antibody treatment reduced the engraftment suggesting that the engraftment of $CD34^{+}$ was dependent on these integrins (Peled et al., 2000). Similarly, homing of human $CD34^{+}CD38^{-/low}CXCR4^{+}$ stem and progenitor cells to the BM and spleen of NOD/SCID mice was also regulated by these integrins and protein kinase C signaling pathway (Kollet et al., 2001). Homing of $CD34^+CD38^+$ cells was enhanced by pretreatment with SCF and repressed by pretreatment with neutralizing antibodies for CXCR4 and integrins and protein kinase C inhibition. Rolling and tethering at endothelium are followed by subsequent HSPC transmigration through endothelium (Katayama et al., 2003; Mazo et al., 1998) and requires activity of matrix-degrading enzymes MMPs.

2.3 MMPs Activity Facilitates Transendothelial Migration

Activity of matrix-degrading MMPs induces a proteolytic milieu in the BM by digesting the BM niche components and inactivating the factors involved in HSPC retention. MMP-2, MMP-9, membrane-anchored MT1-MMP (MMP-14), and neutrophil collagenase MMP-8 constitute the main MMPs present in the BM. MMP-2 and MMP-9 regulate hematopoietic cell migration (in vivo) and their migration toward SDF-1 in vitro (Janowska-Wieczorek, Marquez, Dobrowsky, Ratajczak, & Cabuhat, 2000). The role of MMPs in homing has been shown by various groups. SCF-induced MMPs expression was found to increase homing and repopulation in NOD/SCID mice (Byk et al., 2005). SDF-1 and G-CSF can also induce activation of MT1-MMP by human $CD34^+$ progenitors (Avigdor et al., 2004). Incubation of CB $CD34^+$ cells in MMP2 and MMP9 rich media resulted in elevated CXCR4 expression and migration, suggesting that besides creating a proteolytic environment, MMPs can also modulate motility via CXCR4 upregulation. Our lab has previously shown that stress-induced SDF-1, MMP-9, and HGF recruit human $CD34^+$ progenitors from the BM to the liver of NOD/SCID mice (Kollet et al., 2003), thus suggesting that hematopoietic cells respond to stress signals as part of tissue repair. BM c-kit^+ cells from MT1-MMP-deficient mice exhibited significantly reduced engraftment levels and inhibition of MT1-MMP by monoclonal antibody attenuated homing of human HSPC in a NOD/SCID mouse model (Vagima et al., 2009). Similarly, a coordinated interaction of CXCR4 and MT1-MMP is crucial for the melanoma metastasis to lungs (Bartolomé, Ferreiro, Miquilena-Colina, et al., 2009). These studies support critical role of MT1-MMP in HSC homing and metastasis. MT1-MMP can activate pro-MMP-2 and this can further activate other proteases including MMP-9 which further degrades ECM barriers and thus facilitates transendothelial migration and subsequent homing.

2.4 Bioactive Lipids Promote Homing by Modulating Motility

Among sphingolipids, bioactive sphingosine-1-phosphate (S1P) levels create a gradient to which HSPC respond (Golan et al., 2012; Juarez et al., 2012; Massberg et al., 2007; Ratajczak et al., 2010). The levels of S1P are much higher in the circulation (Bendall & Basnett, 2013; Venkataraman et al., 2008) where mature red blood cells, activated platelets, and endothelial cells contribute to its production (Fukuhara et al., 2012; Pappu et al., 2007). S1P and activation of its receptors increase CXCR4 signaling, trans-endothelial migration, and in vivo homing along with influencing cell proliferation and angiogenesis regulation (Alvarez, Milstien, & Spiegel, 2007; Jo et al., 2005; Kimura et al., 2004). S1P and ceramide-1-phosphate (C1P) were found to be upregulated in conditioned BM for transplantation (Kim et al., 2012). Therefore, despite reduced SDF-1 in response to proteolytic microenvironment, S1P and C1P were proposed to support homing. Activation of S1P receptor on human $CD34^+$ progenitor cells has been shown to increase homing in vivo in xenograft model (Adamiak et al., 2015; Kimura et al., 2004). On the contrary, BM-expressed sphingosine kinase 1 ($Sphk1^{-/-}$)-deficient mice transplanted with hematopoietic cells from normal control and $CXCR4^{-/-}$ mice revealed homing and engraftment defects (Adamiak et al., 2015) further strengthening the involvement of sphingolipids in mediating homing.

Prostaglandins belong to a large family of bioactive lipids. A rare population of αSMA^+ macrophage in the BM was shown to express COX-2 enzyme (Ludin et al., 2012) which can synthesize Prostaglandin E2 (PGE2). In vitro treatment of PGE2 increases the CXCR4 expression among both murine and human HSPC which enhances homing and engraftment (Broxmeyer & Pelus, 2014; Goichberg et al., 2006; Hoggatt, Singh, Sampath, & Pelus, 2009). Stimulation of $CD34^+$ hematopoietic progenitors by PGE2 and cAMP significantly elevated their functional CXCR4 levels which were mediated by PKCζ activity (Goichberg et al., 2006). Inhibition of cAMP effector-Rap1 or interference with the activation of Rac1 reduced cAMP-induced CXCR4 expression and PKCζ activation thus PGE2-induced homing was suggested to be mediated through CXCR4 upregulation. Similarly, ex vivo treatment of umbilical CB with dmPGE2 was found to be safe and enhanced the efficacy of long-term engraftment with accelerated neutrophil recovery (Cutler et al., 2013). They found that PGE2 upregulated the genes like *CXCR4*, *cyclin D1*, and *survivin* which are responsible for homing, proliferation, and cell survival. Taken together, PGE2 regulates hematopoietic stem cell homeostasis (North et al., 2007), increases expression of surface CXCR4 on the HSPCs, facilitates their responsiveness to SDF-1 gradients, enhances homing, and maintains the stem cells in the BM.

2.5 CD44 and HA Interactions Modulate Expression of MMP and Binding to Selectins

CD44 is expressed on the surface of HSPC (Dimitroff, Lee, Rafii, Fuhlbrigge, & Sackstein, 2001) and mature myeloid cells (Katayama, Hidalgo, Chang, Peired, & Frenette, 2005), whereas its ligand hyaluronic acid (HA) is expressed in BM sinusoid endothelium and the endosteum region. CD44 and its ligand HA interaction are considered important for trafficking of human HPCs including homing and repopulation (Avigdor et al., 2004). Treatment with anti-CD44 monoclonal antibodies or intravenous injection of HA prevents the homing of human CB and mobilized peripheral blood $CD34^+$ cells (Avigdor et al., 2004). CD44 is involved in the SDF-1-induced migration of human HPCs, and CD44 expression was suggested to be important for the long-term repopulation of NOD/SCID mice by human $CD34^+$ HSCs. HA improves homing and engraftment of CB HSPC in MT1-MMP-dependent manner. HA-mediated increased MT1-MMP expression in CB HSPC enhances chemoinvasion of $CD34^+$ cells toward low SDF-1 gradient (Shirvaikar, Marquez-Curtis, Ratajczak, & Janowska-Wieczorek, 2011). HA and thrombin elevate MAPK, PI3K, and Rac1 signaling, whereas the inhibition of PI3K and Rac-1-attenuated MT1-MMP expression, reduced pro-MMP-2 activation and chemoinvasion toward SDF-1. Pretreatment of CB HSPC with HA or thrombin before transplantation was suggested to improve their homing and engraftment (Shirvaikar et al., 2011). CD44 can also bind to selectins when properly fucosylated which can be endowed by enforced fucosylation (Sackstein et al., 2008). Thus, the CD44- and HA-regulated homing is dependent on their ability to modulate expression of MMP and binding to selectins and thereby increasing the chemotaxis toward SDF-1.

2.6 Can We Improve Homing?

Under clinical situations, successful homing is relevant only if it is able to result in engraftment and repopulation in the host BM. As discussed before, various chemotactic factors including SDF-1/CXCR4, S1P, PGE2, selectins, integrins, adhesion molecules, CD44/HA interaction, and MMPs facilitate this process of homing and which ultimately induce BM repopulation (Lapidot & Kollet, 2010; Lévesque et al., 2010; Ratajczak, 2015). The cells for transplantation, which are usually administered intravenously, need to travel through circulation, cross the endothelial barrier, and find their way back to the BM niches. The trafficking of mature and immature cell trafficking including adhesion, rolling, and transendothelial migration happens exclusively in sinusoidal blood vessels (Itkin et al., 2016). Exposure of BM HSPC to blood plasma compromised long-term repopulation and survival of HSPC. Homed HSCs are maintained in the periarterial niches of BM in a quiescent mode following metabolic reactive oxygen species (ROS) inhibition. Owing to the fact that not all the administered HSPCs home to the stem cell niches in BM, homing and repopulation efficiency in different people using equal number of infused cells might vary. Therefore, developing efficient strategies to improve the seeding efficiency of HSPCs is very important and one of such approaches is transplanting them directly to the BM microenvironment (Pineault & Abu-Khader, 2015). Such intrabone marrow transplantation was found to enhance allogeneic hematopoietic and stromal cells engraftment (Chen et al., 2015). Following this approach more HSC home to spleen and BM showing higher PB chimerism which was impaired by administration of the CXCR4 antagonist, AMD3100. As mentioned earlier, HSPCs homing to the BM are mediated through a crucial SDF-1–CXCR4 axis (Lapidot & Kollet, 2010; Lévesque et al., 2010; Ratajczak, 2015). Relatively low levels of SDF-1 after preconditioning creates a shallow gradient; however, the factors like complement cascade cleavage fragments (C3a and $_{desArg}$C3a), cathelicidin (LL-37), β2-defensin, HA, fibronectin, soluble VCAM-1, ICAM-1, and uPAR may significantly enhance the responsiveness of HSPCs to even a shallow SDF-1 gradient (Avigdor et al., 2004; Ratajczak, 2015; Ratajczak & Adamiak, 2015). Overexpression of CXCR4 on human CD34$^+$ progenitors was found to enhance SDF-1-mediated chemotaxis, rendering them less sensitive to receptor internalization in presence of high levels of SDF-1 and increased the repopulation of CD34$^+$/CD38$^{(-/low)}$ cells in NOD/SCID mice (Kahn et al., 2004). Campbell, Hangoc, Liu, Pollok, and Broxmeyer (2007) have shown that expression of membrane-bound dipeptidyl peptidase-4 (CD26) on donor cells negatively regulates homing and engraftment. CD26 can cleave SDF-1 and this truncated form of SDF-1 has depleted chemotactic activity. Inhibition of CD26 peptidase was found to increase the engraftment of human CB CD34$^+$ cells. Manipulation of fibroblast growth factor (FGF) signaling was suggested to improve HSPC mobilization and transplantation (Itkin et al., 2012). FGF-2-enhanced durable short- and long-term repopulation potential of HSPC was mediated by reduction of ROS, c-Kit receptor activation, and STAT5 phosphorylation.

An interesting way of enhancing the homing and engraftment is recently reported as mild heat treatment. Ex vivo mild heat treatment of human $CD34^+$ cells (Capitano, Hangoc, Cooper, & Broxmeyer, 2015) increased their CXCR4 expression and colocalization of CXCR4 within lipid raft domains which can better interact with Rac1. Also, exposure of HSPCs to the histone deacetylase inhibitor valporic acid induces CXCR4 expression (Gul et al., 2009) and the expansion of UCB-derived Oct-4^+ primitive HSCs (Chaurasia, Gajzer, Schaniel, D'Souza, & Hoffman, 2014). Besides, other factors including ceramide-1-phosphate (C1P) (Ratajczak et al., 2010), extracellular nucleotides such as ATP or UTP (Rossi et al., 2007), as well as certain ions, such as Ca^{2+} and H^+ (Adams et al., 2005; Okajima, 2013), calcium-sensing receptor (CaR) were shown to influence the efficiency of HSC homing to endosteal regions (Adams et al., 2005). CaR-deficient mice had increased LSK in the blood and HSPC from these mice could not lodge properly within endosteum of WT mice. Exposure of HSPC to hypoxic environment (Mantel et al., 2015) and inhibition of Heme oxygenase-1 (HO-1) activity in HSPCs by small molecule inhibitors (Adamiak et al., 2016; Wysoczynski et al., 2015), and inhibition of GSK1 (Ko et al., 2011) have been suggested to improve the homing. Interestingly, incubation of human mononuclear cells with 10% DMSO which significantly increased the percentage of $CXCR4^+$, $CD38^+$, and $CD34^+$cells, enhanced their chemotactic responsiveness, and moreover increased homing of human $CD45^+$ and $CD45^+CD34^+$ cells to the mouse BM (Jarocha, Zuba-Surma, & Majka, 2016). Strategies based on a balanced levels of the factors influencing HSPC motility and navigation could improve the homing. However, we should keep in mind that each donor and the recipient is unique in themselves with respect to their levels of response. Therefore, the strategy which is good for one cohort of patients may not be universally applicable. This highlights the need of personalized therapy based on patients' responses.

3. GETTING HSPC OUT OF THE BM NICHE (HSPC MOBILIZATION)

HSPC residing in the BM are anchored by adhesive interactions provided by stromal niche cells and ECM in the microenvironment provides anchorage sites for HSCs. Different niche-forming cells by providing biochemical and biophysical cues including chemokine SDF-1, SCF, E-selectin, and VCAM-1 maintain and regulate HSPC functions including their self-renewal, differentiation, retention, and mobilization. In order to replenish the daily needs of mature functional immune and blood cells or during alarmed/stress situations like irradiation, chemotherapy, inflammation, or ischemic insult, adhesion interactions between HSPC and its niche components need to be disrupted. These processes result in secretion of SDF-1 and robust mobilization of stem cells from the BM to replenish the compromised hematopoietic cells. Recently, work from our lab has shown that sinusoidal blood vessels are more permeable and serve as exclusive sites for immature and mature leukocyte trafficking, whereas less permeable arterial blood vessels maintain the HSC in ROS low state (Itkin et al., 2016). Reduced endothelial integrity, increased sinusoidal vascular permeability, and transient increase in ROS levels are required for HSPC trafficking via sinusoidal sites. Exposure of BM HSPC to blood plasma increased their ROS levels transiently and also enhanced their migration and differentiation (Itkin et al., 2016). HSPC for transplantation of patients suffering from hematological disorders can be obtained from different sources including umbilical CB, bone marrow, and mobilized peripheral blood. Nowadays, peripheral blood transplants have become of much practice compared to the BM transplants owing to its less invasive collection, rapid neutrophil and platelet recovery (reviewed by Arai & Klingemann, 2003), faster immune reconstitution, and eventually faster engraftment. Reduced morbidity, reduced total cost, and shorter hospitalization time also make this more favored approach. During steady state, number of circulating HSPC in the peripheral blood is very low, while sufficient number of circulating HSPC is a prerequisite for the transplantation. This can be achieved by use of mobilization agents (chemokines and cytokines) that enforce the HSPC to migrate from the BM to PB. Thus, understanding the mechanisms underlying their mobilization would be useful for clinically improving the cell yields for collection from the donors and ultimately benefitting the patients as well as to combat during alarmed situations.

3.1 Commonly Used Mobilization Agents and Their Mechanisms

3.1.1 Chemotherapy

Chemotherapy was reported to induce HSC numbers in the peripheral blood of patients who underwent chemotherapy (Richman, Weiner, &

Yankee, 1976) and was routinely used for HSPC mobilization protocols until the development of cytokine stimulation protocols (Stiff, Murgo, Wittes, DeRisi, & Clarkson, 1983). Later, chemotherapy was shown to enhance endogenous G-CSF levels (Liu, Poursine-Laurent, & Link, 1997) which promoted mobilization. Cyclophosphamide and etoposide are commonly used chemotherapeutic agents. Use of chemotherapy together with cytokines like G-CSF or GM-CSF (a process called chemomobilization) generally provides higher stem cells (Desikan et al., 2001; Pusic et al., 2008). Chemomobilization results in higher HSPC yield possibly by inducing protease release and disrupting the ability of stroma to support stem cells. It provides improved engraftment and lower failure rates but the negative side of this approach is its higher cost and increased toxicity (Hamadani et al., 2012; Hiwase et al., 2007; Tuchman et al., 2015). Chemotherapy (vincristine and cisplatin) treatment to mice damaged sympathetic neurons and resulted in impaired bone marrow function. These mice mobilized poorly with G-CSF. This chemotherapy-mediated loss of bone marrow niche cells including Nestin^{+}MSC and endothelial cells was reduced by providing pharmacological protection (using 4-methylcatechol) to neurons (Lucas et al., 2013). The mobilization defect was also prevented by protecting the nervous system, suggesting use of drugs to prevent or reduce nerve damage during chemotherapy regimes.

3.1.2 Granulocyte-Colony Stimulating Factor

G-CSF is one of the most commonly used mobilization agent in the clinical practice. Knowing that on the 5th day of G-CSF treatment, CD34^{+} cells reach peak levels, G-CSF is injected subcutaneously at least 4 days before the day of apheresis. G-CSF is a natural glycoprotein, produced by the macrophages, endothelium, and other immune cells. G-CSF-mobilized CD34^{+} cells had more neutrophil and mononuclear phagocyte precursors (Donahue et al., 2009). Evidences suggest that G-CSF can induce stem cell mobilization via both protease-dependent as well as protease-independent manner. G-CSF stimulates the BM to produce more myeloid cells in particular granulocytes leading to increased secretion of serine proteases including Cathepsin G and MMP-9 (Heissig et al., 2002). Enhanced proteases activity in turn cleaves adhesion interactions like VLA4/VCAM-1 (Levesque, Takamatsu, Nilsson, Haylock, & Simmons, 2001) and CXCR4/SDF-1, thus creating a niche which is now less supportive for stem cell maintenance in the BM. We have reported previously that G-CSF reduces levels of chemokine SDF-1 which was induced by degradation of neutrophil elastase and these effects correlated with the stem cell mobilization. Inhibiting the elastase activity declined both SDF-1 reduction and stem cell mobilization. Furthermore, G-CSF-induced mobilization of human and murine stem cells was abrogated by neutralizing CXCR4 or SDF-1. G-CSF-induced mobilization upregulated the expression of CXCR4 in the bone marrow to facilitate the mobilization (Petit et al., 2002). Thus, stem cell mobilization correlated with CXCR4 upregulation, reduction in BM SDF-1 levels which in turn was caused by activity of neutrophil elastase.

Katayama et al. (2006) have reported the involvement of signals from sympathetic nervous system (SNS) in mediating G-CSF-induced mobilization. They suggested that signals from the SNS regulate hematopoietic stem cell egress from BM (Katayama et al., 2006). They showed that mice having aberrant nerve conduction failed to mobilize HSPC from BM in response to G-CSF. G-CSF receptor CSF3R has been reported to be present on neurons (Schneider et al., 2005), indicating the direct modulation of nervous system activity by G-CSF. The involvement of beta-adrenergic receptors in stem cell mobilization was reported. Furthermore, regulation of SDF-1 levels by β3-adrenergic receptor signaling (Mendez-Ferrer, Lucas, Battista, & Frenette, 2008) supports the role of nervous system in stem cell mobilization. Chronic stress in mice increased β3-adrenergic signaling in the bone marrow, released more noradrenaline by sympathetic nerves which reduced stromal SDF-1 expression. Following these signals, HSPC proliferation was elevated leading to increased leukocyte mobilization. Similarly, treatment of β-adrenergic antagonist propranolol reduced G-CSF-induced HSPC mobilization, whereas treatment with an agonist clenbuterol enhanced mobilization (Katayama et al., 2006). SDF-1 mRNA and protein levels were found to oscillate with circadian periodicity and their levels correlated inversely with HSPC egress. Further, these SDF-1 oscillations were abrogated by antagonizing the β3-adrenergic receptors (Mendez-Ferrer et al., 2008). β2- and β3-adrenergic receptors knockout mice exhibited HSPC mobilization in response to G-CSF as compared to the controls (Mendez-Ferrer, Battista, & Frenette, 2010). These results indicated that both of these receptors are involved in regulating HSPC mobilization; however, the fact that they are not the sole factors involved further highlights the complexity of mechanism of stem cell mobilization. Taken together, G-CSF enhances stem cell mobilization by reducing the SDF-1 levels and enhancing the protease activity (Heissig et al., 2002; Lévesque et al., 2002; Petit et al., 2002) where nervous system and innate immunity components also influence the mobilization process (Katayama et al., 2006). The exact mechanism/s of G-CSF-induced stem cell mobilization is complex owing to the involvement of more than one factor and still remains elusive. GM-CSF was found to be less potent as compared to G-CSF in terms of mobilized stem cell number and hematopoietic recovery (Arora et al., 2004).

3.1.3 CXCR4 Antagonists as Mobilization Agents

Owing to expression of CXCR4 by mostly all hematopoietic stem cells, later on modulations in CXCR4, is considered alternative approach to directly mobilize HSPC. Following this CXCR4 antagonists including CTCE-0021, CTCE0214, and Plerixafor (AMD3100) were developed. Among these AMD 3100 which is a reversible inhibitor of CXCR4-SDF-1 binding is most commonly used nowadays for stem cell mobilization. It was developed as a potential treatment for HIV, as it blocks the HIV entry into $CD4^+$ T cells by binding to the HIV coreceptor CXCR4 (Donzella et al., 1998). CXCR4 antagonists abrogate CXCR4/SDF-1 binding and effectively induce HSC mobilization without inducing myeloid hyperplasia or proteolysis.

AMD3100 treatment provides a rapid, less toxic, and convenient approach for normal donors (Devine et al., 2008) as well as for poor mobilizers. It mobilizes stem cells within an hour. AMD3100-mobilized $CD34^+$ cells include more B-, T-, and mast cell precursors. It antagonizes CXCR4 resulting in loss of sensitivity to SDF-1 thus preventing interaction of CXCR4 and its chemokine SDF-1. This triggers rapid mobilization of stem cells into the peripheral blood through a positive signal most likely through a sphingolipid S1P gradient. Other suggested mechanisms for AMD3100-mediated mobilization are that it reduces the levels of SDF-1 produced from BM stromal cells. AMD3100 induce homeostatic release of SDF-1 to the circulation in mice and nonhuman primates (Dar et al., 2011). AMD3100-mediated SDF-1 release from $CXCR4^+$ human bone marrow osteoblasts and endothelial cells in CXCR4/JNK-dependent manner. Neutralization of CXCR4 or SDF-1 inhibited SDF-1 secretion and reduced the egress of murine progenitor cells, whereas intrabone injection of biotinylated SDF-1 induced SDF-1 release and progenitor cell mobilization. Bone marrow endothelial cells (BMECs) have been reported to regulate trafficking and maintenance of HSPC in the BM (Itkin et al., 2016). Conditional deletion of CXCR4 from ECs increased the permeability of

BM blood vessel enhancing the HSPC egress, whereas FGF-2 treatment in vivo reduced endothelial barrier permeability and increased adherens junction molecules expression by BMECs thus reducing HSPC egress of LTR-HSCs into the peripheral blood. Similarly, AMD3100 antagonized BM endothelial CXCR4 signaling, reduced SDF-1 and VE-cadherin expression on sinusoidal BM endothelial cells. All these events enhance permeability of endothelial barrier (Itkin et al., 2016) and favor rapid mobilization of HSPC. AMD3100-mobilized cells including primitive HSPC, B-, T-, dendritic, and natural killer cells have increased expression of adhesion molecule VLA-4 and CXCR4 which provide them advantage to repopulate and reconstitute the immune system efficiently in the BM.

3.1.4 Combination of Mobilization Agents Provides Better Output

In order to improve mobilization efficiency an approach combining more than one mobilizing agents is employed. Likewise, combined treatment of G-CSF and AMD3100 is safe, effective, allow retrieval of more HSPC with fewer apheresis sessions and turned out to be a rescuing protocol for poor G-CSF mobilizers. Combined G-CSF and AMD3100 treatment induces more release of B and T cells. Furthermore, stem cells mobilized with this approach have higher CD49d expression, less expression of CD62L (L-selectin), increased expression of antiapoptotic genes, and those of regulating cell cycle, DNA repair, and oxygen transport genes all of which support the acquisition of an "engrafting" phenotype.

Besides, parathyroid hormone (PTH) (Adams et al., 2007), GRO-β (*CXCL2*), and GROβT are other factors which are known to act synergistically with G-CSF, enhance stem cell mobilization and engraftment (Adams et al., 2007; Dawson et al., 2005; Fukuda, Bian, King, & Pelus, 2007). GRO-β is a member of CXC chemokine family. Mechanistically, GRO-β- and GROβT-induced mobilization is mediated by neutrophil-dependent upregulation of MMP-9 activation. GRO-β alone or in combination with G-CSF elevated the mobilization of LSK and $CD34^-$ LSK as compared to G-CSF alone (Fukuda et al., 2007). Cells mobilized with this approach showed increased homing and superior engraftment efficiency. In addition, several cytokines including IL-1, IL-3, IL-6, IL-7, IL-8, IL-11, IL-12, GM-CSF, SCF, and MIP-1α (reviewed in Cottler-Fox et al., 2003) were suggested to increase mobilization of progenitors in blood.

Lenalidomide is another mobilizing agent used in multiple myeloma patients, but has negative impact on stem cell collection (Paripati et al., 2008). Although, combined Lenalidomide and AMD3100 mobilization strategies resulted in successful stem cell collections (Malard et al., 2012; Mark et al., 2008). S1P receptor agonist SEW2871 facilitated AMD3100 and the combination of AMD3100 and G-CSF-induced mobilization but not with G-CSF alone (Juarez et al., 2012). S1P elevation induced ROS levels which elevated HSPC migration whereas enhanced SDF-1 secretion from the stromal cells (Golan et al., 2012). Most of S1P agonists mediate their effects by inducing internalization and loss of function of S1P receptors except SEW2871 which work without inducing internalization of receptors (Jo et al., 2005). However, use of S1P receptor agonist on the negative side might induce transient bradycardia therefore use of these agonists should be considered very carefully.

Me6TREN is a newly discovered compound for HSPC mobilization when used alone or in combination with ADM3100 or G-CSF (Zhang et al., 2014). It was found to be safe and efficient in mobilizing the HSPC from the BM to peripheral blood in mice. This compound antagonized SDF-1-induced chemotaxis of murine and human hematopoietic progenitors. Mechanistically, Me6TREN activated AKT phosphorylation, mitogen-activated protein kinase, and resulted in induction of MMP9. The successful use of this compound in humans warrants further studies.

In summary, different factors affect the mobilization of HSPC, still not all the patients respond equally to a particular mobilization agent, thus affecting the efficacy of these approaches. The combined use of more than one factor has provided a way to overcome the inability of mobilization in poor mobilizers. Therefore, after the discovery of new mobilization inducing agents poses new challenges itself in order to implement them in correct group of patients. Owing to the fact that all the patients cannot be treated using same approach, this highlights the need for personalized medical treatments, which sometimes could be genetically related.

3.2 Suggestive Mechanistic Insights of Mobilization

During steady state, HSPC resides mainly in the BM with exceptions of a few which are found to be circulating. As mentioned before, during alarmed situations or under clinical mobilization, HSPC–niche interactions in particular VLA-4/VCAM1 and SDF-1/CXCR4 needs to be cleaved. Neutrophil-released proteases were shown to be crucial for cleaving HSPC–niche interactions (Levesque, Hendy, Takamatsu, Simmons, & Bendall, 2003) and their inhibition reduced the G-CSF-induced mobilization (Pelus, Bian, King, Fukuda, et al., 2004). Deletion of dipeptidylpeptidase 4 (also known as CD26) in mice attenuated mobilization of HSPC (Christopherson, Cooper, Hangoc, & Broxmeyer, 2003). They detected significantly reduced number of HPC in the peripheral blood of CD26$^{(-/-)}$ mouse as compared to WT mouse following 2- or 4-day G-CSF regimen. Recently, CD26 was shown to decrease the activity of GM-CSF, G-CSF, IL-3, and erythropoietin by cleaving within their N-termini. They found that inhibition of DPP4 or its knockout enhanced activities of colony stimulating factors (Broxmeyer et al., 2012). Here, we will discuss about different factors which have been suggested to regulate the HSPC mobilization.

3.2.1 Bioactive Lipids Provide Chemoattractant Gradient for HSPC Trafficking

The bioactive lipid S1P is chemoattractant for blood-forming stem cells. Increased S1P levels in the plasma compared to the BM direct the egress of HSPC from the BM to circulation during mobilization (Golan et al., 2012; Juarez et al., 2012). Inactivation of bioactive lipids in the plasma abolished the stem cell chemotactic activity of the plasma (Ratajczak et al., 2010). Mice deficient in S1P production showed reduced mobilization in response to G-CSF and AMD3100 (Golan et al., 2012; Juarez et al., 2012; Ratajczak et al., 2010). Further, desensitization of (in vivo) S1P receptors by FTY720 and disruption of S1P gradient resulted in reduced steady-state egress of immature progenitors and primitive LSK cells. S1P-induced HSPC egress was mediated via activation of ROS signaling. Furthermore, S1P-induced SDF-1 secretion from Nestin-GFP^{+} MSPCs, following its release to the circulation thus S1P forms an important regulatory gradient which attracts HSPC (Golan et al., 2012). Our results with regard to the involvement of S1P in AMD3100-induced HSPC mobilization are supported by others (Juarez et al., 2012; Ratajczak et al., 2010). However, its role in G-CSF-induced mobilization needs to be further explored.

3.2.2 Microenvironmental Disruption Facilitates HSPC Egress

HSPC and its niche interactions mediate HSPC retention in the bone marrow and need to be cleaved in order to drive mobilization. The commonly used mobilizing agent G-CSF increases the myeloid cell numbers in the BM and simultaneously exert effects on the microenvironment. It induces loss of osteoblasts and increases osteoclast activity (Kollet et al., 2006; Semerad et al., 2005). Osteoblast together with endothelial cells contributes as source of bone marrow SDF-1. Semerad et al. (2005) have shown that

G-CSF-mediated downregulated osteoblast activity resulted in reduced SDF-1 levels which in turn correlated to the mobilization of hematopoietic progenitor cells. Conditional deletion of SDF-1 from stroma or inhibition of CXCR4 results in robust HSC mobilization (Broxmeyer et al., 2005; Ding, Saunders, Enikolopov, & Morrison, 2012). We have reported previously that osteoclast activation in response to stress induces secretion of proteolytic enzymes which degrade endosteal components and thus causing mobilization of progenitors (Kollet et al., 2006). The mobilization of immature progenitor cells was enhanced by stimulation of osteoclast with receptor activator of NF-κB ligand (RANKL) in a CXCR4 and MMP-9-dependent manner. Enhanced mobilization was accompanied by reduced SDF-1, SCF, and osteopontin levels along the endosteum. In contrast, inhibition of osteoclasts function by genetic manipulations (gene deletion of *protein tyrosine phosphatase, receptor type, E*) or calcitonin administration reduced egress of progenitors during homeostasis, stress, and G-CSF-induced mobilization (Kollet et al., 2006). Thus, it has been suggested that osteoclast produce proteolytic enzymes, including MMP-9 and cathepsin K disrupt HSPC–niche adhesion interactions and subsequently promoting the egress of HSPC from the BM. G-CSF administration induces osteoclasts appearance. However, some studies showed that osteoclast activity is not essential for G-CSF-induced HSC mobilization. Besides, loss of osteomacs and osteocytes within the bone has been reported during G-CSF-mediated mobilization (Asada et al., 2013; Winkler et al., 2010). This reduction in osteoblasts and osteocytes was shown to be influenced by signals from SNS (Asada et al., 2013; Katayama et al., 2006). In human patients mobilized with G-CSF and +/− chemotherapy, the levels of soluble RANKL, and osteoprotegerin (OPG) were found to be elevated between premobilization and HSC collection period. The HSPC mobilization is associated with increased osteoblastic activity and endothelial vessel destabilization (Angelopoulou, Tsirkinidis, Boutsikas, Vassilakopoulos, & Tsirigotis, 2014), although BM MSCs do not express the receptor for G-CSF (Mendez-Ferrer et al., 2010).

G-CSF induces BM HSPC to secrete both MMP-2 and MMP-9 which further increase migration through reconstituted basement membrane. MMPs enhances pericellular proteolysis by processing ECM components (Sato & Takino, 2010) creating a conducible environment for mobilization. G-CSF-mediated activation of MT1-MMP can activate pro-MMP-2 and this can further activate other proteases including MMP-9. Levels of MMPs increase in plasma after mobilization with G-CSF. During G-CSF treatment, higher MT1-MMP expression in HSPC enhances egress into PB and reduced retention in BM (Vagima et al., 2009). Increased MT1-MMP and reduced expression of the MT1-MMP inhibitor RECK was observed on circulating human $CD34^+$ progenitor cells compared with immature BM cells. G-CSF treatment elevated MT1-MMP activity in murine and human hematopoietic cells in a P13K/Akt-dependent manner. The expression of MT1-MMP correlates with clinical mobilization of $CD34^+$ cells in healthy donors and patients with lymphoid malignancies. Inhibition of MT1-MMP by monoclonal antibody inhibited the G-CSF-induced mobilization whereas RECK neutralization enhanced the motility and egress of $CD34^+$ cells. Proteases induce degradation of adhesion molecules like integrin-α4β1 (VLA-4) and VCAM which in turn degrade amino terminal of CXCR4 and SDF-1. Thus, G-CSF-induced MT1-MMP upregulation and redistribution into lipid rafts contributes to create proteolytic milieu in the BM which facilitates HSPC mobilization.

Previous work from our lab has reported that HSC motility and their microenvironment are regulated through CD45 (Shivtiel et al., 2008). Treatment with G-CSF increased CD45 expression on BM mononuclear cells before their egress whereas, CD45 expression was reduced on the mobilized cells found in the circulation (Shivtiel et al., 2008). These results suggested involvement of dynamic CD45 expression in G-CSF-induced mobilization. Furthermore, CD45KO mice had reduced numbers of LSK progenitor cells in the BM and exhibited a delayed and reduced response to G-CSF in terms of mobilized cells. G-CSF upregulated MT1-MMP levels in WT osteoclasts whereas not in CD45-KO mice. Lower secretion of MMP-9 by CD45-KO BM leukocytes after G-CSF stimulation implies that CD45 regulates MMP activation. During G-CSF-induced mobilization, the bone marrow undergoes active remodeling as osteoclasts secrete increased levels of cathepsin K and MMP-9 which is associated with bone turnover (Kollet et al., 2006; Shivtiel et al., 2008). Similarly, MT1-MMP has also been associated with bone turnover (Holmbeck, Bianco, Yamada, & Birkedal-Hansen, 2004) where mice lacking MT1-MMP had severe skeletal defects.

BM macrophages have been suggested to affect G-CSF-induced mobilization. Loss of BM macrophages has been shown to induce HSPC mobilization (Chow et al., 2011; Winkler et al., 2010). Loss of macrophages decreases SDF-1 expression by $Nestin^+$ MSC which attenuates CXCR4/SDF-1 binding and thus facilitates mobilization. Similarly, BM monocytes also play an important role in G-CSF-induced mobilization (Christopher, Rao, Liu, Woloszynek, & Link, 2011). They showed that monocytic cells inhibit the production of factors required for growth and/or survival of osteoblast lineage cell maintenance, ultimately leading to robust HSPC mobilization. They also showed that G-CSF receptor signals in monocytic cells are sufficient for inducing HSPC mobilization. Treatment of G-CSF induced pronounced loss of monocytic cells in the bone marrow. Recently, we identified BM-resident myeloid niche cells (αSMA$^+$ monocytes/macrophages) which regulate maintenance of primitive murine HSPCs in the BM. These αSMA$^+$ monocytes/macrophages located near small blood sinuses produce PGE2 which limits ROS levels in HSPC via inhibition of AKT signaling. COX-2 expression by α-SMA$^+$ monocytes/macrophages preserves HSPCs in the BM and prevents the HSPC exhaustion (Ludin et al., 2012).

Robust HSPC mobilization is also associated with reduced expression of other niche factors, including kit ligand SCF and angiopoietin-1 (Chang et al., 2008; Winkler et al., 2010). SCF/c-Kit axis has been shown to be important for trafficking of HSPC to the circulation. Tyrosine kinase c-Kit is expressed among all HSPCs, whereas its ligand SCF by BM endothelial cells and stromal cells. G-CSF-induced mobilization was found to be associated with cleavage of membrane-bound SCF (Heissig et al., 2002). Neutralizing c-Kit with antibody led to HSC migration from BM to peripheral blood and efficient transplantation. AMD3100 inhibits c-Kit phosphorylation and thus leads to HSPC mobilization (Cheng et al., 2010). AMD3100-induced HSPC mobilization was inhibited in mice with defective c-kit activity where c-Kit deficiency in mice resulted in poor migration and adhesion capacities (Kimura et al., 2011). VLA-4/VCAM-1-based adhesion interactions are involved in the retention of HSPC in the BM. Both G-CSF- and AMD3100-induced mobilization disrupts VLA-4/VCAM-1 axis. Inhibition of VLA-4/VCAM-1 binding results in HSC mobilization. Similarly, antiadhesion therapy against α4 integrins was found to increase circulating progenitors in a clinical trial. Administration of BIO5192 (a small molecule inhibitor of VLA-4 binding) and anti-VLA-4 monoclonal antibody (natalizumab) enhanced the mobilization induced by G-CSF in mice. As discussed in this section, various factors from BM microenvironment contribute to mediate HSPC–niche interactions and the approaches based on balanced alterations in microenvironment can be utilized to induce adequate amount of HSPC mobilization.

3.2.3 Direct and Indirect Involvement of the Nervous System in HSPC Mobilization

Involvement of the nervous system in HSPC mobilization has been documented through different studies. Neural signals together with stimuli from immune system, bone, BM, and stromal cues/factors synchronize to mediate HSPC migration and their development to maintain bone remodeling (Lapidot & Kollet, 2010). Expression of the neuronal receptors among BM cells indicates that signals from the nervous system can directly affect their functions. The nervous system can send the signals directly through the nerve endings or indirectly sending them via circulation. BM is innervated with nerve fibers which serve as carriers of direct signals/neurotransmitters from SNS. Light and darkness cues stimulate nerves in the suprachiasmatic nucleus, which leads to release of norepinephrine (NE) in the BM from sympathetic nerves (Mendez-Ferrer et al., 2008). The levels of HSPC are at peak 5 h after the onset of light and this is accompanied by lowest SDF-1 expression in the BM. In correlation, the levels of HSPCs harvested at this time from mice mobilized with G-CSF or AMD3100 were higher. Interestingly, in contrast to optimal HSPC mobilization in mice occurring in the morning, HSC mobilization peak occurs late in the evening in humans. Therefore, harvesting HSPC based on the circadian rhythms can be used as an approach to get maximum cell yield (Lucas, Battista, Shi, Isola, & Frenette, 2008) in donors.

Dopamine and β2-adrenergic receptors were found be expressed among immature human $CD34^+$ cells (Spiegel et al., 2007) where higher expression of these receptors was in the primitive $CD34^+CD38^{(low)}$ cells. The expression of these receptors was increased by G-CSF stimulation. Catecholamine's treatment increased cell mobilization through Wnt signaling activation (Spiegel et al., 2007). The HSPC mobilization was enhanced by β2-adrenergic agonist administration. Beta-adrenergic receptors β2 and β3 work cooperatively during progenitor mobilization and deficiency of both these receptors abrogates G-CSF-induced mobilization (Mendez-Ferrer et al., 2010). Signals from peripheral SNS were shown to regulate enforced HSPC mobilization. SNS through NE signaling regulate G-CSF-induced osteoblast suppression, SDF-1 levels, and HSPC mobilization (Katayama et al., 2006). NE treatment rapidly induced SDF-1 release and progenitor cell mobilization (Dar et al., 2011). Stimulation of β2-adrenergic receptors by NE in mice increased HSPC mobilization, whereas antagonists of β2-adrenergic antagonists reduced PB numbers (Dar et al., 2011; Spiegel et al., 2007). Glycogen synthase kinase-3β (GSK3β) was found to regulate SDF-1-induced migration and steady-state egress of murine HSPCs (Lapid et al., 2013). The expression of GSK3β in HSPCs was found to correlate with their physiological egress. Its expression was highest during the peak HSPC egress (Mendez-Ferrer et al., 2008) at zeitgeber time 5 (ZT5) hours after initiation of light, whereas GSK3β expression was at low at ZT13, following darkness. GSK3β induced cytoskeletal rearrangement and mediated HSPC egress which was regulated by adrenergic signals. Expression of vitamin D receptor is crucial for the osteoblast responses to SNS signals and this expression increases by β2-adrenergic receptor stimulation (Kawamori et al., 2010). Peripheral sympathetic neurons show expression of G-CSF receptor and their stimulation with G-CSF reduced NE uptake and thus induced signaling from SNS by Frenette's group. BM corticosterone levels were found to have circadian oscillations in wild type mice. However, these oscillations were not observed in corticotropin-releasing factor receptor 1 (CRFR1) mutant mice. These mutant mice had very low corticosterone levels and had higher HSPC numbers, stromal cells, and increased number of circulating HSPCs. On the contrary, elevated corticosterone levels resulted in reduced long-term repopulation and lower stromal progenitors and increased HSPC apoptosis (Kollet et al., 2013). Proliferation of stromal progenitors was mediated by Notch1 signaling, and increased SDF-1 production by reticular Nestin MSPCs.

3.2.4 Complement Cascades and NO Signaling Affect HSPC Mobilization

Cleavage fragment of the fifth component of the activated complement cascade (C5a)-deficient mice were poor mobilizers suggesting important role for C5a and $_{desArg}$C5a in the egress of HSPCs from the BM to circulation (Lee et al., 2009). On the other hand, previous work from Ratajczak showed that C3- and C3a-deficient mice were more sensitive to G-CSF mobilization. In response to plasma C5a granulocytes undergo degranulation and secrete cationic peptides including cathelicidin and beta-defensin which increase HSPCs response to plasma SDF-1 gradient. Complement activation releases erythrocytes sphingosine-1-phosphate (S1P) into the plasma which makes a gradient for HSPC chemoattraction as discussed earlier. C5 cleavage fragment C5a increases expression of MMP-9 and MT1-MMP thus induce a highly proteolytic microenvironment promoting granulocyte egress to peripheral blood and facilitates subsequent HSPC egress (Jalili et al., 2010). Two proteolytic enzyme cascades, the coagulation cascade (CoaC) and the fibrynolytic cascade (FibC) get activated during G-CSF- and AMD3100-induced mobilization and provide C5 convertase activity (Borkowska et al., 2014). C5a is known to interact with the C5a receptor on granulocytes and activate intracellular hematopoietic cell-specific phospholipase C-β2 (PLC-β2)-mediated signaling. This PLC-β2 was found to be important in pharmacological mobilization of HSPCs (Adamiak et al., 2016). HSPC are maintained in the BM niche due to interaction of membrane lipid raft-associated receptors (CXCR4 and VLA-4) with their respective specific ligands (SDF-1 and VCAM-1). The integrity of the lipid rafts maintained by the glycolipid glycosylphosphatidylinositol anchor (GPI-A) is disrupted by PLC-β2, enabling granulocytes to release proteolytic enzymes. PLC-β2-deficient mice were poor G-CSF and AMD3100 mobilizers. As mentioned before, during steady-state HSPC mobilize to the peripheral blood following circadian oscillations. Borkowska, Suszynska, Ratajczak, and Ratajczak (2016) explored the involvement of distal part of the complement cascade in the diurnal release of HSC into peripheral blood. They found that C5-deficient mice lacked diurnal changes in the number of circulating HSPC in PB and their work suggested that activation of the distal part of the ComC, C5 cleavage, and release of C5a and $_{desArg}$C5a are required for the circadian release of HSPCs to circulation.

Besides, nitric oxide (NO) is an important player in regulation of stem cell localization and function. The most primitive noncycling and nonmotile EPCR-expressing mouse BM HSC contain very low levels of NO, while the more mature cycling and motile progenitors lack EPCR expression and contain high levels of NO (Gur-Cohen et al., 2015). Mice lacking endothelial NO synthase (eNOS) showed reduced $CD34^+$ mobilization (Ozüyaman et al., 2005). Genetic thrombomodulin mutation carrying mice (TMpro/pro) and reduced activated protein C (aPC) generation exhibited reduced BM stem cell retention and increased migrating HSCs in the blood (Gur-Cohen et al., 2015). Binding of aPC and EPCR activates PAR1 signaling which inhibits NO and cdc42 activity while promoting VLA-4 affinity. On the contrary, activation of PAR1 by thrombin increases NO generation, cdc42 activity, reduced VLA-4 affinity in HSPC, and SDF-1 secretion. Therefore, aPC-PAR1 signaling induced LTR-HSC retention whereas thrombin-PAR1 signaling promotes mobilization of LTR-HSC.

3.3 Involvement of Other Factors

Studies using genetic and pharmacologic models of PTH receptor activation have suggested PTH a useful candidate to study activation of HSC niche (Calvi et al., 2003; Itkin, Kaufmann, Gur-Cohen, Ludin, & Lapidot, 2013). Combining PTH with G-CSF facilitated the HSPC mobilization in the patients who failed to mobilize before. PTH increased progenitor cells egress in mice was suggested to be dependent on elevated endogenous

G-CSF release. Besides, stabilization of HIF-1 is suggested another approach to modulate HSPC motility. On the same line FG-4497 that stabilizes HIF-1 through inhibition of its hydroxylation is being tested. HIF-1a has been shown to increase SDF-1α (Ceradini et al., 2004), CXCR4 expression (Staller et al., 2003), and prevent ROS-induced cell damage (Kirito, Hu, & Komatsu, 2009), suggesting that hypoxia regulate important signaling axis responsible for HSPC retention in the BM. Exercise is suggested as an adjuvant therapy for hematopoietic stem cell mobilization and has been discussed in detail by Emmons, Niemiro, and De Lisio (2016).

4. SUMMARY AND CONCLUDING REMARKS

This chapter discusses the key steps, the different molecules, and the mechanistic pathways involved in bidirectional trafficking of HSPC, i.e., homing to and mobilization from the BM. As mentioned before, during homeostasis most HSPC occupy their dynamic BM niches (Fig. 1A) with very less in circulation, however, recruited there in large numbers in response to the external stimuli like stress or inflammation (Fig. 1B). Such migration of HSPC is a multifacet process which is mediated by intrinsic motility mechanisms as well as complex extrinsic microenvironmental dynamics. Various extrinsic/microenvironmental factors comprised of BM resident osteoblasts, osteoclasts, macrophages, neutrophils, immune system, complement cascades, and signals from SNS regulate the HSPC trafficking under physiological conditions and stress-induced recruitment to the circulation. The disrupted chemotactic and adhesion interactions (including CXCR4/SDF-A, VLA-4/VCAM-1, and SCF/c-Kit), activity of MMPs, and disruption of ECM exert a proteolytic environment in the BM which induce HSPC mobilization following the S1P gradient to the circulation. Different approaches/factors employed for HSPC mobilization have been discussed in the chapter including chemotherapy, G-CSF, CXCR4 antagonists. Most HSPC mobilization studies aim to develop beneficial and novel therapeutic strategies for optimal HSPC collection with the ultimate aim of improving hematopoietic recovery after transplantation. Knowledge about the basic mechanisms of HSPC retention in the BM niche/microenvironment can be exploited for clinical stem cell transplantation which requires harvesting large number of HSPC. Various newer studies suggest many factors regulating the HSPC mobilization, however, very few have been implicated in humans successfully. AMD3100 is one such example where basic knowledge of inhibiting the CXCR4/SDF-1 interaction was used and

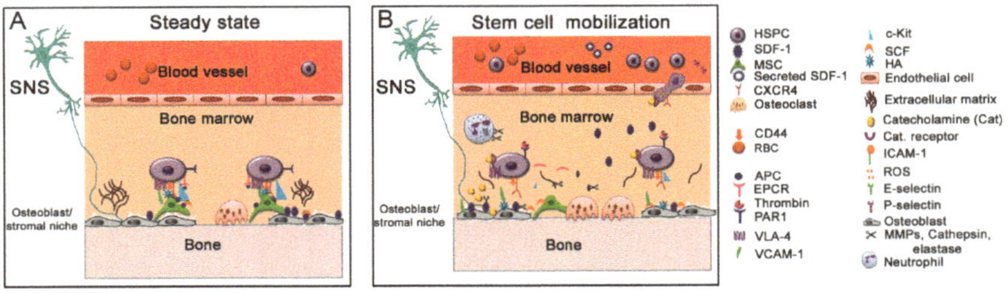

Fig. 1 (A) Steady-state retention of hematopoietic stem and progenitor cells (HSPC). During steady state, HSPC are maintained in the nonmotile quiescent mode in the bone marrow via chemotactic and adhesion interactions mediated by CXCR4/SDF-1, VLA-4/VCAM-1, c-Kit/SCF, CD44/HA, and EPCR/aPC axis. (B) Mobilization of HSPC. In response to the mobilization agents or under alarmed situations, the activity of osteoclasts and proteases including elastase, cathepsin, and matrix metalloproteinases (MMPs) is upregulated which disrupt the chemotactic and adhesion interactions of HSPC and the microenvironment. The consequent cleavage of ligands including CD44, SDF-1, SCF, VCAM-1, and EPCR, upregulation of CXCR4 and thrombin/PAR1 signaling and cues from nervous system direct HSPC to proliferate, differentiate, and mobilize to the circulation.

now this is successfully being used to treat many patients. As we know that not all patients/donors mobilize to the same extent in response to a particular mobilizing agent. To address the same, HSPC mobilization currently employ use of more than one mobilizing agents to overcome the inability of donors to respond to single mobilizing agent or improve their mobilized HSPC yield. Involvement of multiple factors in driving homing and mobilization could be one of the underlying causes for different responses of donors or recipients to a particular mobilization and homing regimes. Under the circumstances where the patients fail to mobilize with G-CSF were cotreated with AMD3100 to get the desired outcomes. Such approach was also found to be safe and effective. Besides AMD3100, various other factors including PTH, GRO-β (*CXCL2*), and GROβT were found to synergize with G-CSF. Similarly, Lenalidomide and AMD3100-based mobilization strategies are effective for treating multiple myeloma patients. We also discussed about a newly identified Me6TREN compound which induced efficient HSPC mobilization in mice when used alone or in combination with ADM3100 or G-CSF (Zhang et al., 2014). However, its use in humans warrants further studies.

The mobilized HSPC when transplanted undergo active navigation from the circulation, cross the blood–BM endothelial barrier, and

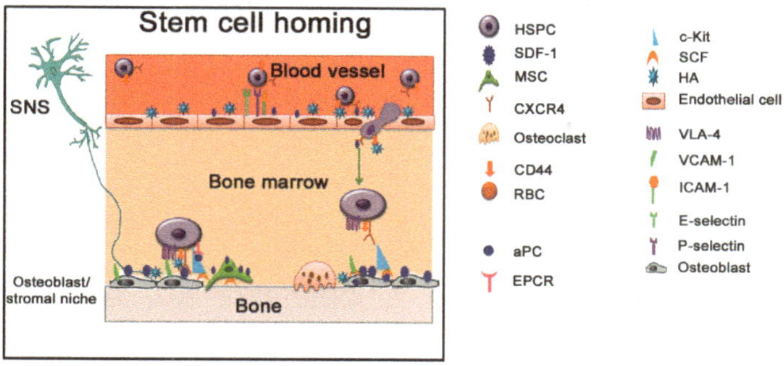

Fig. 2 Homing of transplanted HSPCs to the bone marrow. Transplanted HSPC in the preconditioned recipients follow the path through the circulation and then roll and tether at the endothelial wall with the help of selectins and integrins. Subsequently, $CXCR4^{+}$ HSPC adheres to SDF-1 and HA-expressing endothelial cells, transmigrate through blood-endothelial barrier, and ultimately they reach their specialized niches to repopulate and engraft there.

ultimately reconstitute the recipient BM following their lodgment and repopulation there (Fig. 2). These homing and repopulation events are mediated by orchestrated activity of integrins, selectins, MMPs, adhesion molecules like ICAM-1, chemotactic factors including SDF-1/CXCR4, S1P, PGE2, and CD44/HA interaction. We discuss about some interesting studies focusing on suggestive approaches to improve HSPC mobilization and homing including overexpression of CXCR4, intrabone transplantation, inhibition of CD26 peptidase, manipulation of FGF signaling, mild heat treatment, use of priming agents, and administration of PTH. Newer mechanistic insights of the HSPC trafficking to and from the BM can be employed to improve the current HSPC mobilization protocols to obtain optimal donor cells, directing the homing of transplanted cells efficiently to the recipient BM and ultimately improving the hematopoietic recovery. As we know that not all the transplanted cells reach the BM, many are stuck in between and this is why current clinical protocols rely on use optimal donor cells numbers. However, improving the basic understanding of the homing would surely enhance the chances of optimizing the efficient HSPC homing and achieving of the desired therapeutic effect using even lower number of donor cells. Thus, it will be bliss for patients suffering with hematological disorders as the sources like CB could also be used under such scenario. Thus, we think that newer knowledge in the field could help in designing different strategies for efficient mobilization and homing protocols to combat the alarmed situations and will confer better personalized treatments to the patients as well.

REFERENCES

Adamiak, M., Borkowska, S., Wysoczynski, M., Suszynska, M., Kucia, M., Rokosh, G., et al. (2015). Evidence for the involvement of sphingosine-1-phosphate in the homing and engraftment of hematopoietic stem cells to bone marrow. *Oncotarget*, *6*(22), 18819–18828.

Adamiak, M., Iv, J. B., Zhao, J., Abdelbaset-Ismail, A., Grubczak, K., Borkowska, S., et al. (2016). Downregulation of heme oxygenase 1 (HO-1) activity in hematopoietic cells enhances their engraftment after transplantation. *Cell Transplantation*, *25*(7), 1265–1276.

Adams, G. B., Chabner, K. T., Alley, I. R., Olson, D. P., Szczepiorkowski, Z. M., Poznansky, M. C., et al. (2005). Stem cell engraftment at the endosteal niche is specified by the calcium-sensing receptor. *Nature*, *439*(7076), 599–603.

Adams, G. B., Martin, R. P., Alley, I. R., Chabner, K. T., Cohen, K. S., Calvi, L. M., et al. (2007). Therapeutic targeting of a stem cell niche. *Nature Biotechnology*, *25*(2), 238–243.

Alvarez, S. E., Milstien, S., & Spiegel, S. (2007). Autocrine and paracrine roles of sphingosine-1-phosphate. *Trends in Endocrinology and Metabolism*, *18*(8), 300–307.

Angelopoulou, M. K., Tsirkinidis, P., Boutsikas, G., Vassilakopoulos, T. P., & Tsirigotis, P. (2014). New insights in the mobilization of hematopoietic stem cells in lymphoma and multiple myeloma patients. *BioMed Research International*, *2014*, 835138.

Arai, S., & Klingemann, H. G. (2003). Hematopoietic stem cell transplantation: Bone marrow vs. mobilized peripheral blood. *Archives of Medical Research*, *34*(6), 545–553.

Arora, M., Burns, L. J., Barker, J. N., Miller, J. S., Defor, T. E., Olujohungbe, A. B., et al. (2004). Randomized comparison of granulocyte colony-stimulating factor versus granulocyte macrophage colony-stimulating factor plus intensive chemotherapy for peripheral blood stem cell mobilization and autologous transplantation in multiple myeloma. *Biology of Blood and Marrow Transplantation*, *10*(6), 395–404.

Asada, N., Katayama, Y., Sato, M., Minagawa, K., Wakahashi, K., Kawano, H., et al. (2013). Matrix-embedded osteocytes regulate mobilization of hematopoietic stem/progenitor cells. *Cell Stem Cell*, *12*(6), 737–747.

Avigdor, A., Goichberg, P., Shivtiel, S., Dar, A., Peled, A., Samira, S., et al. (2004). CD44 and hyaluronic acid cooperate with SDF-1 in the trafficking of human CD34+ stem/progenitor cells to bone marrow. *Blood*, *103*(8), 2981–2989.

Bartolomé, R. A., Ferreiro, S., Miquilena-Colina, M. E., et al. (2009). The chemokine receptor CXCR4 and the metalloproteinase MT1-MMP are mutually required during melanoma metastasis to lungs. *American Journal of Pathology*, *174*(2), 602–612.

Bendall, L. J., & Basnett, J. (2013). Role of sphingosine 1-phosphate in trafficking and mobilization of hematopoietic stem cells. *Current Opinion in Hematology*, *20*(4), 281–288.

Bonig, H., Priestley, G. V., Nilsson, L. M., Jiang, Y., & Papayannopoulou, T. (2004). PTX-sensitive signals in bone marrow homing of fetal and adult hematopoietic progenitor cells. *Blood*, *104*(8), 2299–2306.

Borkowska, S., Suszynska, M., Mierzejewska, K., Ismail, A., Budkowska, M., Salata, D., et al. (2014). Novel evidence that crosstalk between the complement, coagulation and fibrinolysis proteolytic cascades is involved in mobilization of hematopoietic stem/progenitor cells (HSPCs). *Leukemia*, *28*(11), 2148–2154.

Borkowska, S., Suszynska, M., Ratajczak, J., & Ratajczak, M. Z. (2016). Evidence of a pivotal role for the distal part of the complement cascade in the diurnal release of hematopoietic stem cells into peripheral blood. *Cell Transplantation*, *25*(2), 275–282.

Broxmeyer, H. E., Hoggatt, J., O'Leary, H. A., Mantel, C., Chitteti, B. R., Cooper, S., et al. (2012). Dipeptidylpeptidase 4 negatively regulates colony-stimulating factor activity and stress hematopoiesis. *Nature Medicine*, *18*(12), 1786–1796.

Broxmeyer, H. E., Orschell, C. M., Clapp, D. W., Hangoc, G., Cooper, S., Plett, P. A., et al. (2005). Rapid mobilization of murine and human hematopoietic stem and progenitor cells with AMD3100, a CXCR4 antagonist. *Journal of Experimental Medicine, 201*(8), 1307–1318.
Broxmeyer, H. E., & Pelus, L. M. (2014). Inhibition of DPP4/CD26 and dmPGE$_2$ treatment enhances engraftment of mouse bone marrow hematopoietic stem cells. *Blood Cells Molecules and Disease, 53*(1–2), 34–38.
Burns, J. M., Summers, B. C., Wang, Y., Melikian, A., Berahovich, R., Miao, Z., et al. (2006). A novel chemokine receptor for SDF-1 and I-TAC involved in cell survival, cell adhesion, and tumor development. *Journal of Experimental Medicine, 203*(9), 2201–2213.
Byk, T., Kahn, J., Kollet, O., Petit, I., Samira, S., Shivtiel, S., et al. (2005). Cycling G1 CD34+/CD38+ cells potentiate the motility and engraftment of quiescent G0 CD34+/CD38-/low severe combined immunodeficiency repopulating cells. *Stem Cells, 23*(4), 561–574.
Calvi, L. M., Adams, G. B., Weibrecht, K. W., Weber, J. M., Olson, D. P., Knight, M. C., et al. (2003). Osteoblastic cells regulate the haematopoietic stem cell niche. *Nature, 425*(6960), 841–846.
Campbell, T. B., Hangoc, G., Liu, Y., Pollok, K., & Broxmeyer, H. E. (2007). Inhibition of CD26 in human cord blood CD34+cells enhances their engraftment of nonobese diabetic/severe combined immunodeficiency mice. *Stem Cells and Development, 16*(3), 347–354.
Cancelas, J. A., Lee, A. W., Prabhakar, R., Stringer, K. F., Zheng, Y., & Williams, D. A. (2005). RacGTPases differentially integrate signals regulating hematopoietic stem cell localization. *Nature Medicine, 11*(8), 886–891.
Capitano, M. L., Hangoc, G., Cooper, S., & Broxmeyer, H. E. (2015). Mild heat treatment primes human CD34(+) cord blood cells for migration toward SDF-1α and enhances engraftment in an NSG mouse model. *Stem Cells, 33*(6), 1975–1984.
Ceradini, D. J., Kulkarni, A. R., Callaghan, M. J., Tepper, O. M., Bastidas, N., Kleinman, M. E., et al. (2004). Progenitor cell trafficking is regulated by hypoxic gradients through HIF-1 induction of SDF-1. *Nature Medicine, 10*(8), 858–864.
Chang, M. K., Raggatt, L. J., Alexander, K. A., Kuliwaba, J. S., Fazzalairi, N. L., Schroder, K., et al. (2008). Osteal tissue macrophages are intercalated throughout human and mouse bone lining tissues and regulate osteoblast function in vitro and in vivo. *Journal of Immunology, 181*(2), 1232–1244.
Chaurasia, P., Gajzer, D. C., Schaniel, C., D'Souza, S., & Hoffman, R. (2014). Epigenetic reprogramming induces the expansion of cord blood stem cells. *The Journal of Clinical Investigation, 124*(6), 2378–2395.
Chen, C., Su, Y., Chen, J., Zhang, D., Song, Y., & Guo, S. (2015). Intra-bone marrow transplantation of endosteal bone marrow cells facilitates allogeneic hematopoietic and stromal cells engraftment dependent on early expression of CXCL-12. *Medical Science Monitor, 21*, 2757–2766.
Cheng, M., Zhou, J., Wu, M., Boriboun, C., Thorne, T., Liu, T., et al. (2010). CXCR4-mediated bone marrow progenitor cell maintenance and mobilization are modulated by c-kit activity. *Circulation Research, 107*(9), 1083–1093.
Chow, A., Lucas, D., Hidalgo, A., Mendez-Ferrer, S., Hashimoto, D., Scheiermann, C., et al. (2011). Bone marrow CD169+ macrophages promote the retention of hematopoietic stem and progenitor cells in the mesenchymal stem cell niche. *The Journal of Experimental Medicine, 208*(2), 261–271.
Christopher, M. J., Rao, M., Liu, F., Woloszynek, J. R., & Link, D. C. (2011). Expression of the G-CSF receptor in monocytic cells is sufficient to mediate hematopoietic progenitor mobilization by G-CSF in mice. *Journal of Experimental Medicine, 208*(2), 251–260.

Christopherson, K. W., Cooper, S., Hangoc, G., & Broxmeyer, H. E. (2003). CD26 is essential for normal G-CSF-induced progenitor cell mobilization as determined by $CD26^{-/-}$ mice. *Experimental Hematology, 31*(11), 1126–1134.

Cottler-Fox, M. H., Lapidot, T., Petit, I., Kollet, O., DiPersio, J. F., Link, D., et al. (2003). Stem cell mobilization. *American Society Hematology Education Program, 2003*, 419–437.

Cutler, C., Multani, P., Robbins, D., Kim, H. T., Le, T., Hoggatt, J., et al. (2013). Prostaglandin modulated umbilical cord blood hematopoietic stem cell transplantation. *Blood, 122*(17), 3074–3081.

Dar, A., Goichberg, P., Shinder, V., Kalinkovich, A., Kollet, O., Netzer, N., et al. (2005). Chemokine receptor CXCR4-dependent internalization and resecretion of functional chemokine SDF-1 by bone marrow endothelial and stromal cells. *Nature Immunology, 6*(10), 1038–1046.

Dar, A., Kollet, O., & Lapidot, T. (2006). Mutual, reciprocal SDF-1/CXCR4 interactions between hematopoietic and bone marrow stromal cells regulate human stem cell migration and development in NOD/SCID chimeric mice. *Experimental Hematology, 34*(8), 967–975.

Dar, A., Schajnovitz, A., Lapid, K., Kalinkovich, A., Itkin, T., Ludin, A., et al. (2011). Rapid mobilization of hematopoietic progenitors by AMD3100 and catecholamines is mediated by CXCR4-dependent SDF-1 release from bone marrow stromal cells. *Leukemia, 25*(8), 1286–1296.

Dawson, M. A., Schwarer, A. P., Muirhead, J. L., Bailey, M. J., Bollard, G. M., & Spencer, A. (2005). Successful mobilization of peripheral blood stem cells using recombinant human stem cell factor in heavily pretreated patients who have failed a previous attempt with a granulocyte colony-stimulating factor-based regimen. *Bone Marrow Transplantation, 36*(5), 389–396.

Desikan, K. R., Tricot, G., Munshi, N. C., Anaissie, E., Spoon, D., Fassas, A., et al. (2001). Preceding chemotherapy, tumour load and age influence engraftment in multiple myeloma patients mobilized with granulocyte colony-stimulating factor alone. *British Journal of Haematology, 112*(1), 242–247.

Devine, S. M., Vij, R., Rettig, M., Todt, L., McGlauchlen, K., Fisher, N., et al. (2008). Rapid mobilization of functional donor hematopoietic cells without G-CSF using AMD3100, an antagonist of the CXCR4/SDF-1 interaction. *Blood, 112*(4), 990–998.

Dimitroff, C. J., Lee, J. Y., Rafii, S., Fuhlbrigge, R. C., & Sackstein, R. (2001). CD44 is a major E-selectin ligand on human hematopoietic progenitor cells. *The Journal of Cell Biology, 153*(6), 1277–1286.

Ding, L., Saunders, T. L., Enikolopov, G., & Morrison, S. J. (2012). Endothelial and perivascular cells maintain haematopoietic stem cells. *Nature, 481*(7382), 457–462.

Dominici, M., Rasini, V., Bussolari, R., Chen, X., Hofmann, T. J., Spano, C., et al. (2009). Restoration and reversible expansion of the osteoblastic hematopoietic stem cell niche after marrow radioablation. *Blood, 114*(11), 2333–2343.

Donahue, R. E., Jin, P., Bonifacino, A. C., Metzger, M. E., Ren, J., Wang, E., et al. (2009). Plerixafor (AMD3100) and granulocyte colony-stimulating factor (GCSF) mobilize different CD34+ cell populations based on global gene and microRNA expression signatures. *Blood, 114*(12), 2530–2541.

Donzella, G. A., Schols, D., Lin, S. W., Este, J. A., Nagashima, K. A., Maddon, P. J., et al. (1998). AMD3100, a small molecule inhibitor of HIV-1 entry via the CXCR4 coreceptor. *Nature Medicine, 4*(1), 72–77.

Dorrance, A. M., De Vita, S., Radu, M., Reddy, P. N., McGuinness, M. K., Harris, C. E., et al. (2013). The RacGTPase effector p21-activated kinase is essential for hematopoietic stem/progenitor cell migration and engraftment. *Blood, 121*(13), 2474–2482.

Ellis, S. L., Grassinger, J., Jones, A., Borg, J., Camenisch, T., Haylock, D., et al. (2011). The relationship between bone, hemopoietic stem cells, and vasculature. *Blood, 118*(6), 1516–1524.

Emmons, R., Niemiro, G. M., & De Lisio, M. (2016). Exercise as an adjuvent therapy for hematopoietic stem cellmobilization. *Stem Cells International, 2016*, 7131359.

Esencay, M., Sarfraz, Y., & Zagzag, D. (2013). CXCR7 is induced by hypoxia and mediates glioma cell migration towards SDF-1alpha. *BMC Cancer, 13*, 347.

Foudi, A., Jarrier, P., Zhang, Y., Wittner, M., Geay, J. F., Lecluse, Y., et al. (2006). Reduced retention of radioprotective hematopoietic cells within the bone marrow microenvironment in CXCR4$^{-/-}$ chimeric mice. *Blood, 107*(6), 2243–2251.

Fukuda, S., Bian, H., King, A. G., & Pelus, A. M. (2007). The chemokine GRObeta mobilizes early hematopoietic stem cells characterized by enhanced homing and engraftment. *Blood, 110*(3), 860–869.

Fukuhara, S., Simmons, S., Kawamura, S., Inoue, A., Orba, Y., Tokudome, T., et al. (2012). The sphingosine-1-phosphate transporter spns2 expressed on endothelial cells regulates lymphocyte trafficking in mice. *The Journal of Clinical Investigation, 122*(4), 1416–1426.

Goichberg, P., Kalinkovich, A., Borodovsky, N., Tesio, M., Petit, I., Nagler, A., et al. (2005). cAMP-induced PKCzeta activation increases functional CXCR4 expression on human CD34+ hematopoietic progenitors. *Blood, 107*(3), 870–879.

Goichberg, P., Kalinkovich, A., Borodovsky, N., Tesio, M., Petit, I., Nagler, A., et al. (2006). cAMP-induced PKCzeta activation increases functional CXCR4 expression on human CD34+ hematopoietic progenitors. *Blood, 107*(3), 870–879.

Golan, K., Vagima, Y., Ludin, A., Itkin, T., Cohen-Gur, S., Kalinkovich, A., et al. (2012). S1P promotes murine progenitor cell egress and mobilization via S1P1-mediated ROS signaling and SDF-1 release. *Blood, 119*(11), 2478–2488.

Gu, Y., Filippi, M. D., Cancelas, J. A., Siefring, J. E., Williams, E. P., Jasti, A. C., et al. (2003). Hematopoietic cell regulation by Rac1 and Rac2 guanosine triphosphatases. *Science, 302*(5644), 445–449.

Gul, H., Marquez-Curtis, L. A., Jahroudi, N., Lo, J., Turner, A. R., & Janowska-Wieczorek, A. (2009). Valproic acid increases CXCR4 expression in hematopoietic stem/progenitor cells by chromatin remodeling. *Stem Cells and Development, 18*(6), 831–838.

Gur-Cohen, S., Itkin, T., Chakrabarty, S., Graf, C., Kollet, O., Ludin, A., et al. (2015). PAR1 signaling regulates the retention and recruitment of EPCR-expressing bone marrow hematopoietic stem cells. *Nature Medicine, 21*(11), 1307–1317.

Hamadani, M., Kochuparambil, S. T., Osman, S., Cumpston, A., Leadmon, S., Bunner, P., et al. (2012). Intermediate-dose versus low-dose cyclophosphamide and granulocyte colonystimulating factor for peripheral blood stem cell mobilization in patients with multiple myeloma treated with novel induction therapies. *Biology of Blood and Marrow Transplantation, 18*(7), 1128–1135.

Hartmann, T. N., Grabovsky, V., Pasvolsky, R., Shulman, Z., Buss, E. C., Spiegel, A., et al. (2008). A crosstalk between intracellular CXCR7 and CXCR4 involved in rapid CXCL12-triggered integrin activation but not in chemokine-triggered motility of human T lymphocytes and CD34+ cells. *Journal of Leukocyte Biology, 84*(4), 1130–1140.

Heissig, B., Hattori, K., Dias, S., Friedrich, M., Ferris, B., Hackett, N. R., et al. (2002). Recruitment of stem and progenitor cells from the bone marrow niche requires MMP-9 mediated release of kit-ligand. *Cell, 109*(5), 625–637.

Hiwase, D. K., Bollard, G., Hiwase, S., Bailey, M., Muirhead, J., & Schwarer, A. P. (2007). Intermediate-dose CY and GCSF more efficiently mobilize adequate numbers of PBSC for tandem autologous PBSC transplantation compared with low-dose CY in patients with multiple myeloma. *Cytotherapy, 9*(6), 539–547.

Hoggatt, J., Singh, P., Sampath, J., & Pelus, L. M. (2009). Prostaglandin E2 enhances hematopoietic stem cell homing, survival, and proliferation. *Blood*, *113*(22), 5444–5455.
Holmbeck, K., Bianco, P., Yamada, S., & Birkedal-Hansen, H. (2004). MT1-MMP: A tethered collagenase. *Journal of Cellular Physiology*, *200*(1), 11–19.
Infantino, S., Moepps, B., & Thelen, M. (2006). Expression and regulation of the orphan receptor RDC1 and its putative ligand in human dendritic and B cells. *Journal of Immunology*, *176*(4), 2197–2207.
Itkin, T., Gur-Cohen, S., Spencer, J. A., Schajnovitz, A., Ramasamy, S. K., Kusumbe, A. P., et al. (2016). Distinct bone marrow blood vessels differentially regulate haematopoiesis. *Nature*, *532*, 323–328. http://dx.doi.org/10.1038/nature17624.
Itkin, T., Kaufmann, K. B., Gur-Cohen, S., Ludin, A., & Lapidot, T. (2013). Fibroblast growth factor signaling promotes physiological bone remodeling and stem cell self-renewal. *Current Opinion in Hematology*, *20*(3), 237–244.
Itkin, T., Ludin, A., Gradus, B., Gur-Cohen, S., Kalinkovich, A., Schajnovitz, A., et al. (2012). FGF-2 expands murine hematopoietic stem and progenitor cells via proliferation of stromal cells, c-Kit activation, and CXCL12 down-regulation. *Blood*, *120*(9), 1843–1855.
Jalili, A., Shirvaikar, N., Marquez-Curtis, L., Qiu, Y., Korol, C., Lee, H., et al. (2010). Fifth complement cascade protein (C5) cleavage fragments disrupt the SDF-1/CXCR4 axis: Further evidence that innate immunity orchestrates the mobilization of hematopoietic stem/progenitor cells. *Experimental Hematology*, *38*(4), 321–332.
Janowska-Wieczorek, A., Marquez, L. A., Dobrowsky, A., Ratajczak, M. Z., & Cabuhat, M. L. (2000). Differential MMP and TIMP production by human marrow and peripheral blood CD34(+) cells in response to chemokines. *Experimental Hematology*, *28*(11), 1274–1285.
Jarocha, D., Zuba-Surma, E., & Majka, M. (2016). Dimethyl sulfoxide (DMSO) increases percentage of CXCR4(+) hematopoietic stem/progenitor cells, their responsiveness to an SDF-1 gradient, homing capacities, and survival. *Cell Transplantation*, *25*(7), 1247–1257.
Jo, E., Sanna, M. G., Gonzalez-Cabrera, P. J., Thangada, S., Tigyi, G., Osborne, D. A., et al. (2005). S1P1-selective in vivo-active agonists from high-throughput screening: Off-the-shelf chemical probes of receptor interactions, signaling, and fate. *Chemical Biology*, *12*(6), 703–715.
Juarez, J. G., Harun, N., Thien, M., Welschinger, R., Baraz, R., Pena, A. D., et al. (2012). Sphingosine-1-phosphate facilitates trafficking of hematopoietic stem cells and their mobilization by CXCR4 antagonists in mice. *Blood*, *119*(3), 707–716.
Kahn, J., Byk, T., Jansson-Sjostrand, L., Petit, I., Shivtiel, S., Nagler, A., et al. (2004). Overexpression of CXCR4 on human CD34 (+) progenitors increases their proliferation, migration, and NOD/SCID repopulation. *Blood*, *103*(8), 2942–2949.
Katayama, Y., Battista, M., Kao, W. M., Hidalgo, A., Peired, A. J., Thomas, S. A., et al. (2006). Signals from the sympathetic nervous system regulate hematopoietic stem cell egress from bone marrow. *Cell*, *124*(2), 407–421.
Katayama, Y., Hidalgo, A., Chang, J., Peired, A., & Frenette, P. S. (2005). CD44 is a physiological E-selectin ligand on neutrophils. *The Journal of Experimental Medicine*, *201*(8), 1183–1189.
Katayama, Y., Hidalgo, A., Furie, B. C., Vestweber, D., Furie, B., & Frenette, P. S. (2003). PSGL-1 participates in E-selectin-mediated progenitor homing to bone marrow: Evidence for cooperation between E-selectin ligands and a4 integrin. *Blood*, *102*(6), 2060–2067.
Kawamori, Y., Katayama, Y., Asada, N., Minagawa, K., Sato, M., Okamura, A., et al. (2010). Role for vitamin D receptor in the neuronal control of the hematopoietic stem cell niche. *Blood*, *116*(25), 5528–5535.

Kiel, M. J., & Morrison, S. J. (2008). Uncertainty in the niches that maintain haematopoietic stem cells. *Nature Reviews. Immunology*, *8*(4), 290–301.

Kim, C. H., Wu, W., Wysoczynski, M., Abdel-Latif, A., Sunkara, M., Morris, A., et al. (2012). Conditioning for hematopoietic transplantation activates the complement cascade and induces a proteolytic environment in bone marrow: A novel role for bioactive lipids and soluble C5b-C9 as homing factors. *Leukemia*, *26*(1), 106–116.

Kimura, T., Boehmler, A. M., Seitz, G., Kuci, S., Wiesner, T., Brinkmann, V., et al. (2004). The sphingosine 1-phosphate receptor agonist fty720 supports cxcr4-dependent migration and bone marrow homing of human cd34+ progenitor cells. *Blood*, *103*(12), 4478–4486.

Kimura, Y., Ding, B., Imai, N., Nolan, D. J., Butler, J. M., & Rafii, S. (2011). c-Kit-mediated functional positioning of stem cells to their niches is essential for maintenance and regeneration of adult hematopoiesis. *PLoS One*, *6*(10), e26918.

Kirito, K., Hu, Y., & Komatsu, N. (2009). HIF-1 prevents the overproduction of mitochondrial ROS after cytokine stimulation through induction of PDK-1. *Cell Cycle*, *8*(17), 2844–2849.

Ko, K. H., Holmes, T., Palladinetti, P., Song, E., Nordon, R., O'Brien, T. A., et al. (2011). GSK-3β inhibition promotes engraftment of ex vivo-expanded hematopoietic stem cells and modulates gene expression. *Stem Cells*, *29*, 108–118.

Kollet, O., Dar, A., Shivtiel, S., Kalinkovich, A., Lapid, K., Sztainberg, Y., et al. (2006). Osteoclasts degrade endosteal components and promote mobilization of hematopoietic progenitor cells. *Nature Medicine*, *12*(6), 657–664.

Kollet, O., Shivtiel, S., Chen, Y. Q., Suriawinata, J., Thung, Sn, Dabeva, M. D., et al. (2003). Hgf, Sdf-1, and Mmp-9 are involved in stress-induced human Cd34(+) stem cell recruitment to the liver. *The Journal of Clinical Investigation*, *112*(2), 160–169.

Kollet, O., Spiegel, A., Peled, A., Petit, I., Byk, T., Hershkoviz, R., et al. (2001). Rapid and efficient homing of human Cd34(+)Cd38(-/low)Cxcr4(+) stem and progenitor cells to the bone marrow and spleen of Nod/Scid and Nod/Scid/B2m(null) mice. *Blood*, *97*(10), 3283–3291.

Kollet, O., Vagima, Y., D'Uva, G., Golan, K., Canaani, J., Itkin, T., et al. (2013). Physiologic corticosterone oscillations regulate murine hematopoietic stem/progenitor cell proliferation and CXCL12 expression by bone marrow stromal progenitors. *Leukemia*, *27*(10), 2006–2015.

Laird, D. J., von Andrian, U. H., & Wagers, A. J. (2008). Stem cell trafficking in tissue development, growth, and disease. *Cell*, *132*(4), 612–630.

Lapid, K., Itkin, T., D'Uva, G., Ovadya, Y., Ludin, A., Caglio, G., et al. (2013). GSK3β regulates physiological migration of stem/progenitor cells via cytoskeletal rearrangement. *The Journal of Clinical Investigation*, *123*(4), 1705–1717.

Lapidot, T., Dar, A., & Kollet, O. (2005). How do stem cells find their way home? *Blood*, *106*(6), 1901–1910.

Lapidot, T., & Kollet, O. (2010). The brain-bone-blood triad: Traffic lights for stem-cell homing and mobilization. *Hematology American Society Hematology Education Program*, *2010*, 1–6.

Lee, H. M., Wu, W., Wysoczynski, M., Liu, R., Zuba-Surma, E. K., Kucia, M., et al. (2009). Impaired mobilization of hematopoietic stem/progenitor cells in C5-deficient mice supports the pivotal involvement of innate immunity in this process and reveals novel promobilization effects of granulocytes. *Leukemia*, *23*(11), 2052–2062.

Lévesque, J. P., Helwani, F. M., & Winkler, I. G. (2010). The endosteal 'osteoblastic' niche and its role in hematopoietic stem cell homing and mobilization. *Leukemia*, *24*(12), 1979–1992.

Levesque, J., Hendy, J., Takamatsu, Y., Simmons, P., & Bendall, L. (2003). Disruption of the CXCR4/CXCL12 chemotactic interaction during hematopoietic stem cell mobilization induced by GCSF or cyclophosphamide. *The Journal of Clinical Investigation, 111*(2), 187–196.

Lévesque, J. P., Hendy, J., Takamatsu, Y., Williams, B., Winkler, I. G., & Simmons, P. J. (2002). Mobilization by either cyclophosphamide or granulocyte colony-stimulating factor transforms the bone marrow into a highly proteolytic environment. *Experimental Hematology, 30*(5), 440–449.

Levesque, J., Takamatsu, Y., Nilsson, S., Haylock, D., & Simmons, P. (2001). Vascular cell adhesion molecule-1 (CD106) is cleaved by neutrophil proteases in the bone marrow following hematopoietic progenitor cell mobilization by granulocyte colony-stimulating factor. *Blood, 98*(5), 1289–1297.

Levoye, A., Balabanian, K., Baleux, F., Bachelerie, F., & Lagane, B. (2009). CXCR7 heterodimerizes with CXCR4 and regulates CXCL12-mediated G protein signaling. *Blood, 113*(24), 6085–6093.

Liu, F., Poursine-Laurent, J., & Link, D. C. (1997). The granulocyte colony-stimulating factor receptor is required for the mobilization of murine hematopoietic progenitors into peripheral blood by cyclophosphamide or interleukin-8 but not Flt-3 ligand. *Blood, 90*(7), 2522–2528.

Lucas, D., Battista, M., Shi, P. A., Isola, L., & Frenette, P. S. (2008). Mobilized hematopoietic stem cell yield depends on species-specific circadian timing. *Cell Stem Cell, 3*(4), 364–366.

Lucas, D., Scheiermann, C., Chow, A., Kunisaki, Y., Bruns, I., Barrick, C., et al. (2013). Chemotherapy-induced bone marrow nerve injury impairs hematopoietic regeneration. *Nature Medicine, 19*(6), 695–703.

Ludin, A., Itkin, T., Gur-Cohen, S., Mildner, A., Shezen, E., Golan, K., et al. (2012). Monocytes-macrophages that express α-smooth muscle actin preserve primitive hematopoietic cells in the bone marrow. *Nature Immunology, 13*(11), 1072–1082.

Malard, F., Kröger, N., Gabriel, I. H., Hübel, K., Apperley, J. F., Basak, G. W., et al. (2012). Plerixafor for autologous peripheral blood stem cell mobilization in patients previously treated with fludarabine or lenalidomide. *Biology of Blood and Marrow Transplantation, 18*(2), 314–317.

Mantel, C. R., O'Leary, H. A., Chitteti, B. R., Huang, X., Cooper, S., Hangoc, G., et al. (2015). Enhancing hematopoietic stem cell transplantation efficacy by mitigating oxygen shock. *Cell, 161*(7), 1553–1565.

Mark, T., Stern, J., Furst, J. R., Jayabalan, D., Zafar, F., LaRow, A., et al. (2008). Stem cell mobilization with cyclophosphamide overcomes the suppressive effect of lenalidomide therapy on stem cell collection in multiple myeloma. *Biology of Blood and Marrow Transplantation, 14*(7), 795–798.

Massberg, S., Schaerli, P., Knezevic-Maramica, I., Kollnberger, M., Tubo, N., Moseman, E. A., et al. (2007). Immunosurveillance by hematopoietic progenitor cells trafficking through blood, lymph, and peripheral tissues. *Cell, 131*(5), 994–1008.

Mazo, I. B., Gutierrez-Ramos, J. C., Frenette, P. S., Hynes, R. O., Wagner, D. D., & von Andrian, U. H. (1998). Hematopoietic progenitor cell rolling in bone marrow microvessels: Parallel contributions by endothelial selectins and vascular cell adhesion molecule 1. *The Journal of Experimental Medicine, 188*(3), 465–474.

Mendez-Ferrer, S., Battista, M., & Frenette, P. S. (2010). Cooperation of beta (2)- and beta(3)-adrenergic receptors in hematopoietic progenitor cell mobilization. *Annals of the New York Academy of Sciences, 1192*, 139–144.

Mendez-Ferrer, S., Lucas, D., Battista, M., & Frenette, P. S. (2008). Haematopoietic stem cell release is regulated by circadian oscillations. *Nature, 452*(7186), 442–447.

Morrison, S. J., & Scadden, D. T. (2014). The bone marrow niche for haematopoietic stem cells. *Nature, 505*(7483), 327–334.

Naumann, U., Cameroni, E., Pruenster, M., Mahabaleshwar, H., Raz, E., Zerwes, H. G., et al. (2010). CXCR7 functions as a scavenger for CXCL12 and CXCL11. *PLoS One, 5*(2), e9175.

Neusser, M. A., Kraus, A. K., Regele, H., Cohen, C. D., Fehr, T., Kerjaschki, D., et al. (2010). The chemokine receptor CXCR7 is expressed on lymphatic endothelial cells during renal allograft rejection. *Kidney International*, *77*(9), 801–808.

Nilsson, S. K., Johnston, H. M., & Coverdale, J. A. (2001). Spatial localization of transplanted hemopoietic stem cells: Inferences for the localization of stem cell niches. *Blood*, *97*(8), 2293–2299.

North, T. E., Goessling, W., Walkley, C. R., Lengerke, C., Kopani, K. R., Lord, A. M., et al. (2007). Prostaglandin E2 regulates vertebrate hematopoietic stem cell homeostasis. *Nature*, *447*(7147), 1007–1011.

Okajima, F. (2013). Regulation of inflammation by extracellular acidification and proton-sensing GPCRs. *Cellular Signalling*, *25*(11), 2263–2271.

Olson, T. S., Caselli, A., Otsuru, S., Hofmann, T. J., Williams, R., Paolucci, P., et al. (2013). Megakaryocytes promote murine osteoblastic HSC niche expansion and stem cell engraftment after radioablative conditioning. *Blood*, *121*(26), 5238–5249.

Ozüyaman, B., Ebner, P., Niesler, U., Ziemann, J., Kleinbongard, P., Jax, T., et al. (2005). Nitric oxide differentially regulates proliferation and mobilization of endothelial progenitor cells but not of hematopoietic stem cells. *Thrombosis and Haemostasis*, *94*(4), 770–772.

Pappu, R., Schwab, S. R., Cornelissen, I., Pereira, J. P., Regard, J. B., Xu, Y., et al. (2007). Promotion of lymphocyte egress into blood and lymph by distinct sources of sphingosine-1-phosphate. *Science*, *316*(5822), 295–298.

Paripati, H., Stewart, A. K., Cabou, S., Dueck, A., Zepeda, V. J., Pirooz, N., et al. (2008). Compromised stem cell mobilization following induction therapy with lenalidomide in myeloma. *Leukemia*, *22*(6), 1282–1284.

Peled, A., Grabovsky, V., Habler, L., Sandbank, J., Arenzana-Seisdedos, F., Petit, I., et al. (1999). The chemokine SDF-1 stimulates integrin-mediated arrest of CD34(+) cells on vascular endothelium under shear flow. *The Journal of Clinical Investigation*, *104*(9), 1199–1211.

Peled, A., Kollet, O., Ponomaryov, T., Petit, I., Franitza, S., Grabovsky, V., et al. (2000). The chemokine SDF-1 activates the integrins LFA-1, VLA-4, and VLA-5 on immature human CD341 cells: Role in transendothelial/stromal migration and engraftment of NOD/SCID mice. *Blood*, *95*(11), 3289–3296.

Pelus, L. M., Bian, H., King, A. G., Fukuda, S., et al. (2004). Neutrophil-derived MMP-9 mediates synergistic mobilization of hematopoietic stem and progenitor cells by the combination of G-CSF and the chemokines GRObeta/CXCL2 and GRObetaT/CXCL2delta4. *Blood*, *103*(1), 110–119.

Petit, I., Goichberg, P., Spiegel, A., Peled, A., Brodie, C., Seger, R., et al. (2005). Atypical Pkc-zeta regulates Sdf-1-mediated migration and development of human Cd34 + progenitor cells. *The Journal of Clinical Investigation*, *115*(1), 168–176.

Petit, I., Szyper-Kravitz, M., Nagler, A., Lahav, M., Peled, A., Habler, L., et al. (2002). G-CSF induces stem cell mobilization by decreasing bone marrow Sdf-1 and up-regulating Cxcr4. *Nature Immunology*, *3*(7), 687–694.

Pineault, N., & Abu-Khader, A. (2015). Advances in umbilical cord blood stem cell expansion and clinical translation. *Experimental Hematology*, *43*(7), 498–513.

Plett, P. A., Frankovitz, S. M., Wolber, F. M., Abonour, R., & Orschell-Traycoff, C. M. (2002). Treatment of circulating CD34(+) cells with SDF-1alpha or anti-CXCR4 antibody enhances migration and NOD/SCID repopulating potential. *Experimental Hematology*, *30*(9), 1061–1069.

Ponomaryov, T., Peled, A., Petit, I., Taichman, R. S., Habler, L., Sandbank, J., et al. (2000). Induction of the chemokine stromal-derived factor-1 following DNA damage improves human stem cell function. *The Journal of Clinical Investigation*, *106*(11), 1331–1339.

Pusic, I., Jiang, S. Y., Landua, S., Uy, G. L., Rettig, M. P., Cashen, A. F., et al. (2008). Impact of mobilization and remobilization strategies on achieving sufficient stem cell yields for autologous transplantation. *Biology of Blood and Marrow Transplantation, 14*(9), 1045–1056.

Ratajczak, M. Z. (2015). A novel view of the adult bone marrow stem cell hierarchy and stem cell trafficking. *Leukemia, 29*(4), 776–782.

Ratajczak, M. Z., & Adamiak, M. (2015). Membrane lipid rafts, master regulators of hematopoietic stem cell retention in bone marrow and their trafficking. *Leukemia, 29*(7), 1452–1457.

Ratajczak, M. Z., Lee, H., Wysoczynski, M., Wan, W., Marlicz, W., Laughlin, M. J., et al. (2010). Novel insight into stem cell mobilization-plasma sphingosine-1-phosphate is a major chemoattractant that directs the egress of hematopoietic stem progenitor cells from the bone marrow and its level in peripheral blood increases during mobilization due to activation of complement cascade/membrane attack complex. *Leukemia, 24*(5), 976–985.

Reddy, P. N., Radu, M., Xu, K., Wood, J., Harris, C. E., Chernoff, J., et al. (2016). p21-activated kinase 2 regulates HSPC cytoskeleton, migration, and homing via CDC42 activation and interaction with β-Pix. *Blood, 127*(16), 1967–1975.

Richman, C. M., Weiner, R. S., & Yankee, R. A. (1976). Increase in circulating stem cells following chemotherapy in man. *Blood, 47*(6), 1031–1039.

Rossi, L., Manfredini, R., Bertolini, F., Ferrari, D., Fogli, M., Zini, R., et al. (2007). The extracellular nucleotide UTP is a potent inducer of hematopoietic stem cell migration. *Blood, 109*(2), 533–542.

Sackstein, R., Merzaban, J. S., Cain, D. W., Dagia, N. M., Spencer, J. A., Lin, C. P., et al. (2008). Ex vivo glycan engineering of CD44 programs human multipotent mesenchymal stromal cell trafficking to bone. *Nature Medicine, 14*(2), 181–187.

Sato, H., & Takino, T. (2010). Coordinate action of membranetype matrix metalloproteinase-1 (MT1-MMP) and MMP-2 enhances pericellular proteolysis and invasion. *Cancer Science, 101*(4), 843–847.

Schneider, A., Krüger, C., Steigleder, T., Weber, D., Pitzer, C., Laage, R., et al. (2005). The hematopoietic factor G-CSF is a neuronal ligand that counteracts programmed cell death and drives neurogenesis. *The Journal of Clinical Investigation, 115*(8), 2083–2098.

Semerad, C. L., Christopher, M. J., Liu, F., Short, B., Simmons, P. J., Winkler, I., et al. (2005). G-CSF potently inhibits osteoblast activity and CXCL12 mRNA expression in the bone marrow. *Blood, 106*(9), 3020–3027.

Shirvaikar, N., Marquez-Curtis, L. A., Ratajczak, M. Z., & Janowska-Wieczorek, A. (2011). Hyaluronic acid and thrombin upregulate MT1-MMP through PI3K and Rac-1 signaling and prime the homing-related responses of cord blood hematopoietic stem/progenitor cells. *Stem Cells and Development, 20*(1), 19–30.

Shivtiel, S., Kollet, O., Lapid, K., Schajnovitz, A., Goichberg, P., Kalinkovich, A., et al. (2008). CD45 regulates retention, motility, and numbers of hematopoietic progenitors, and affects osteoclast remodeling of metaphyseal trabecules. *Journal of Experimental Medicine, 205*(10), 2381–2395.

Sipkins, D. A., Wei, X., Wu, J. W., Runnels, J. M., Côté, D., Means, T. K., et al. (2005). In vivo imaging of specialized bone marrow endothelial microdomains for tumour engraftment. *Nature, 435*(7044), 969–973.

Spiegel, A., Shivtiel, S., Kalinkovich, A., Ludin, A., Netzer, N., Goichberg, P., et al. (2007). Catecholaminergic neurotransmitters regulate migration and repopulation of immature human CD34+ cells through Wnt signaling. *Nature Immunology, 8*(10), 1123–1131.

Staller, P., Sulitkova, J., Lisztwan, J., Moch, H., Oakeley, E. J., & Krek, W. (2003). Chemokine receptor CXCR4 downregulated by von Hippel–Lindau tumour suppressor pVHL. *Nature, 425*(6955), 307–311.

Stiff, P. J., Murgo, A. J., Wittes, R. E., DeRisi, M. F., & Clarkson, B. D. (1983). Quantification of the peripheral blood colony forming unit-culture rise following chemotherapy. Could leukocytaphereses replace bone marrow for autologous transplantation? *Transfusion, 23*(6), 500–503.

Sugiyama, T., Kohara, H., Noda, M., & Nagasawa, T. (2006). Maintenance of the hematopoietic stem cell pool by CXCL12–CXCR4 chemokine signaling in bone marrow stromal cell niches. *Immunity*, *25*(6), 977–988.

Tarnowski, M., Liu, R., Wysoczynski, M., Ratajczak, J., Kucia, M., & Ratajczak, M. Z. (2010). CXCR7: A new SDF-1-binding receptor in contrast to normal CD34(+) progenitors is functional and is expressed at higher level in human malignant hematopoietic cells. *European Journal of Haematology*, *85*(6), 472–483.

Thelen, M., & Thelen, S. (2008). CXCR7, CXCR4 and CXCL12: An eccentric trio? *Journal of Neuroimmunology*, *198*(1–2), 9–13.

Tuchman, S. A., Bacon, W. A., Huang, L. W., Long, G., Rizzieri, D., Horwitz, M., et al. (2015). Cyclophosphamide based hematopoietic stem cell mobilization before autologous stem cell transplantation in newly diagnosed multiple myeloma. *Journal of Clinical Apheresis*, *30*(3), 176–182.

Uto-Konomi, A., McKibben, B., Wirtz, J., Sato, Y., Takano, A., Nanki, T., et al. (2013). CXCR7 agonists inhibit the function of CXCL12 by down-regulation of CXCR4. *Biochemical and Biophysical Research Communications*, *431*(4), 772–776.

Vagima, Y., Avigdor, A., Goichberg, P., Shivtiel, S., Tesio, M., Kalinkovich, A., et al. (2009). MT1-MMP and RECK are involved in human CD34+ progenitor cell retention, egress, and mobilization. *The Journal of Clinical Investigation*, *119*(3), 492–503.

Venkataraman, K., Lee, Y. M., Michaud, J., Thangada, S., Ai, Y., Bonkovsky, H. L., et al. (2008). Vascular endothelium as a contributor of plasma sphingosine 1-phosphate. *Circulation Research*, *102*(6), 669–676.

Williams, B., & Nilsson, S. K. (2009). Investigating the interactions between haemopoietic stem cells and their niche: Methods for the analysis of stem cell homing and distribution within the marrow following transplantation. *Methods in Molecular Biology*, *482*, 93–107.

Williams, D. A., Zheng, Y., & Cancelas, J. A. (2008). Rho GTPases and regulation of hematopoietic stem cell localization. *Methods in Enzymology*, *439*, 365–393.

Winkler, I. G., Sims, N. A., Pettit, A. R., Barbier, V., Nowlan, B., Helwani, F., et al. (2010). Bone marrow macrophages maintain hematopoietic stem cell (HSC) niches and their depletion mobilizes HSCs. *Blood*, *116*(23), 4815–4828.

Wysoczynski, M., Ratajczak, J., Pedziwiatr, D., Rokosh, G., Bolli, R., & Ratajczak, M. Z. (2015). Identification of heme oxygenase 1 (HO-1) as a novel negative regulator of mobilization of hematopoietic stem/progenitor cells. *Stem Cell Reviews*, *11*(1), 110–118.

Wysoczynski, M., Reca, R., Ratajczak, J., Kucia, M., Shirvaikar, N., Honczarenko, M., et al. (2005). Incorporation of CXCR4 into membrane lipid rafts primes homing-related responses of hematopoietic stem/progenitor cells to an SDF-1 gradient. *Blood*, *105*(1), 40–48.

Xie, Y., Yin, T., Wiegraebe, W., He, X. C., Miller, D., Stark, D., et al. (2009). Detection of functional haematopoietic stem cell niche using real-time imaging. *Nature*, *457*(7225), 97–101.

Zeng, Y., Broxmeyer, H. E., Staser, K., Chitteti, B. R., Park, S. J., Hahn, S., et al. (2015). Pak2 regulates hematopoietic progenitor cell proliferation, survival, and differentiation. *Stem Cells*, *33*(5), 1630–1641.

Zhang, J., Ren, X., Shi, W., Wang, S., Chen, H., Zhang, B., et al. (2014). Small molecule Me6TREN mobilizes hematopoietic stem/progenitor cells by activating MMP-9 expression and disrupting SDF-1/CXCR4 axis. *Blood*, *123*(3), 428–441.

Zhao, Y., Zhan, Y., Burke, K. A., & Anderson, W. F. (2005). Soluble factor(s) from bone marrow cells can rescue lethally irradiated mice by protecting endogenous hematopoietic stem cells. *Experimental Hematology*, *33*(4), 428–434.

CHAPTER SIX

Alterations of HSC Niche in Myeloid Malignancies

L. Han, M. Konopleva[1]
The University of Texas MD Anderson Cancer Center, Houston, TX, United States
[1]Corresponding author: e-mail address: mkonople@mdanderson.org

Contents

The normal hematopoietic system provides lifelong production of blood cells to sustain oxygen delivery, hemostasis, and immune functions. This requires tight control of hematopoietic stem cells (HSCs), which possess the potential for self-renewal to maintain the stem cell pool and differentiate into functional lineage cells. The fate of HSCs is partially determined by intrinsic clues involving chromatin remodeling and the transcription factor network (Enver, Pera, Peterson, & Andrews, 2009; Rizo, Vellenga, de Haan, & Schuringa, 2006). However, emerging data indicate that stem cells' potential are largely regulated extrinsically by their microenvironment (or niche). The concept of stem cell niche in which stem cell behaviors are regulated by their associated cells was initially postulated by Schofield (1978) and first demonstrated in *Drosophila* (Kiger, Jones, Schulz, Rogers, & Fuller, 2001; Kiger, White-Cooper, & Fuller, 2000). Over a decade ago, murine models were manipulated to investigate the composition and

Advances in Stem Cells and their Niches, Volume 1
ISSN 2468-5097
http://dx.doi.org/10.1016/bs.asn.2017.01.003

molecular determinants of the bone marrow (BM) niche, as well as its roles in regulating stem cells and developing hematopoietic malignancies (Arai et al., 2004; Calvi et al., 2003). In this chapter, we focus on dissecting the components of the HSC niche and the role of altered niche function in the setting of myeloid malignancies. Understanding the differences between physiologic and pathologic modulations of the stem cell niche may provide the key to developing therapeutics that restore normal hematopoiesis through abrogation of malignant niches.

1. LOCATION AND CELLULAR COMPONENTS OF NORMAL STEM CELL NICHE

The BM niche refers to a specialized marrow microenvironment composed of a variety of cell types in unique anatomic regions with defined functions (Carter et al., 2011; Morrison & Scadden, 2014; Reagan & Rosen, 2016). It provides essential endocrine, autocrine, and paracrine signaling, as well as physical interactions to maintain hematopoiesis and modulate skeletal remodeling. The cellular components of the BM niche comprise two types with distinct functions: one type that provides direct signals to HSCs, including mesenchymal stem cells (MSCs), endothelial cells, and megakaryocytes; and a second type that indirectly facilitates HSC functions, such as osteoblasts and nerve cells (Fig. 1) (Schepers, Campbell, & Passegue, 2015).

1.1 Location of the HSC Niche

1.1.1 The Endosteal Niche

The organization of the BM niche and its association with stem cells have been identified in murine BM using imaging technologies coupled with a series of functional assays. The endosteal niche was initially defined as the major location for HSCs (Nilsson, Johnston, & Coverdale, 2001; Xie et al., 2009). The osteoblast lineage cells that are progenies of pluripotent MSCs line the surface of the endosteum and play an indispensable role in normal hematopoiesis. These cells regulate HSC function through production of a variety of cytokines, including C-X-C motif chemokine ligand 12 (CXCL12), angiopoietin-1, and granulocyte colony-stimulating factor (G-CSF) (Calvi & Link, 2015). Calvi et al. observed that the HSC pool was expanded by increasing the osteoblast lineage cell number through stimulation of the osteoblast-specific parathyroid hormone receptor (Calvi et al., 2003). In a transgenic Col2.3 Delta TK mouse model, genetic ablation of osteoblasts severely impaired normal hematopoiesis (Visnjic et al., 2004).

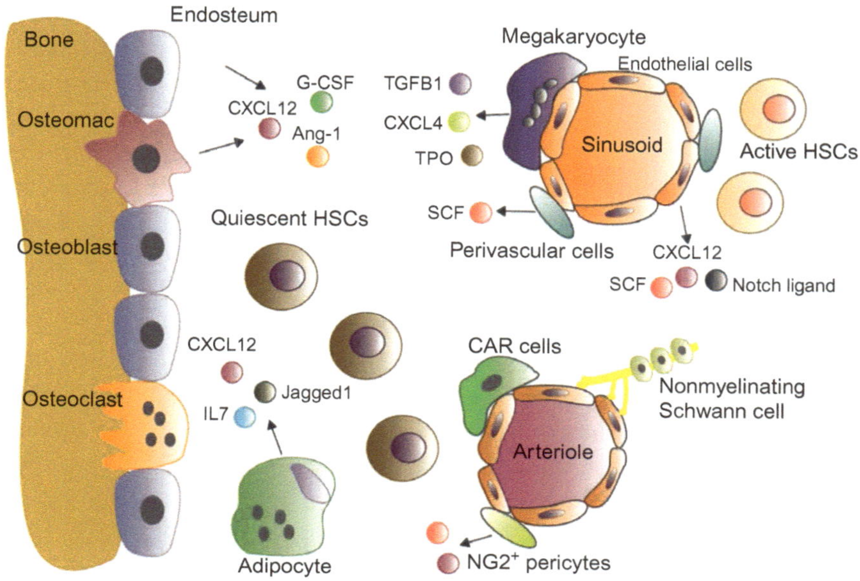

Fig. 1 Bone marrow niche structures of normal hematopoietic stem cells. The endosteal niche is composed of the osteoblast lineage cells derived from pluripotent MSCs. The sinusoidal-megakaryocyte niche contains endothelial cells, megakaryocytes, and perivascular cells. The arteriolar niche includes arteriolar endothelial cells, NG2^{+} pericytes, and sympathetic nerve fibers ensheathed by nonmyelinating Schwann cells. Osteoblasts, osteomacs, adipocytes, endothelial cells, and perivascular cells regulate HSC fate by secreting soluble factors.

Reduction of HSC pool size induced by osteoblastic deficiency is mediated by loss of stem cell quiescence and suppression of self-renewal (Bowers et al., 2015). The primitive osteolineage cells, not the differentiated cells, may be responsible for HSC maintenance, is suggested by the finding that immature osteolineage cells in C57BL/6 mice expressed higher levels of niche-regulating factors stem cell factor (SCF) and CXCL12 (Nakamura et al., 2010).

1.1.2 The Vascular Niche

Several studies have emphasized the role of vascular niches in supporting HSC maintenance. Kiel et al. using immunofluorescence to visualize the location of HSCs found that about 60% of HSCs were associated with sinusoidal endothelium in both spleen and BM, although some HSCs were present in the osteoblast-enriched endosteal location (Kiel et al., 2005). An imaging cytometry platform utilized to investigate the HSC distribution in BM cavities revealed that most HSCs interact closely with microvessels in the endosteal zones (Nombela-Arrieta et al., 2013). A recent study using high-resolution in vivo imaging confirmed that HSCs preferentially interacted with the vascular niche (Koechlein et al., 2016). However, their differentiated progenies displayed dramatically reduced affinity for the vascular area. Notch signaling in endothelial cells was demonstrated to be the critical regulator of the HSC niche, by increasing $CD31^+$ vessels, platelet-derived growth factor receptor beta (PDGFRB)-positive perivascular cells, production of SCF, and maintaining the niche-forming vessels in aging mice (Kusumbe et al., 2016).

Emerging data support distinct vascular niches in BM, the arteriolar and sinusoidal-megakaryocyte niches. Kunisaki et al. demonstrated that quiescent HSCs specifically localized to the arterioles in the endosteal space (Kunisaki et al., 2013). This arteriolar niche included $Nestin^{high}$ $NG2^+LEPR^-$ pericytes, CXCL12-abundant reticular (CAR) cells, endothelial cells, nonmyelinating Schwann cells, and sympathetic nerves, as well as CXCL12 and SCF produced by pericytes. In turn, the $NG2^-$ MSCs formed the sinusoidal niche that supported actively cycling HSCs. Nestin, a marker for precursor cells that gives rise to both endothelial and mesenchymal lineages (Ono et al., 2014) was expressed by $Sca\text{-}1^+CD31^{high}$ arterial endothelial cells, but not by $Sca\text{-}1^-CD31^+$ sinusoid endothelial cells (Itkin et al., 2016). A subset of MSCs was termed CAR cells because of high expression of CXCL12, which plays an important role in regulation of HSC proliferation (Omatsu et al., 2010). A study by Itkin et al. demonstrated that the lower permeability of arterial vessels provides a relatively metabolically inactive metabolically microenvironment and maintains HSC at a quiescent state with low levels of reactive oxygen species (ROS) (Itkin et al., 2016). Their imaging data also showed that ROS^{low} HSCs reside in close proximity to megakaryocytes.

1.2 Cellular Components

1.2.1 MSCs: Phenotypes and Functions

MSCs are of the mesenchymal lineage origin. The International Society for Cellular Therapy position statement suggests that only the fibroblast-like plastic-adherent cells that meet specified stem cell criteria can be termed MSCs (Horwitz et al., 2005). MSCs have the capacities for self-renewal and multilineage differentiation to osteoblasts, adipocytes, and chondroblasts in vitro and in vivo (Dominici et al., 2006; Mo, Wang, Zhou, Li, & Wu, 2016; Pittenger et al., 1999). BM-derived MSCs are heterogeneous populations, maintaining an osteoblast pool through dynamic response to tissue stress (Park et al., 2012). Functionally, MSCs form the bone structure and organize the supportive microenvironment for hematopoiesis (Mendez-Ferrer et al., 2010; Sacchetti et al., 2007).

It has been proposed that the minimum criteria to define the phenotype of human MSCs comprise positivity for CD105, CD73, and CD90, and negativity for hematopoietic markers CD45, CD34, CD14, or CD11b; CD79alpha or CD19; and HLA-DR surface molecules (Dominici et al., 2006). However, increasing lines of evidence indicate that MSCs involve a mixture of diverse subpopulations with distinct surface antigens and biological functions. Furthermore, their phenotype can be altered by in vitro culture (Busser et al., 2015). Bensidhoum et al. demonstrated that Stro-1^{+} MSCs represented 6% of adherent BM mononuclear cells (Bensidhoum et al., 2004). More Stro-1^{+} MSCs than Stro-1^{-} MSCs were detected in BM, spleen, muscles, liver, and kidneys, indicating that the Stro-1^{+} MSCs have superior migration capacity. The MSCs with high-growth capacity highly expressed Stro-1 and PDGFRA (Samsonraj et al., 2015). Battula et al. identified heterogeneous clones presenting distinct phenotypes and functions by isolating and culturing MSC antigen-1 (MSCA-1)$^{+}$CD56$^{+/-}$ MSCs (Battula et al., 2009). MSCA-1^{+}CD56$^{+/-}$ MSCs coexpressed CD166 were enriched in colony-forming unit fibroblasts (CFU-F)-generating cells and differentiated to chondrocytes and pancreas-like islets. However, MSCA-1^{+}CD56^{-} MSCs expressed CD10, CD26, CD106 (vascular cell adhesion molecule 1 [VCAM1]), and CD146 and differentiated to adipocytes. Nestin^{+} MSCs were spatially associated with HSCs and adrenergic nerve fibers, playing a key role in supporting HSC expansion (Isern et al., 2013; Mendez-Ferrer et al., 2010). Besides these, CD271, SCF, CD29 (β1 integrin), CD324 (E-cadherin) TWIST, and SSEA-4 were also reported to be expressed on MSCs with self-renewing capacity (Pleyer, Valent, & Greil, 2016; Samsonraj et al., 2015). Phenotypic diversity of MSCs in various studies is mostly related to the differences in MSC origins, cell culture conditions, and cell isolation procedures.

MSCs localize adjacent to arterioles and sinusoid vessels and are in direct contact with endothelial cells, forming pericytes. MSCs regulate the fate of HSCs via adhesion molecule-mediated cell–cell interactions, production of cytokines, and modulation of regulatory signals (Pleyer et al., 2016). Single-cell RNA sequencing on isolated MSCs with distinct proximity to HSCs revealed that HSC proximal MSCs regulated stem cell quiescence by expressing higher levels of secreted RNase angiogenin, cytokine interleukin 18 (IL18), and the adhesion molecule embigin, as well as putative regulators such as CXCL12 and VCAM1 (Silberstein et al., 2016).

1.2.2 Vascular Endothelial Cells

Endothelial cells line the blood vessels that deliver nutrients and oxygen to cells and establish a vascular niche to regulate HSCs through angiocrine factors and direct cellular contacts (Butler, Kobayashi, & Rafii, 2010; Rafii, Butler, & Ding, 2016). Endothelial cells were initially identified as an HSC niche-supporting component through their capacity to expand cord blood-derived stem/progenitor cells ex vivo (Rafii et al., 1995). Endothelium-specific markers, including CD31, VE-cadherin, vascular endothelial growth factor receptors 2 and 3 (VEGFR2/3), and TIE2, recognize endothelial cells (Avecilla et al., 2004; Hooper et al., 2009; Poulos et al., 2013). The endothelial cells secrete angiocrine factors, including SCF, CXCL12, VEGFA, fibroblast growth factor 2 (FGF2), angiopoietin, thrombospondin-1 (TSP1), and Notch ligands, to activate HSC self-renewal (Butler, Nolan, et al., 2010; Kobayashi et al., 2010; Rafii et al., 2016). Further evidence suggests that endothelial cells are also required in multilineage reconstitution of HSCs, modulating angiocrine factors induced by signaling pathways activated within endothelial cells (Avecilla et al., 2004; Hooper et al., 2009; Kobayashi et al., 2010). Kobayashi et al. demonstrated that the activation status of endothelial cells determines the fate of HSCs. At steady state, through activation of the AKT pathway, endothelial cells produce angiocrine factors, including SCF, CXCL12, and Notch ligands, to maintain the stem cell pool and regulate self-renewal of HSCs. In the setting of myeloablative stress, coactivation of the AKT and MAPK pathways in endothelial cells results in production of lineage-commitment factors, such as G-CSF, granulocyte-macrophage colony-stimulating factor (GM-CSF) and IL6 (Kobayashi et al., 2010). The mature hematopoietic cells (megakaryocytes) can in turn halt the angiogenesis by producing antiangiogenic factor TSP1 to maintain a homeostasis (Kopp et al., 2006). Endothelial cells also establish a perisinusoidal niche in the spleen for extramedullary hematopoiesis through production of SCF (Inra et al., 2015).

1.2.3 Adipocytes

MSCs give rise to adipocytes when the adipogenic-lineage transcription factor PPARG is activated (Kersten, Desvergne, & Wahli, 2000).

The connective tissue growth factor has been shown to direct human MSCs to adipocyte differentiation (Battula et al., 2013). BM adipocytes reside close to the endosteal surface and have been in the past considered simply as "filler" for the fibrotic or empty marrow spaces (Reagan & Rosen, 2016; Rosen, Ackert-Bicknell, Rodriguez, & Pino, 2009). Recent studies, however, demonstrated BM adipocytes to be critical niche components in regulating HSC homeostasis. Marrow adipose tissue was reported to exist in two types, regulated and constitutive (Scheller & Rosen, 2014). The constitutive marrow adipose tissue is formed shortly after birth, while the regulated marrow adipose tissue is located in close proximity to hematopoietic cells in trabecular bone. The adipocytes act as negative regulators of hematopoiesis by preventing progenitor expansion while preserving the stem cell pool (Naveiras et al., 2009). Radiation-damaged BM adipocytes enhanced progenitor expansion and posttransplant recovery, and pharmacologic suppression of adipogenesis promotes BM engraftment. Metabolic stress induced by a high-fat diet activated PPARG, enhancing adipogenesis at the price of osteoblastogenesis (Luo et al., 2015). The adipocyte-rich niche produces less Jagged-1, CXCL12, and IL7, which compromises HSC support. Recent data show that the enhanced adipogenesis and reduced osteoblastogenesis are regulated by leptin/leptin receptor signaling through activation of JAK2/STAT3 signaling in MSCs (Yue, Zhou, Shimada, Zhao, & Morrison, 2016).

1.2.4 Mature Hematopoietic Cells

Some mature hematopoietic cells have been revealed to serve as niche components. Megakaryocytes are identified by their specific markers, including CD41 and von Willenbrand factor. Recent studies highlight the regulation of HSC cycling by megakaryocytes in the sinusoid through production of CXCL4 (Bruns et al., 2014), transforming growth factor (TGF)-beta 1 (Zhao et al., 2014), and thrombopoietin (TPO) (Nakamura-Ishizu, Takubo, Fujioka, & Suda, 2014). The membrane protein C-type lectin-like receptor-2 (CLEC2) accounted for the production of TPO and other factors (Nakamura-Ishizu, Takubo, Kobayashi, Suzuki-Inoue, & Suda, 2015). Deletion of CLEC2 drove BM HSCs to cell cycle entry and reduced the stem cell repopulation potential. Overall, megakaryocytes are critical components of the HSC niche, regulating HSC cell cycle activity (Bruns et al., 2014; Kiel et al., 2005; Zhao et al., 2014). Megakaryocytes contributed to expansion of the endosteal niche after total body radiation, through TPO signaling and CD41 integrin-mediated adhesion (Olson et al., 2013).

BM-resident macrophages including osteomacs (osteal macrophages) and osteoclasts have been recently recognized as HSC niche components. Osteomacs maintain the endosteal HSC niche by supporting osteoblast function, and loss of osteomacs leads to the mobilization of HSCs into blood (Winkler et al., 2010). Deletion of macrophages reduced HSC-retention factor CXCL12 and promoted HSC mobilization (Chow et al., 2011). Maintenance of the HSC by macrophages was recently found to rely partially on the tumor necrosis factor expression in the macrophages (Lu et al., 2016). In addition, CD234-expressing macrophages in the BM promoted stem cell quiescence via binding of CD82 that is predominantly expressed on long-term HSCs (Hur et al., 2016). Osteoclasts resorb bone tissues and regulate osteoblast development that is crucial for HSC function. In the absence of osteoclasts, mesenchymal progenitors accumulated and blocked osteoblastic differentiation, leading to impairment of the HSC niche (Mansour et al., 2012). These studies demonstrated that the cells of the hematologic origin support their HSC ancestors in the niche.

1.2.5 Sympathetic Nervous System

The sympathetic nervous system regulates niche components through binding of the neurotransmitter norepinephrine to the β3- and β2-adrenergic receptors expressed on MSCs and osteoblasts (Asada et al., 2013; Mendez-Ferrer, Lucas, Battista, & Frenette, 2008). Trafficking of HSCs from the BM to the bloodstream is a consequence of rhythmic downregulation of CXCL12 expression regulated by the molecular clock genes through circadian noradrenaline secretion by the sympathetic nervous system (Mendez-Ferrer et al., 2008). Casanova-Acebes et al. demonstrated that this circadian release is regulated by rhythmic elimination of neutrophils by BM-resident macrophages (Casanova-Acebes et al., 2013). The migration of aged neutrophils is CXCR4-dependent and disrupts the HSC niche by reducing CAR cells, thereby reducing CXCL12 protein expression. Nonmyelinating Schwann cells that ensheath and protect sympathetic nerve fibers are positive for Nestin, like some perivascular MSCs. These glial cells maintain HSC dormancy by releasing active TGFB1, which triggers signaling pathways required for HSC quiescence (Yamazaki et al., 2009, 2011). Therefore, the sympathetic nervous system together with its glial cells regulates HSC trafficking and proliferation through modulation of niche components.

2. STEM CELL NICHE ALTERATION IN MYELOID DISORDERS

Myeloid malignancies include acute myeloid leukemia (AML), chronic myeloid leukemia (CML), myelodysplastic syndrome (MDS), and myeloproliferative neoplasms (MPNs); their pathogenesis involves recurrent genomic alterations. Recent studies demonstrated that the malignant residents may shape the niche to their benefit, while suppressing normal hematopoiesis. The remodeled niche (i.e., malignant niche) subsequently contributes to tumor progression. An alternative theory, supported by a series of studies in genetically defined mouse models, postulates that genetic alteration within the niche itself may lead to the malignant transformation of hematopoiesis (Fig. 2).

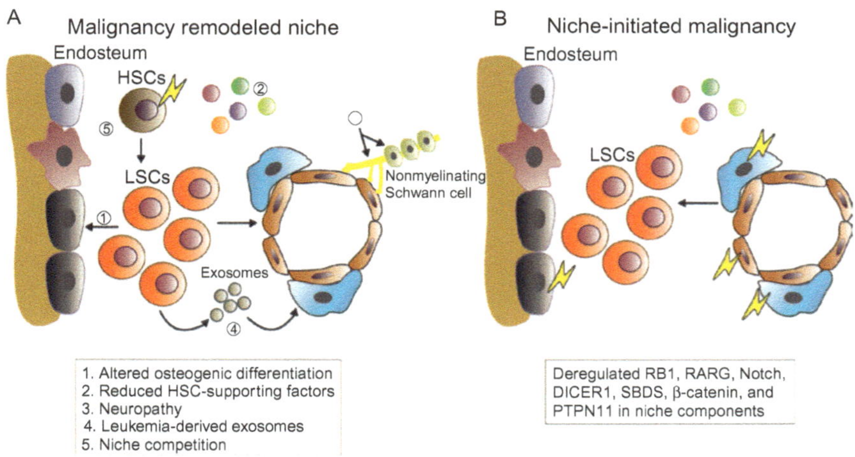

Fig. 2 Cross talk between niche components and leukemia stem cells (LSCs) in hematological malignancies. (A) Myeloid malignancies caused by genetic alterations occurring in HSCs trigger niche remodeling. The mechanisms involve alterations of osteogenic differentiation, reduced production of HSC-supporting soluble factors, leukemia-induced neuropathy, altered MSC function by leukemia-derived exosomes, and niche competition between normal and malignant cell. (B) Genetic alterations occurring in the niches may lead to the malignant transformation of hematopoiesis. Selected genetic alterations of nice cells shown to facilitate development of preleukemia or leukemia are listed in the *box*.

2.1 Stromal Changes in Patients With Myeloid Disorders

2.1.1 Alterations in MSC Phenotype in AML and MDS

Several studies have embarked on identification of altered phenotype or functional properties of MSCs isolated from leukemic BM compared to their healthy controls. AML-derived MSCs express established stromal cell markers CD73, CD90, and CD105 (Despeaux et al., 2011; Geyh et al., 2016). An extensive panel of antibodies against surface molecules on MSCs revealed that expression levels of CD29, CD44, CD73, CD90, and CD324 were comparable between AML MSCs and normal MSCs. CD105 and CD146 were overexpressed on AML MSCs (Huang et al., 2015). In contrast, Le et al. found that AML-derived MSCs displayed a normal immunophenotype (defined as $CD14^-CD19^-CD34^-CD73^+CD90^+CD105^+CD146^+$) but expressed lower levels of CD146 than normal MSCs (Le et al., 2016). Through its function as a cell adhesion molecule, CD146 on MSCs in the BM supports self-renewal of HSCs through release of soluble factors and mediation of cell–cell interaction (Sacchetti et al., 2007; Sorrentino et al., 2008). It has been reported that expression of cell surface molecules involved in interaction with HSCs is decreased in MDS stromal elements (Geyh et al., 2013). The elevated fraction of $CD271^+$ MSCs has been reported as higher in MDS and AML than in normal BM (Flores-Figueroa, Varma, Montgomery, Greenberg, & Gratzinger, 2012). These $CD271^+$ MSCs favor expansion of leukemia blasts via upregulation of cross talk molecule CXCL12 (Flores-Figueroa et al., 2012; Kim, Shim, et al., 2015).

2.1.2 Structural and Functional Alterations of Malignant MSCs

Geyh et al. demonstrated that isolated MSCs from AML patients displayed disorganized architecture and reduced CFU-F activity. In this study, osteogenic differentiation was significantly impaired in AML MSCs, and this impairment was accompanied by diminished levels of osterix and osteocalcin mRNA (Geyh et al., 2016). AML-derived MSCs display a distinct methylation signature with alteration of pathways involved in cell differentiation, proliferation, and skeletal development. Assessment of the frequency of long-term culture-initiating healthy $CD34^+$ HSCs in AML showed that stromal hematopoietic support was severely impaired. The underlying mechanisms involved downregulation of SCF and overexpression of the Notch ligand Jagged-1. It remains unclear whether MSC alteration is the primary contributor to leukemia initiation or a consequence of malignant hematopoiesis. Geyh et al. reported the instructor role of leukemia cells on MSC functionality, which recovered to near normal when leukemia cells were therapeutically removed. Similarly, MDS MSCs demonstrated elevated levels of osteopontin and Jagged-1, while production of SCF, angiopoietin-1, and chemokines was reduced. These MSCs showed an MDS-dependent hypermethylation pattern and a compromised capacity to support HSCs (Geyh et al., 2013). Through MDS patient-derived mouse models and transcriptional profiling, Medyouf et al. demonstrated that MDS cells reprogrammed MSCs by genetic or epigenetic alterations, whereby MSCs maintained a strong hypoxia signature under normoxic conditions (Medyouf et al., 2014). These patient-derived MSCs overproduced N-cadherin, IGFBP2, VEGFA, and leukemia-inhibitory factor to promote expansion of MDS cells.

Kim et al. investigated the mesenchymal cellular composition of BM stromal cells collected from patients with newly diagnosed AML and found that primitive MSCs ($CD146^{+}CD166^{-}$) were present at a lower frequency than normal (Kim, Shim, et al., 2015). Transcriptome analysis revealed that leukemia cells reprogrammed MSCs by suppressing cell cycle-related genes and upregulating cytokine-related genes. Leukemia cells remodeled the niche by regulating cross talk molecules Jagged-1 and CXCL12 to favor leukemogenesis and suppress normal hematopoiesis. Kim et al. further showed that niche composition correlated with clinical outcome; patients who displayed large numbers of primitive MSCs experienced earlier relapse.

Osteoblastic lineage cells derived from MSCs play a critical role in maintaining normal hematopoiesis. Several studies have demonstrated that leukemia alters niche components, which further enables leukemia progression. Krevvata et al. showed that osteoblast number was reduced by 55% in patients with MDS/AML as compared to healthy controls (Krevvata et al., 2014). This finding was confirmed in DTA_{osb} mouse models in which genetic depletion of osteoblasts yielded a higher tumor burden and shorter survival. Pharmacologic tryptophan hydroxylase inhibitor LP533401 stimulated the normal hematopoiesis and reduced tumor burden by maintaining the osteoblast pool through inhibition of the synthesis of duodenal serotonin. Functional inhibition of osteoblasts was also observed in a murine AML model (Frisch et al., 2012).

The regulation of stromal components differed, however, in different hematological malignancies. Schepers et al. described increased numbers of osteoblasts in an inducible BCR-ABL transgenic mouse model of CML (Schepers et al., 2013). This expansion of osteoblasts was reversible by removing BCL-ABL expression and was mediated by TPO and C-C chemokine (CCL3, MIP1A). However, the expanded osteoblasts functioned abnormally, with reduced expression of normal HSC-retaining factors, such as CXCL12, SCF, and leptin.

2.1.3 Cytogenetic and Molecular Alterations of MSCs in Myeloid Disorders

Although the critical role of the BM microenvironment in determining the fate of leukemia cells has been shown by many studies, it remains controversial whether MSCs undergo cytogenetic and genetic alterations along with leukemic transformation. Cytogenetic aberrations in BM MSCs have been reported in patients with AML, MDS, or MPNs. Genomic changes in MDS MSCs have been detected by fluorescence in situ hybridization and array-comparative genomic hybridization (CGH) (Lopez-Villar et al., 2009). Array-CGH demonstrated minimal regions with gains localized in 19p13.3, 11q13.1, and 20q13.33. An unsupervised hierarchical cluster analysis identified one cluster linked to MDS with 5q- syndrome (Lopez-Villar et al., 2009). Blau et al. demonstrated structural chromosomal aberrations in a large number of patients with MDS (44%) or AML (54%) (Blau et al., 2007). Most abnormalities were observed in chromosome 1, 7, or 10. Another study from the same group showed both structural and numerical cytogenetic alterations in patients with MDS/AML (Blau et al., 2011). Patients with aberrations in MSCs were more likely to experience leukemia-related mortality. However, a more recent study reported chromosomal abnormalities in only 5% of patients with MDS/AML (3 of 79), involving an extra chromosome 5 or 7, and a balanced translocation (Kim, Jekarl, et al., 2015). A large proportion of random loss of chromosomal material in MSCs in MDS suggests that chromosomal instability of MDS MSCs may facilitate the growth (and possibly the transformation) of the malignant clones (Song et al., 2012). Several studies found that MSCs displayed chromosomal abnormalities distinct from those in their corresponding leukemia blasts, indicating a different origin for MSCs (Blau et al., 2007, 2011; Klaus et al., 2010; Lopez-Villar et al., 2009). In one study of samples from only four AML patients, abnormal cytogenetics was both overlapping with and distinct from leukemia blasts (Huang et al., 2015). This may argue the existence of a common origin for both hematopoietic cells and MSCs. In these studies, no clonal abnormalities were detected in BM MSCs from healthy donors.

In CML, the contribution of MSCs to malignant hematopoiesis remains to be investigated. Previous studies proved that MSCs isolated from patient with BCR-ABL$^+$ CML do not express *t*(9;22) translocation, indicating that the MSCs are not related to leukemia clone (Carrara et al., 2007; Jootar et al., 2006; Wohrer et al., 2007). BCR-ABL-negative MPNs typically harbor the JAK2^{V617F} mutation. Avanzini et al. demonstrated that MSCs at early passages from 20 patients with an MPN carrying the JAK2^{V617F} mutation in their granulocytes did not show the same mutation in MSC by PCR analysis (Avanzini et al., 2014), consistent with other reports (Bacher et al., 2010; Mercier, Monczak, Francois, Prchal, & Galipeau, 2009; Pieri et al., 2008; Schneider et al., 2014). However, other genetic abnormalities involving different chromosomes were identified in 17% of patients with BCR-ABL-negative MPN (Avanzini et al., 2014).

AML cells with IDH1 and IDH2 mutations produced oncometabolite R-2-hydroxyglutarate (R-2HG), which not only alters epigenetics and differentiation in leukemia cells but also activates the NF-κB pathway in BM stromal cells via ROS/MAPK-dependent phosphorylation (Chen, Lai, et al., 2016). R-2HG upregulates an NF-κB-dependent gene signature and contributes to the enhanced proliferation and chemoresistance of AML cells.

In high-risk MDS patients, MSCs displayed upregulated levels of CDKN2B (p15INK4B) and altered global DNA methylation status, which accounted for impaired stromal support for normal HSCs and may have contributed to ineffective hematopoiesis (Poloni et al., 2014). DNA demethylating therapy was reported to restore MSCs derived from high-risk MDS patients to a normal phenotype (Maurizi et al., 2016). Massive parallel RNA sequencing revealed that MSCs from low-risk MDS patients displayed transcriptomic profiles distinct from their normal counterparts by enriching genes in cellular stress and inflammation (Chen, Zambetti, et al., 2016). Abnormal PI3K/AKT and WNT signaling pathways were identified in MDS patient-derived MSCs, correlating with abnormal phenotypes, impaired proliferation, and altered expression of early osteogenesis-related genes (Falconi et al., 2016; Pavlaki et al., 2014).

A recent study applied whole exome sequencing of BM MSCs collected from AML patients at serial time points (diagnosis, remission, and relapse), and identified a persistent mutation in the PLEC gene in one patient, which encodes key cytoskeleton player pectin (von der Heide et al., 2016). RNA sequencing revealed deregulation of proteoglycans, adhesion molecules, and cytokines in BM MSCs collected from AML patients at diagnosis. KEGG pathway enrichment analysis of AML MSCs identified alterations involved in metabolic pathways, endocytosis, and transcriptional and epigenetic signatures (von der Heide et al., 2016).

2.2 Other Malignant Niche Components

Recent studies suggest that endothelial cells, megakaryocytes, and adipocytes are likewise altered by their malignant hematopoietic cells. Compared to normal endothelial cells, AML-derived endothelial cells present similar functions and phenotypes but demonstrate a distinct gene signature characterized by upregulation of c-FOS and senescence-related genes including interferon signaling and COL3A1 (Pizzo et al., 2016). Strikingly Cogle et al. found in a patient-derived xenograft model that AML cells not only localized to the vascular but that some of those cells integrated into the endothelium and contributed to the vasculature niche by producing endothelium-like cells. The vessel-associated AMLs were more quiescent and responded poorly to cytarabine, indicating that endothelial cells may serve as a sanctuary site for AML (Cogle et al., 2014). Dissociating AML from endothelial cells with tubulin-binding combretastatins caused degradation of endothelial cells adhesion molecules that tethered AML cells and rendered AML sensitive to chemotherapy (Bosse et al., 2016).

A recent study highlighted the interplay between leukemic stem cells (LSCs) and adipose tissue. Gonadal adipose tissue-resident LSCs induced lipolysis in adipose tissue, which promoted fatty acid metabolism of leukemia cells and conferred chemoresistance on LSCs (Ye et al., 2016).

Another study has demonstrated that hematopoietic clonal cells can directly generate altered niche components in myelofibrosis models. Primary myelofibrosis is characterized by increased myeloid proliferation and progressive BM fibrosis. In the study of Verstovsek et al., the neoplastic monocyte-derived fibrocytes were shown to produce collagen and fibronectin that contribute directly to the BM fibrosis, which is critical for disease progression (Verstovsek et al., 2016).

2.3 Neuropathy Induced by Malignant Cells

HSCs harboring the $JAK2^{V617}$ mutation produce IL1β, which causes damage to sympathetic nerve fibers and death of Schwann cells, leading to loss of $Nestin^{+}$ MSCs and decreased CXCL12 production (Arranz et al., 2014). The impaired niche favors expansion of mutant HSCs and further accelerates MPN progression. As already mentioned, β3-adrenergic signaling mediates sympathetic nervous system regulation of MSCs. Application of β3-adrenergic agonists restores the regulation and prevents loss of MSCs. Hanoun et al., using a syngeneic MLL-AF9 AML model, demonstrated that infiltration of AML cells into the BM induced sympathetic neuropathy and remodeled the HSC niche by driving abnormal $Nestin^+$ MSC expansion while losing mature bone-forming osteoblasts through β2-adrenergic signaling (Hanoun et al., 2014). The impaired niche lost $NG2^+$ periarteriolar niche cells and downregulated expression of HSC-retention factors, including CXCL12, SCF, angiopoietin-1, and VCAM1. BM neuropathy was also demonstrated in patients with diabetes and tumor patients receiving chemotherapy. In a rat model of type 2 diabetes, BM neuropathy caused reduction in CLOCK gene expression and loss of circulating endothelial progenitor cells (Busik et al., 2009). Chemotherapy induced cytotoxic damage to BM neuron fibers, which resulted in loss of quiescence of MSCs (Lucas et al., 2013). Subsequent chemotherapy killed the cycling MSCs and damaged niches, causing HSC exhaustion and compromising HSC engraftment. A recent study revealed differential requirements between HSCs and LSCs for core circadian transcription factors (Puram et al., 2016). As mentioned in the discussion of Section 1.2.5, circadian rhythms regulate HSC mobilization via extrinsic adrenergic signaling. In AML, circadian rhythm transcription factors CLOCK and BMAL1 were found to be prerequisites for LSC proliferation and self-renewal (Puram et al., 2016).

2.4 Remodeling of the BM Niche by Tumor-Derived Exosomes

Exosomes are microvesicles (30–100 nm) of endocytic origin that are secreted by most cells in culture and are present inside large multivesicular endosomes (Thery, Zitvogel, & Amigorena, 2002). Exosomes serve as paracrine regulators of the malignant niche and contain complex cargo. Tumor cell-derived exosomes were found to enable BM MSCs to enhance tumor cell growth through production of CCL2 and CCL7, which promoted macrophage recruitment (Lin et al., 2016). BM stromal cell uptake of AML-derived exosomes with RNA enrichment displayed distinct functional effects on secretion of growth factors (Huan et al., 2013). Huan et al. demonstrated that AML exosomes suppress normal HSC functions by downregulation of HSC-retention factors SCF and CXCL12 in stromal cells (Huan et al., 2015). AML exosome-directed microRNAs trafficking to HSCs can also directly suppress HSC proliferation and differentiation via regulation of c-MYB in a stroma-independent manner (Hornick et al., 2016). Exosomes released from CML cells induced IL8 production from BM stromal cells (Corrado et al., 2016). Exosomes/microvesicles released from BM MSCs collected from MDS patients were demonstrated to transfer miR-10a and miR-15a to CD34+ cells and modify expression of MDM2 and p53 genes (Muntion et al., 2016). Paggetti et al. demonstrated that chronic lymphocytic leukemia-derived exosomes and their molecular cargo are rapidly uptaken by surrounding stromal cells, including MSCs and endothelial cells. The reprogrammed stromal cells in turn contribute a tumor-supportive microenvironment by enhancing proliferation, releasing cytokines, and promoting angiogenesis (Paggetti et al., 2015). These studies indicate that exosomes are critical mediators of intercellular communication.

2.5 Metabolic Alterations in Stem Cell Niche

The stem cell niche has been characterized as hypoxic by either direct measurement of oxygen tension in the BM or indirectly by expression of hypoxia-inducible factor-1α (HIF-1α) and surrogate hypoxic markers (Parmar, Mauch, Vergilio, Sackstein, & Down, 2007; Spencer et al., 2014; Takubo et al., 2010). HIFs act as critical transcriptional regulators of cellular responses to low oxygen tension. Knockdown of HIF-2α markedly decreased the engraftment potential of AML LSCs, indicating that HIFs are required for LSC maintenance and may represent a therapeutic target in AML (Rouault-Pierre et al., 2013). Hypoxia-activated prodrug TH-302 has been proven to efficiently deplete hypoxic leukemia cells and prolong survival of AML xenograft models (Benito et al., 2016). However, the importance of HIFs was challenged by results in a conditional genetic model in which HIF-1α and HIF-2α were deleted in the setting of MEIS1/HOXA9- or MLL-AF9-driven leukemogenesis; the results indicated HIF is a tumor suppressor for AML initiation (Vukovic et al., 2015). These conflicting findings require further validation in other models and in patient-derived xenografts.

Normal HSCs utilize glycolysis for their energy generation, via which HSCs are maintained in a quiescent state by reducing oxidative stress and production of ROS (Simsek et al., 2010). However, AML is highly dependent on oxidative phosphorylation for survival (Lagadinou et al., 2013). Moschoi et al. demonstrated that AML LSCs and blasts take up functional mitochondria from BM stromal cells through an endocytic pathway and that this uptake was enhanced by chemotherapeutic agents

(Moschoi et al., 2016). This mitochondrial transfer provides a survival advantage following chemotherapy and may contribute to chemoresistance.

2.6 Niche Competition Between Normal and Malignant Cells

It remains undetermined whether leukemia cells create their specific malignant niche or outcompete normal HSCs for the native HSC niche. Colmone et al. demonstrated that leukemia cells disrupt the HSC niche by overproducing SCF, which attracts HSCs to migrate into the niche with high leukemia burden (Colmone et al., 2008). The sequestration of HSCs was relieved by neutralizing the SCF. Several recent studies have demonstrated that leukemia cells compete with HSCs for the niche, thereby inhibiting hematopoiesis by displacing HSCs. In the MLL-AF9 murine AML model, increasing doses of HSCs with a fixed number of LSCs significantly prolonged survival and delayed leukemia progression (Glait-Santar et al., 2015). Confocal microscopy demonstrated that the endosteal BM niche allows occupation by either leukemia cells or normal cells, but not both. Boyd et al. proved, in patient-derived mouse models, that AML LSCs shared the niche with their normal counterparts (Boyd et al., 2014). Cotransplantation of HSCs with AML cells could outcompete LSCs and delay leukemia progression, while displacing leukemia cells from the niche by mobilizing agents facilitated reconstitution of normal HSCs. This study suggests that predisruption of the LSC–niche interaction could improve HSC replacement in the setting of allogeneic HSC transplantation.

2.7 Niche-Initiated Hematologic Malignancies

It has been demonstrated that, by manipulating either the cytokines or the recipient strain of mouse, human $CD34^{+}$ cells overexpressing MLL-AF9 were directed to generate AML, acute lymphoid leukemia, or biphenotypic leukemia, indicating the importance of microenvironmental cues in determining lineage outcome (Wei et al., 2008). There is experimental evidence supporting the idea that abnormalities in the niche components can drive leukemogenesis. Walkley et al. published two reports demonstrating that disruption of the niche leads to myeloproliferative disease (Walkley, Olsen, et al., 2007; Walkley, Shea, Sims, Purton, & Orkin, 2007). Loss of the retinoblastoma (RB1) protein in the myeloid-specific lineage caused only mild defects. The myeloproliferative disease was fully developed only when RB1 was deleted from both the hematopoietic cells and the microenvironment (Walkley, Shea, et al., 2007). Loss of RB1 increased osteoclasts, which resulted in depletion of osteoblasts and loss of HSC quiescence, causing extramedullary hematopoiesis, and HSC mobilization. In these authors' complementary study, deficiency of retinoic acid receptor γ (RARG) in the niche was sufficient to induce MPN-like disease, which was partially due to elevated levels in RARG-deficient mice (Walkley, Olsen, et al., 2007). Myeloproliferative disease was also developed in mice with a Notch-defective microenvironment by inactivating Notch ligand regulator Mind bomb-1 (Kim et al., 2008). Raaijmakers et al. found that deletion of DICER1, an RNase III endonuclease regulating microRNA biogenesis and RNA processing, specifically in mouse osteoprogenitors but not in mature osteoblasts, led to MDS and AML (Raaijmakers et al., 2010). The *Shwachman–Bodian–Diamond syndrome* (SBDS) gene was inactivated in a preleukemic disorder, Shwachman–Diamond syndrome. Expression of the SBDS gene was reduced in MSCs with loss of DICER1 (Raaijmakers et al., 2010). Most recently, Raaijmakers and colleagues reported hematopoietic malignant transformation from MSCs lacking the SBDS gene, which induced mitochondrial dysfunction, oxidative stress, and DNA damage responses to HSCs via p53-S100A8/9-TLS inflammatory signaling (Zambetti et al., 2016). Transcriptome sequencing of $CD271^{+}$ niche cells revealed that high expression of S100A8/9 in low-risk MDS patients significantly predicted higher risk for leukemic evolution. Reduced expression of DICER1 and SBDS in MSCs has been also detected in patients with MDS (Santamaria et al., 2012). Kode et al. demonstrated that osteoblasts carrying an oncogenic mutation were capable of inducing leukemia (Kode et al., 2014). In genetically engineered mice, the $Ctnnb1^{CAosb}$ mutation, which resulted in constitutively active mutation of β-catenin in osteoblasts, led to AML/MDS, supporting evidence for niche-driven leukemogenesis (Kode et al., 2014). Gene expression microarray revealed that Notch ligand Jagged-1 was upregulated in osteoblasts of mice carrying the $Ctnnb1^{CAosb}$ mutation. Studies using BM biopsy samples revealed that 38% of patients with AML/MDS displayed nuclear (activated) β-catenin in osteoblasts and elevated Notch signaling in hematopoietic cells, supporting potential relevance in human disease. The follow-up study into the underlying molecular mechanisms revealed that overexpression of the NOTCH ligand was due to interaction between FOXO1 and β-catenin in osteoblasts (Kode et al., 2016). Dong et al. recently found that mice carrying the $PTPN11^{E76K/+}$ mutation in $Nestin^{+}$ MSCs but not in differentiated osteoblasts or endothelial cells developed MPN-like disease characterized by splenomegaly and increased numbers of myeloid cells (Dong et al., 2016). PTPN11-mutant MSCs overproduced CCL3, which recruited monocytes to the HSC niche, and HSCs were subsequently hyperactivated by monocyte-produced IL1β and other proinflammatory cytokines to develop MPN. Activation of parathyroid hormone receptor in osteoblasts reduced tumor burden in BCR-ABL1 oncogene-induced CML-like MPN but enhanced MLL-AF9 oncogene-induced AML in mouse models, indicating that distinct niches are requisite for CML and AML (Krause et al., 2013).

2.8 Targeting the LSC Niche

Uncoupling LSCs from their protective BM niche can enhance drug-induced apoptotic cell death and drive LSCs to enter the cell cycle, making them more responsive to chemotherapy. CXCL12/CXCR4 signaling promotes LSC anchoring in the BM niche and maintains the LSCs in a quiescent state, causing resistance to chemotherapy (Nervi et al., 2009; Zeng et al., 2009). The microRNA let-7a was downregulated by CXCR4 activation linked by a transcription factor YY1 and acted as a key player mediating CXCR4-driven chemoresistance. Targeting the CXCL12/CXCR4 axis has been proven to be effective in treating AML, mobilizing leukemia cells out of their niche and suppressing multiple survival pathways via CXCR4 antagonist, AMD3100 (plerixafor) or LY2510924 (Cho et al., 2015; Uy et al., 2012). The hyaluronic acid receptor CD44 was demonstrated to mediate LSC lodging in the BM niche, making CD44 a promising target to dissociate LSCs from their niche (Jin, Hope, Zhai, Smadja-Joffe, & Dick, 2006; Krause, Lazarides, von Andrian, & Van Etten, 2006). Jacamo and colleagues found that the cross talk between leukemia cells and the BM niche was mediated by VCAM1/very late antigen-4 (VLA-4)-induced activation of the NF-κB signaling pathway (Jacamo et al., 2014). Small molecules targeting NF-κB (Guzman et al., 2007) or VCAM1 (Hsieh et al., 2014) may be useful in targeting diseased microenvironment in myeloid malignancies. CD98, which controls adhesion and amino acid transport, was discovered recently as a key regulator of interactions between AML LSCs and vasculature (Bajaj et al., 2016). Antibody blockade of CD98 disrupted adhesion of LSCs to vasculature and impaired propagation of AML in mouse models.

3. FUTURE PERSPECTIVES

Emerging studies convincingly demonstrate that BM microenvironmental cues regulate normal and malignant hematopoiesis. Modulating the malignant BM niche may provide novel strategies that are applicable for malignancies of diverse genetic backgrounds. Drugs that disrupt aberrant adhesion can be applied to dissociate LSCs from their protective niche. Antagonizing key niche factors, such as CXCL12 and Jagged-1, may reduce malignant niche formation. Novel agents are needed that block MSC remodeling, including reversing aberrant differentiation into fibrotic tissue and modulating β3-adrenergic signals to restore niche damage. Furthermore, a humanized niche in mouse models improves leukemia engraftment (Antonelli et al., 2016; Chen et al., 2012; Reinisch et al., 2016), which facilitates understanding of malignant transformation and testing of various compounds in mouse models that introduce humanized microenvironment and capture the heterogeneity of human diseases. Restoring niche function is also critical in the setting of stem cell transplantation to improve normal HSC engraftment. Targeting the deregulated niche is a promising therapeutic rationale for a wide range of myeloid malignancies.

REFERENCES

Antonelli, A., Noort, W. A., Jaques, J., de Boer, B., de Jong-Korlaar, R., Brouwers-Vos, A. Z., ... Schuringa, J. J. (2016). Establishing human leukemia xenograft mouse models by implanting human bone marrow-like scaffold-based niches. *Blood, 128*, 2949–2959. http://dx.doi.org/10.1182/blood-2016-05-719021.

Arai, F., Hirao, A., Ohmura, M., Sato, H., Matsuoka, S., Takubo, K., ... Suda, T. (2004). Tie2/angiopoietin-1 signaling regulates hematopoietic stem cell quiescence in the bone marrow niche. *Cell, 118*(2), 149–161. http://dx.doi.org/10.1016/j.cell.2004.07.004.

Arranz, L., Sanchez-Aguilera, A., Martin-Perez, D., Isern, J., Langa, X., Tzankov, A., ... Mendez-Ferrer, S. (2014). Neuropathy of haematopoietic stem cell niche is essential for myeloproliferative neoplasms. *Nature, 512*(7512), 78–81. http://dx.doi.org/10.1038/nature13383.

Asada, N., Katayama, Y., Sato, M., Minagawa, K., Wakahashi, K., Kawano, H., ... Tanimoto, M. (2013). Matrix-embedded osteocytes regulate mobilization of hematopoietic stem/progenitor cells. *Cell Stem Cell, 12*(6), 737–747. http://dx.doi.org/10.1016/j.stem.2013.05.001.

Avanzini, M. A., Bernardo, M. E., Novara, F., Mantelli, M., Poletto, V., Villani, L., ... Investigators, A. (2014). Functional and genetic aberrations of in vitro-cultured marrow-derived mesenchymal stromal cells of patients with classical Philadelphia-negative myeloproliferative neoplasms. *Leukemia, 28*(8), 1742–1745. http://dx.doi.org/10.1038/leu.2014.97.

Avecilla, S. T., Hattori, K., Heissig, B., Tejada, R., Liao, F., Shido, K., ... Rafii, S. (2004). Chemokine-mediated interaction of hematopoietic progenitors with the bone marrow vascular niche is required for thrombopoiesis. *Nature Medicine, 10*(1), 64–71. http://dx.doi.org/10.1038/nm973.

Bacher, U., Asenova, S., Badbaran, A., Zander, A. R., Alchalby, H., Fehse, B., ... Ayuk, F. (2010). Bone marrow mesenchymal stromal cells remain of recipient origin after allogeneic SCT and do not harbor the JAK2V617F mutation in patients with myelofibrosis. *Clinical and Experimental Medicine, 10*(3), 205–208. http://dx.doi.org/10.1007/s10238-009-0058-9.

Bajaj, J., Konuma, T., Lytle, N. K., Kwon, H. Y., Ablack, J. N., Cantor, J. M., ... Reya, T. (2016). CD98-mediated adhesive signaling enables the establishment and propagation of acute myelogenous leukemia. *Cancer Cell*, *30*(5), 792–805. http://dx.doi.org/10.1016/j.ccell.2016.10.003.

Battula, V. L., Chen, Y., Cabreira Mda, G., Ruvolo, V., Wang, Z., Ma, W., ... Andreeff, M. (2013). Connective tissue growth factor regulates adipocyte differentiation of mesenchymal stromal cells and facilitates leukemia bone marrow engraftment. *Blood*, *122*(3), 357–366. http://dx.doi.org/10.1182/blood-2012-06-437988.

Battula, V. L., Treml, S., Bareiss, P. M., Gieseke, F., Roelofs, H., de Zwart, P., ... Buhring, H. J. (2009). Isolation of functionally distinct mesenchymal stem cell subsets using antibodies against CD56, CD271, and mesenchymal stem cell antigen-1. *Haematologica*, *94*(2), 173–184. http://dx.doi.org/10.3324/haematol.13740.

Benito, J., Ramirez, M. S., Millward, N. Z., Velez, J., Harutyunyan, K. G., Lu, H., ... Konopleva, M. (2016). Hypoxia-activated prodrug TH-302 targets hypoxic bone marrow niches in preclinical leukemia models. *Clinical Cancer Research*, *22*(7), 1687–1698. http://dx.doi.org/10.1158/1078-0432.CCR-14-3378.

Bensidhoum, M., Chapel, A., Francois, S., Demarquay, C., Mazurier, C., Fouillard, L., ... Lopez, M. (2004). Homing of in vitro expanded Stro-1- or Stro-1+ human mesenchymal stem cells into the NOD/SCID mouse and their role in supporting human CD34 cell engraftment. *Blood*, *103*(9), 3313–3319. http://dx.doi.org/10.1182/blood-2003-04-1121.

Blau, O., Baldus, C. D., Hofmann, W. K., Thiel, G., Nolte, F., Burmeister, T., ... Blau, I. W. (2011). Mesenchymal stromal cells of myelodysplastic syndrome and acute myeloid leukemia patients have distinct genetic abnormalities compared with leukemic blasts. *Blood*, *118*(20), 5583–5592. http://dx.doi.org/10.1182/blood-2011-03-343467.

Blau, O., Hofmann, W. K., Baldus, C. D., Thiel, G., Serbent, V., Schumann, E., ... Blau, I. W. (2007). Chromosomal aberrations in bone marrow mesenchymal stroma cells from patients with myelodysplastic syndrome and acute myeloblastic leukemia. *Experimental Hematology*, *35*(2), 221–229. http://dx.doi.org/10.1016/j.exphem.2006.10.012.

Bosse, R. C., Wasserstrom, B., Meacham, A., Wise, E., Drusbosky, L., Walter, G. A., ... Cogle, C. R. (2016). Chemosensitizing AML cells by targeting bone marrow endothelial cells. *Experimental Hematology*, *44*(5), 363–377. http://dx.doi.org/10.1016/j.exphem.2016.02.003. e365.

Bowers, M., Zhang, B., Ho, Y., Agarwal, P., Chen, C. C., & Bhatia, R. (2015). Osteoblast ablation reduces normal long-term hematopoietic stem cell self-renewal but accelerates leukemia development. *Blood*, *125*(17), 2678–2688. http://dx.doi.org/10.1182/blood-2014-06-582924.

Boyd, A. L., Campbell, C. J., Hopkins, C. I., Fiebig-Comyn, A., Russell, J., Ulemek, J., ... Bhatia, M. (2014). Niche displacement of human leukemic stem cells uniquely allows their competitive replacement with healthy HSPCs. *The Journal of Experimental Medicine*, *211*(10), 1925–1935. http://dx.doi.org/10.1084/jem.20140131.

Bruns, I., Lucas, D., Pinho, S., Ahmed, J., Lambert, M. P., Kunisaki, Y., ... Frenette, P. S. (2014). Megakaryocytes regulate hematopoietic stem cell quiescence through CXCL4 secretion. *Nature Medicine*, *20*(11), 1315–1320. http://dx.doi.org/10.1038/nm.3707.

Busik, J. V., Tikhonenko, M., Bhatwadekar, A., Opreanu, M., Yakubova, N., Caballero, S., ... Grant, M. B. (2009). Diabetic retinopathy is associated with bone marrow neuropathy and a depressed peripheral clock. *The Journal of Experimental Medicine*, *206*(13), 2897–2906. http://dx.doi.org/10.1084/jem.20090889.

Busser, H., Najar, M., Raicevic, G., Pieters, K., Velez Pombo, R., Philippart, P., ... Lagneaux, L. (2015). Isolation and characterization of human mesenchymal stromal cell

subpopulations: Comparison of bone marrow and adipose tissue. *Stem Cells and Development, 24*(18), 2142–2157. http://dx.doi.org/10.1089/scd.2015.0172.
Butler, J. M., Kobayashi, H., & Rafii, S. (2010). Instructive role of the vascular niche in promoting tumour growth and tissue repair by angiocrine factors. *Nature Reviews. Cancer, 10*(2), 138–146. http://dx.doi.org/10.1038/nrc2791.
Butler, J. M., Nolan, D. J., Vertes, E. L., Varnum-Finney, B., Kobayashi, H., Hooper, A. T., ... Rafii, S. (2010). Endothelial cells are essential for the self-renewal and repopulation of Notch-dependent hematopoietic stem cells. *Cell Stem Cell, 6*(3), 251–264. http://dx.doi.org/10.1016/j.stem.2010.02.001.
Calvi, L. M., Adams, G. B., Weibrecht, K. W., Weber, J. M., Olson, D. P., Knight, M. C., ... Scadden, D. T. (2003). Osteoblastic cells regulate the haematopoietic stem cell niche. *Nature, 425*(6960), 841–846. http://dx.doi.org/10.1038/nature02040.
Calvi, L. M., & Link, D. C. (2015). The hematopoietic stem cell niche in homeostasis and disease. *Blood, 126*(22), 2443–2451. http://dx.doi.org/10.1182/blood-2015-07-533588.
Carrara, R. C., Orellana, M. D., Fontes, A. M., Palma, P. V., Kashima, S., Mendes, M. R., ... Covas, D. T. (2007). Mesenchymal stem cells from patients with chronic myeloid leukemia do not express BCR-ABL and have absence of chimerism after allogeneic bone marrow transplant. *Brazilian Journal of Medical and Biological Research, 40*(1), 57–67.
Carter, B. Z., Mak, D. H., Morris, S. J., Borthakur, G., Estey, E., Byrd, A. L., ... Andreeff, M. (2011). XIAP antisense oligonucleotide (AEG35156) achieves target knockdown and induces apoptosis preferentially in CD34+ 38− cells in a phase 1/2 study of patients with relapsed/refractory AML. *Apoptosis: An International Journal on Programmed Cell Death, 16*(1), 67–74. http://dx.doi.org/10.1007/s10495-010-0545-1.
Casanova-Acebes, M., Pitaval, C., Weiss, L. A., Nombela-Arrieta, C., Chevre, R., A-González, N., ... Hidalgo, A. (2013). Rhythmic modulation of the hematopoietic niche through neutrophil clearance. *Cell, 153*(5), 1025–1035. http://dx.doi.org/10.1016/j.cell.2013.04.040.
Chen, Y., Jacamo, R., Shi, Y. X., Wang, R. Y., Battula, V. L., Konoplev, S., ... Andreeff, M. (2012). Human extramedullary bone marrow in mice: A novel in vivo model of genetically controlled hematopoietic microenvironment. *Blood, 119*(21), 4971–4980. http://dx.doi.org/10.1182/blood-2011-11-389957.
Chen, J. Y., Lai, Y. S., Tsai, H. J., Kuo, C. C., Yen, B. L., Yeh, S. P., ... Hung, W. C. (2016). The oncometabolite R-2-hydroxyglutarate activates NF-kappaB-dependent tumor-promoting stromal niche for acute myeloid leukemia cells. *Scientific Reports, 6*, 32428. http://dx.doi.org/10.1038/srep32428.
Chen, S., Zambetti, N. A., Bindels, E. M., Kenswill, K., Mylona, A. M., Adisty, N. M., ... Raaijmakers, M. H. (2016). Massive parallel RNA sequencing of highly purified mesenchymal elements in low-risk MDS reveals tissue-context-dependent activation of inflammatory programs. *Leukemia, 30*(9), 1938–1942. http://dx.doi.org/10.1038/leu.2016.91.
Cho, B. S., Zeng, Z., Mu, H., Wang, Z., Konoplev, S., McQueen, T., ... Konopleva, M. (2015). Antileukemia activity of the novel peptidic CXCR4 antagonist LY2510924 as monotherapy and in combination with chemotherapy. *Blood, 126*(2), 222–232. http://dx.doi.org/10.1182/blood-2015-02-628677.
Chow, A., Lucas, D., Hidalgo, A., Mendez-Ferrer, S., Hashimoto, D., Scheiermann, C., ... Frenette, P. S. (2011). Bone marrow CD169+ macrophages promote the retention of hematopoietic stem and progenitor cells in the mesenchymal stem cell niche. *The Journal of Experimental Medicine, 208*(2), 261–271. http://dx.doi.org/10.1084/jem.20101688.
Cogle, C. R., Goldman, D. C., Madlambayan, G. J., Leon, R. P., Al Masri, A., Clark, H. A., ... Fleming, W. H. (2014). Functional integration of acute myeloid leukemia into the vascular niche. *Leukemia, 28*(10), 1978–1987. http://dx.doi.org/10.1038/leu.2014.109.

Colmone, A., Amorim, M., Pontier, A. L., Wang, S., Jablonski, E., & Sipkins, D. A. (2008). Leukemic cells create bone marrow niches that disrupt the behavior of normal hematopoietic progenitor cells. *Science*, *322*(5909), 1861–1865. http://dx.doi.org/10.1126/science.1164390.

Corrado, C., Saieva, L., Raimondo, S., Santoro, A., De Leo, G., & Alessandro, R. (2016). Chronic myelogenous leukaemia exosomes modulate bone marrow microenvironment through activation of epidermal growth factor receptor. *Journal of Cellular and Molecular Medicine*, *20*(10), 1829–1839. http://dx.doi.org/10.1111/jcmm.12873.

Despeaux, M., Labat, E., Gadelorge, M., Prade, N., Bertrand, J., Demur, C., ... Racaud-Sultan, C. (2011). Critical features of FAK-expressing AML bone marrow microenvironment through leukemia stem cell hijacking of mesenchymal stromal cells. *Leukemia*, *25*(11), 1789–1793. http://dx.doi.org/10.1038/leu.2011.145.

Dominici, M., Le Blanc, K., Mueller, I., Slaper-Cortenbach, I., Marini, F., Krause, D., ... Horwitz, E. (2006). Minimal criteria for defining multipotent mesenchymal stromal cells. The International Society for Cellular Therapy position statement. *Cytotherapy*, *8*(4), 315–317. http://dx.doi.org/10.1080/14653240600855905.

Dong, L., Yu, W. M., Zheng, H., Loh, M. L., Bunting, S. T., Pauly, M., ... Qu, C. K. (2016). Leukaemogenic effects of Ptpn11 activating mutations in the stem cell microenvironment. *Nature*, *539*, 304–308. http://dx.doi.org/10.1038/nature20131.

Enver, T., Pera, M., Peterson, C., & Andrews, P. W. (2009). Stem cell states, fates, and the rules of attraction. *Cell Stem Cell*, *4*(5), 387–397. http://dx.doi.org/10.1016/j.stem.2009.04.011.

Falconi, G., Fabiani, E., Fianchi, L., Criscuolo, M., Raffaelli, C. S., Bellesi, S., ... Leone, G. (2016). Impairment of PI3K/AKT and WNT/beta-catenin pathways in bone marrow mesenchymal stem cells isolated from patients with myelodysplastic syndromes. *Experimental Hematology*, *44*(1), 75–83. http://dx.doi.org/10.1016/j.exphem.2015.10.005. e71–74.

Flores-Figueroa, E., Varma, S., Montgomery, K., Greenberg, P. L., & Gratzinger, D. (2012). Distinctive contact between CD34+ hematopoietic progenitors and CXCL12+ CD271+ mesenchymal stromal cells in benign and myelodysplastic bone marrow. *Laboratory Investigation*, *92*(9), 1330–1341. http://dx.doi.org/10.1038/labinvest.2012.93.

Frisch, B. J., Ashton, J. M., Xing, L., Becker, M. W., Jordan, C. T., & Calvi, L. M. (2012). Functional inhibition of osteoblastic cells in an in vivo mouse model of myeloid leukemia. *Blood*, *119*(2), 540–550. http://dx.doi.org/10.1182/blood-2011-04-348151.

Geyh, S., Oz, S., Cadeddu, R. P., Frobel, J., Bruckner, B., Kundgen, A., ... Schroeder, T. (2013). Insufficient stromal support in MDS results from molecular and functional deficits of mesenchymal stromal cells. *Leukemia*, *27*(9), 1841–1851. http://dx.doi.org/10.1038/leu.2013.193.

Geyh, S., Rodriguez-Paredes, M., Jager, P., Khandanpour, C., Cadeddu, R. P., Gutekunst, J., ... Schroeder, T. (2016). Functional inhibition of mesenchymal stromal cells in acute myeloid leukemia. *Leukemia*, *30*(3), 683–691. http://dx.doi.org/10.1038/leu.2015.325.

Glait-Santar, C., Desmond, R., Feng, X., Bat, T., Chen, J., Heuston, E., ... Dunbar, C. E. (2015). Functional niche competition between normal hematopoietic stem and progenitor cells and myeloid leukemia cells. *Stem Cells*, *33*(12), 3635–3642. http://dx.doi.org/10.1002/stem.2208.

Guzman, M. L., Rossi, R. M., Neelakantan, S., Li, X., Corbett, C. A., Hassane, D. C., ... Jordan, C. T. (2007). An orally bioavailable parthenolide analog selectively eradicates acute myelogenous leukemia stem and progenitor cells. *Blood*, *110*(13), 4427–4435. http://dx.doi.org/10.1182/blood-2007-05-090621.

Hanoun, M., Zhang, D., Mizoguchi, T., Pinho, S., Pierce, H., Kunisaki, Y., ... Frenette, P. S. (2014). Acute myelogenous leukemia-induced sympathetic neuropathy promotes malignancy in an altered hematopoietic stem cell niche. *Cell Stem Cell*, *15*(3), 365–375. http://dx.doi.org/10.1016/j.stem.2014.06.020.

Hooper, A. T., Butler, J. M., Nolan, D. J., Kranz, A., Iida, K., Kobayashi, M., ... Rafii, S. (2009). Engraftment and reconstitution of hematopoiesis is dependent on VEGFR2-mediated regeneration of sinusoidal endothelial cells. *Cell Stem Cell*, *4*(3), 263–274. http://dx.doi.org/10.1016/j.stem.2009.01.006.

Hornick, N. I., Doron, B., Abdelhamed, S., Huan, J., Harrington, C. A., Shen, R., ... Kurre, P. (2016). AML suppresses hematopoiesis by releasing exosomes that contain microRNAs targeting c-MYB. *Science Signaling*, *9*(444), ra88. http://dx.doi.org/10.1126/scisignal.aaf2797.

Horwitz, E. M., Le Blanc, K., Dominici, M., Mueller, I., Slaper-Cortenbach, I., Marini, F. C., ... International Society for Cellular, T. (2005). Clarification of the nomenclature for MSC: The International Society for Cellular Therapy position statement. *Cytotherapy*, 7(5), 393–395. http://dx.doi.org/10.1080/14653240500319234.

Hsieh, Y. T., Gang, E. J., Shishido, S. N., Kim, H. N., Pham, J., Khazal, S., ... Kim, Y. M. (2014). Effects of the small-molecule inhibitor of integrin alpha4, TBC3486, on pre-B-ALL cells. *Leukemia*, *28*(10), 2101–2104. http://dx.doi.org/10.1038/leu.2014.182.

Huan, J., Hornick, N. I., Goloviznina, N. A., Kamimae-Lanning, A. N., David, L. L., Wilmarth, P. A., ... Kurre, P. (2015). Coordinate regulation of residual bone marrow function by paracrine trafficking of AML exosomes. *Leukemia*, *29*(12), 2285–2295. http://dx.doi.org/10.1038/leu.2015.163.

Huan, J., Hornick, N. I., Shurtleff, M. J., Skinner, A. M., Goloviznina, N. A., Roberts, C. T., Jr., & Kurre, P. (2013). RNA trafficking by acute myelogenous leukemia exosomes. *Cancer Research*, *73*(2), 918–929. http://dx.doi.org/10.1158/0008-5472.CAN-12-2184.

Huang, J. C., Basu, S. K., Zhao, X., Chien, S., Fang, M., Oehler, V. G., ... Becker, P. S. (2015). Mesenchymal stromal cells derived from acute myeloid leukemia bone marrow exhibit aberrant cytogenetics and cytokine elaboration. *Blood Cancer Journal*, *5*. e302. http://dx.doi.org/10.1038/bcj.2015.17.

Hur, J., Choi, J. I., Lee, H., Nham, P., Kim, T. W., Chae, C. W., ... Kim, H. S. (2016). CD82/KAI1 maintains the dormancy of long-term hematopoietic stem cells through interaction with DARC-expressing macrophages. *Cell Stem Cell*, *18*(4), 508–521. http://dx.doi.org/10.1016/j.stem.2016.01.013.

Inra, C. N., Zhou, B. O., Acar, M., Murphy, M. M., Richardson, J., Zhao, Z., & Morrison, S. J. (2015). A perisinusoidal niche for extramedullary haematopoiesis in the spleen. *Nature*, *527*(7579), 466–471. http://dx.doi.org/10.1038/nature15530.

Isern, J., Martin-Antonio, B., Ghazanfari, R., Martin, A. M., Lopez, J. A., del Toro, R., ... Mendez-Ferrer, S. (2013). Self-renewing human bone marrow mesenspheres promote hematopoietic stem cell expansion. *Cell Reports*, *3*(5), 1714–1724. http://dx.doi.org/10.1016/j.celrep.2013.03.041.

Itkin, T., Gur-Cohen, S., Spencer, J. A., Schajnovitz, A., Ramasamy, S. K., Kusumbe, A. P., ... Lapidot, T. (2016). Distinct bone marrow blood vessels differentially regulate haematopoiesis. *Nature*, *532*(7599), 323–328. http://dx.doi.org/10.1038/nature17624.

Jacamo, R., Chen, Y., Wang, Z., Ma, W., Zhang, M., Spaeth, E. L., ... Andreeff, M. (2014). Reciprocal leukemia-stroma VCAM-1/VLA-4-dependent activation of NF-kappaB mediates chemoresistance. *Blood*, *123*(17), 2691–2702. http://dx.doi.org/10.1182/blood-2013-06-511527.

Jin, L., Hope, K. J., Zhai, Q., Smadja-Joffe, F., & Dick, J. E. (2006). Targeting of CD44 eradicates human acute myeloid leukemic stem cells. *Nature Medicine*, *12*(10), 1167–1174. http://dx.doi.org/10.1038/nm1483.

Jootar, S., Pornprasertsud, N., Petvises, S., Rerkamnuaychoke, B., Disthabanchong, S., Pakakasama, S., ... Hongeng, S. (2006). Bone marrow derived mesenchymal stem cells

from chronic myeloid leukemia t(9;22) patients are devoid of Philadelphia chromosome and support cord blood stem cell expansion. *Leukemia Research*, *30*(12), 1493–1498. http://dx.doi.org/10.1016/j.leukres.2006.04.013.
Kersten, S., Desvergne, B., & Wahli, W. (2000). Roles of PPARs in health and disease. *Nature*, *405*(6785), 421–424. http://dx.doi.org/10.1038/35013000.
Kiel, M. J., Yilmaz, O. H., Iwashita, T., Yilmaz, O. H., Terhorst, C., & Morrison, S. J. (2005). SLAM family receptors distinguish hematopoietic stem and progenitor cells and reveal endothelial niches for stem cells. *Cell*, *121*(7), 1109–1121. http://dx.doi.org/10.1016/j.cell.2005.05.026.
Kiger, A. A., Jones, D. L., Schulz, C., Rogers, M. B., & Fuller, M. T. (2001). Stem cell self-renewal specified by JAK-STAT activation in response to a support cell cue. *Science*, *294*(5551), 2542–2545. http://dx.doi.org/10.1126/science.1066707.
Kiger, A. A., White-Cooper, H., & Fuller, M. T. (2000). Somatic support cells restrict germline stem cell self-renewal and promote differentiation. *Nature*, *407*(6805), 750–754. http://dx.doi.org/10.1038/35037606.
Kim, Y., Jekarl, D. W., Kim, J., Kwon, A., Choi, H., Lee, S., … Kim, M. (2015). Genetic and epigenetic alterations of bone marrow stromal cells in myelodysplastic syndrome and acute myeloid leukemia patients. *Stem Cell Research*, *14*(2), 177–184. http://dx.doi.org/10.1016/j.scr.2015.01.004.
Kim, Y. W., Koo, B. K., Jeong, H. W., Yoon, M. J., Song, R., Shin, J., … Kong, Y. Y. (2008). Defective Notch activation in microenvironment leads to myeloproliferative disease. *Blood*, *112*(12), 4628–4638. http://dx.doi.org/10.1182/blood-2008-03-148999.
Kim, J. A., Shim, J. S., Lee, G. Y., Yim, H. W., Kim, T. M., Kim, M., … Oh, I. H. (2015). Microenvironmental remodeling as a parameter and prognostic factor of heterogeneous leukemogenesis in acute myelogenous leukemia. *Cancer Research*, *75*(11), 2222–2231. http://dx.doi.org/10.1158/0008-5472.CAN-14-3379.
Klaus, M., Stavroulaki, E., Kastrinaki, M. C., Fragioudaki, P., Giannikou, K., Psyllaki, M., … Papadaki, H. A. (2010). Reserves, functional, immunoregulatory, and cytogenetic properties of bone marrow mesenchymal stem cells in patients with myelodysplastic syndromes. *Stem Cells and Development*, *19*(7), 1043–1054. http://dx.doi.org/10.1089/scd.2009.0286.
Kobayashi, H., Butler, J. M., O'Donnell, R., Kobayashi, M., Ding, B. S., Bonner, B., … Rafii, S. (2010). Angiocrine factors from Akt-activated endothelial cells balance self-renewal and differentiation of haematopoietic stem cells. *Nature Cell Biology*, *12*(11), 1046–1056. http://dx.doi.org/10.1038/ncb2108.
Kode, A., Manavalan, J. S., Mosialou, I., Bhagat, G., Rathinam, C. V., Luo, N., … Kousteni, S. (2014). Leukaemogenesis induced by an activating beta-catenin mutation in osteoblasts. *Nature*, *506*(7487), 240–244. http://dx.doi.org/10.1038/nature12883.
Kode, A., Mosialou, I., Manavalan, S. J., Rathinam, C. V., Friedman, R. A., Teruya-Feldstein, J., … Kousteni, S. (2016). FoxO1-dependent induction of acute myeloid leukemia by osteoblasts in mice. *Leukemia*, *30*(1), 1–13. http://dx.doi.org/10.1038/leu.2015.161.
Koechlein, C. S., Harris, J. R., Lee, T. K., Weeks, J., Fox, R. G., Zimdahl, B., … Reya, T. (2016). High-resolution imaging and computational analysis of haematopoietic cell dynamics in vivo. *Nature Communications*, 7, 12169. http://dx.doi.org/10.1038/ncomms12169.
Kopp, H. G., Hooper, A. T., Broekman, M. J., Avecilla, S. T., Petit, I., Luo, M., … Rafii, S. (2006). Thrombospondins deployed by thrombopoietic cells determine angiogenic switch and extent of revascularization. *The Journal of Clinical Investigation*, *116*(12), 3277–3291. http://dx.doi.org/10.1172/JCI29314.
Krause, D. S., Fulzele, K., Catic, A., Sun, C. C., Dombkowski, D., Hurley, M. P., … Scadden, D. T. (2013). Differential regulation of myeloid leukemias by the bone marrow

microenvironment. *Nature Medicine*, *19*(11), 1513–1517. http://dx.doi.org/10.1038/nm.3364.

Krause, D. S., Lazarides, K., von Andrian, U. H., & Van Etten, R. A. (2006). Requirement for CD44 in homing and engraftment of BCR-ABL-expressing leukemic stem cells. *Nature Medicine*, *12*(10), 1175–1180. http://dx.doi.org/10.1038/nm1489.

Krevvata, M., Silva, B. C., Manavalan, J. S., Galan-Diez, M., Kode, A., Matthews, B. G., … Kousteni, S. (2014). Inhibition of leukemia cell engraftment and disease progression in mice by osteoblasts. *Blood*, *124*(18), 2834–2846. http://dx.doi.org/10.1182/blood-2013-07-517219.

Kunisaki, Y., Bruns, I., Scheiermann, C., Ahmed, J., Pinho, S., Zhang, D., … Frenette, P. S. (2013). Arteriolar niches maintain haematopoietic stem cell quiescence. *Nature*, *502*(7473), 637–643. http://dx.doi.org/10.1038/nature12612.

Kusumbe, A. P., Ramasamy, S. K., Itkin, T., Mae, M. A., Langen, U. H., Betsholtz, C., … Adams, R. H. (2016). Age-dependent modulation of vascular niches for haematopoietic stem cells. *Nature*, *532*(7599), 380–384. http://dx.doi.org/10.1038/nature17638.

Lagadinou, E. D., Sach, A., Callahan, K., Rossi, R. M., Neering, S. J., Minhajuddin, M., … Jordan, C. T. (2013). BCL-2 inhibition targets oxidative phosphorylation and selectively eradicates quiescent human leukemia stem cells. *Cell Stem Cell*, *12*(3), 329–341. http://dx.doi.org/10.1016/j.stem.2012.12.013.

Le, Y., Fraineau, S., Chandran, P., Sabloff, M., Brand, M., Lavoie, J. R., … Allan, D. S. (2016). Adipogenic mesenchymal stromal cells from bone marrow and their hematopoietic supportive role: Towards understanding the permissive marrow microenvironment in acute myeloid leukemia. *Stem Cell Reviews*, *12*(2), 235–244. http://dx.doi.org/10.1007/s12015-015-9639-z.

Lin, L. Y., Du, L. M., Cao, K., Huang, Y., Yu, P. F., Zhang, L. Y., … Shi, Y. F. (2016). Tumour cell-derived exosomes endow mesenchymal stromal cells with tumour-promotion capabilities. *Oncogene*, *35*, 6038–6042. http://dx.doi.org/10.1038/onc.2016.131.

Lopez-Villar, O., Garcia, J. L., Sanchez-Guijo, F. M., Robledo, C., Villaron, E. M., Hernandez-Campo, P., … del Canizo, M. C. (2009). Both expanded and uncultured mesenchymal stem cells from MDS patients are genomically abnormal, showing a specific genetic profile for the 5q- syndrome. *Leukemia*, *23*(4), 664–672. http://dx.doi.org/10.1038/leu.2008.361.

Lu, X. J., Chen, Q., Rong, Y. J., Yang, G. J., Li, C. H., Xu, N. Y., … Chen, J. (2016). LECT2 drives haematopoietic stem cell expansion and mobilization via regulating the macrophages and osteolineage cells. *Nature Communications*, 7, 12719. http://dx.doi.org/10.1038/ncomms12719.

Lucas, D., Scheiermann, C., Chow, A., Kunisaki, Y., Bruns, I., Barrick, C., … Frenette, P. S. (2013). Chemotherapy-induced bone marrow nerve injury impairs hematopoietic regeneration. *Nature Medicine*, *19*(6), 695–703. http://dx.doi.org/10.1038/nm.3155.

Luo, Y., Chen, G. L., Hannemann, N., Ipseiz, N., Kronke, G., Bauerle, T., … Bozec, A. (2015). Microbiota from obese mice regulate hematopoietic stem cell differentiation by altering the bone niche. *Cell Metabolism*, *22*(5), 886–894. http://dx.doi.org/10.1016/j.cmet.2015.08.020.

Mansour, A., Abou-Ezzi, G., Sitnicka, E., Jacobsen, S. E., Wakkach, A., & Blin-Wakkach, C. (2012). Osteoclasts promote the formation of hematopoietic stem cell niches in the bone marrow. *The Journal of Experimental Medicine*, *209*(3), 537–549. http://dx.doi.org/10.1084/jem.20110994.

Maurizi, G., Mattiucci, D., Mariani, M., Ciarlantini, M., Traini, S., Mancini, S., … Poloni, A. (2016). DNA demethylating therapy reverts mesenchymal stromal cells derived from high risk myelodysplastic patients to a normal phenotype. *British Journal of Haematology*. http://dx.doi.org/10.1111/bjh.14115. [Epub ahead of print].

Medyouf, H., Mossner, M., Jann, J. C., Nolte, F., Raffel, S., Herrmann, C., … Nowak, D. (2014). Myelodysplastic cells in patients reprogram mesenchymal stromal cells to establish a transplantable stem cell niche disease unit. *Cell Stem Cell*, *14*(6), 824–837. http://dx.doi.org/10.1016/j.stem.2014.02.014.
Mendez-Ferrer, S., Lucas, D., Battista, M., & Frenette, P. S. (2008). Haematopoietic stem cell release is regulated by circadian oscillations. *Nature*, *452*(7186), 442–447. http://dx.doi.org/10.1038/nature06685.
Mendez-Ferrer, S., Michurina, T. V., Ferraro, F., Mazloom, A. R., Macarthur, B. D., Lira, S. A., … Frenette, P. S. (2010). Mesenchymal and haematopoietic stem cells form a unique bone marrow niche. *Nature*, *466*(7308), 829–834. http://dx.doi.org/10.1038/nature09262.
Mercier, F., Monczak, Y., Francois, M., Prchal, J., & Galipeau, J. (2009). Bone marrow mesenchymal stromal cells of patients with myeloproliferative disorders do not carry the JAK2-V617F mutation. *Experimental Hematology*, *37*(3), 416–420. http://dx.doi.org/10.1016/j.exphem.2008.11.008.
Mo, M., Wang, S., Zhou, Y., Li, H., & Wu, Y. (2016). Mesenchymal stem cell subpopulations: Phenotype, property and therapeutic potential. *Cellular and Molecular Life Sciences*, *73*(17), 3311–3321. http://dx.doi.org/10.1007/s00018-016-2229-7.
Morrison, S. J., & Scadden, D. T. (2014). The bone marrow niche for haematopoietic stem cells. *Nature*, *505*(7483), 327–334. http://dx.doi.org/10.1038/nature12984.
Moschoi, R., Imbert, V., Nebout, M., Chiche, J., Mary, D., Prebet, T., … Griessinger, E. (2016). Protective mitochondrial transfer from bone marrow stromal cells to acute myeloid leukemic cells during chemotherapy. *Blood*, *128*(2), 253–264. http://dx.doi.org/10.1182/blood-2015-07-655860.
Muntion, S., Ramos, T. L., Diez-Campelo, M., Roson, B., Sanchez-Abarca, L. I., Misiewicz-Krzeminska, I., … Del Canizo, M. C. (2016). Microvesicles from mesenchymal stromal cells are involved in HPC-microenvironment crosstalk in myelodysplastic patients. *PloS One*, *11*(2). e0146722http://dx.doi.org/10.1371/journal.pone.0146722
Nakamura, Y., Arai, F., Iwasaki, H., Hosokawa, K., Kobayashi, I., Gomei, Y., … Suda, T. (2010). Isolation and characterization of endosteal niche cell populations that regulate hematopoietic stem cells. *Blood*, *116*(9), 1422–1432. http://dx.doi.org/10.1182/blood-2009-08-239194.
Nakamura-Ishizu, A., Takubo, K., Fujioka, M., & Suda, T. (2014). Megakaryocytes are essential for HSC quiescence through the production of thrombopoietin. *Biochemical and Biophysical Research Communications*, *454*(2), 353–357. http://dx.doi.org/10.1016/j.bbrc.2014.10.095.
Nakamura-Ishizu, A., Takubo, K., Kobayashi, H., Suzuki-Inoue, K., & Suda, T. (2015). CLEC-2 in megakaryocytes is critical for maintenance of hematopoietic stem cells in the bone marrow. *The Journal of Experimental Medicine*, *212*(12), 2133–2146. http://dx.doi.org/10.1084/jem.20150057.
Naveiras, O., Nardi, V., Wenzel, P. L., Hauschka, P. V., Fahey, F., & Daley, G. Q. (2009). Bone-marrow adipocytes as negative regulators of the haematopoietic microenvironment. *Nature*, *460*(7252), 259–263. http://dx.doi.org/10.1038/nature08099.
Nervi, B., Ramirez, P., Rettig, M. P., Uy, G. L., Holt, M. S., Ritchey, J. K., … DiPersio, J. F. (2009). Chemosensitization of acute myeloid leukemia (AML) following mobilization by the CXCR4 antagonist AMD3100. *Blood*, *113*(24), 6206–6214. http://dx.doi.org/10.1182/blood-2008-06-162123.
Nilsson, S. K., Johnston, H. M., & Coverdale, J. A. (2001). Spatial localization of transplanted hemopoietic stem cells: Inferences for the localization of stem cell niches. *Blood*, *97*(8), 2293–2299.
Nombela-Arrieta, C., Pivarnik, G., Winkel, B., Canty, K. J., Harley, B., Mahoney, J. E., … Silberstein, L. E. (2013). Quantitative imaging of haematopoietic stem and progenitor

cell localization and hypoxic status in the bone marrow microenvironment. *Nature Cell Biology*, *15*(5), 533–543. http://dx.doi.org/10.1038/ncb2730.
Olson, T. S., Caselli, A., Otsuru, S., Hofmann, T. J., Williams, R., Paolucci, P., … Horwitz, E. M. (2013). Megakaryocytes promote murine osteoblastic HSC niche expansion and stem cell engraftment after radioablative conditioning. *Blood*, *121*(26), 5238–5249. http://dx.doi.org/10.1182/blood-2012-10-463414.
Omatsu, Y., Sugiyama, T., Kohara, H., Kondoh, G., Fujii, N., Kohno, K., & Nagasawa, T. (2010). The essential functions of adipo-osteogenic progenitors as the hematopoietic stem and progenitor cell niche. *Immunity*, *33*(3), 387–399. http://dx.doi.org/10.1016/j.immuni.2010.08.017.
Ono, N., Ono, W., Mizoguchi, T., Nagasawa, T., Frenette, P. S., & Kronenberg, H. M. (2014). Vasculature-associated cells expressing nestin in developing bones encompass early cells in the osteoblast and endothelial lineage. *Developmental Cell*, *29*(3), 330–339. http://dx.doi.org/10.1016/j.devcel.2014.03.014.
Paggetti, J., Haderk, F., Seiffert, M., Janji, B., Distler, U., Ammerlaan, W., … Moussay, E. (2015). Exosomes released by chronic lymphocytic leukemia cells induce the transition of stromal cells into cancer-associated fibroblasts. *Blood*, *126*(9), 1106–1117. http://dx.doi.org/10.1182/blood-2014-12-618025.
Park, D., Spencer, J. A., Koh, B. I., Kobayashi, T., Fujisaki, J., Clemens, T. L., … Scadden, D. T. (2012). Endogenous bone marrow MSCs are dynamic, fate-restricted participants in bone maintenance and regeneration. *Cell Stem Cell*, *10*(3), 259–272. http://dx.doi.org/10.1016/j.stem.2012.02.003.
Parmar, K., Mauch, P., Vergilio, J. A., Sackstein, R., & Down, J. D. (2007). Distribution of hematopoietic stem cells in the bone marrow according to regional hypoxia. *Proceedings of the National Academy of Sciences of the United States of America*, *104*(13), 5431–5436. http://dx.doi.org/10.1073/pnas.0701152104.
Pavlaki, K., Pontikoglou, C. G., Demetriadou, A., Batsali, A. K., Damianaki, A., Simantirakis, E., … Papadaki, H. A. (2014). Impaired proliferative potential of bone marrow mesenchymal stromal cells in patients with myelodysplastic syndromes is associated with abnormal WNT signaling pathway. *Stem Cells and Development*, *23*(14), 1568–1581. http://dx.doi.org/10.1089/scd.2013.0283.
Pieri, L., Guglielmelli, P., Bogani, C., Bosi, A., Vannucchi, A. M., & Myeloproliferative Disorders Research C. (2008). Mesenchymal stem cells from JAK2(V617F) mutant patients with primary myelofibrosis do not harbor JAK2 mutant allele. *Leukemia Research*, *32*(3), 516–517. http://dx.doi.org/10.1016/j.leukres.2007.07.001.
Pittenger, M. F., Mackay, A. M., Beck, S. C., Jaiswal, R. K., Douglas, R., Mosca, J. D., … Marshak, D. R. (1999). Multilineage potential of adult human mesenchymal stem cells. *Science*, *284*(5411), 143–147.
Pizzo, R. J., Azadniv, M., Guo, N., Acklin, J., Lacagnina, K., Coppage, M., & Liesveld, J. L. (2016). Phenotypic, genotypic, and functional characterization of normal and acute myeloid leukemia-derived marrow endothelial cells. *Experimental Hematology*, *44*(5), 378–389. http://dx.doi.org/10.1016/j.exphem.2016.01.008.
Pleyer, L., Valent, P., & Greil, R. (2016). Mesenchymal stem and progenitor cells in normal and dysplastic hematopoiesis-masters of survival and clonality? *International Journal of Molecular Sciences*, *17*(7), 1009. http://dx.doi.org/10.3390/ijms17071009..
Poloni, A., Maurizi, G., Mattiucci, D., Amatori, S., Fogliardi, B., Costantini, B., … Leoni, P. (2014). Overexpression of CDKN2B (p15INK4B) and altered global DNA methylation status in mesenchymal stem cells of high-risk myelodysplastic syndromes. *Leukemia*, *28*(11), 2241–2244. http://dx.doi.org/10.1038/leu.2014.197.
Poulos, M. G., Guo, P., Kofler, N. M., Pinho, S., Gutkin, M. C., Tikhonova, A., … Butler, J. M. (2013). Endothelial Jagged-1 is necessary for homeostatic and regenerative

hematopoiesis. *Cell Reports*, *4*(5), 1022–1034. http://dx.doi.org/10.1016/j.celrep.2013.07.048.
Puram, R. V., Kowalczyk, M. S., de Boer, C. G., Schneider, R. K., Miller, P. G., McConkey, M., … Ebert, B. L. (2016). Core circadian clock genes regulate leukemia stem cells in AML. *Cell*, *165*(2), 303–316. http://dx.doi.org/10.1016/j.cell.2016.03.015.
Raaijmakers, M. H., Mukherjee, S., Guo, S., Zhang, S., Kobayashi, T., Schoonmaker, J. A., … Scadden, D. T. (2010). Bone progenitor dysfunction induces myelodysplasia and secondary leukaemia. *Nature*, *464*(7290), 852–857. http://dx.doi.org/10.1038/nature08851.
Rafii, S., Butler, J. M., & Ding, B. S. (2016). Angiocrine functions of organ-specific endothelial cells. *Nature*, *529*(7586), 316–325. http://dx.doi.org/10.1038/nature17040.
Rafii, S., Shapiro, F., Pettengell, R., Ferris, B., Nachman, R. L., Moore, M. A., & Asch, A. S. (1995). Human bone marrow microvascular endothelial cells support long-term proliferation and differentiation of myeloid and megakaryocytic progenitors. *Blood*, *86*(9), 3353–3363.
Reagan, M. R., & Rosen, C. J. (2016). Navigating the bone marrow niche: Translational insights and cancer-driven dysfunction. *Nature Reviews. Rheumatology*, *12*(3), 154–168. http://dx.doi.org/10.1038/nrrheum.2015.160.
Reinisch, A., Thomas, D., Corces, M. R., Zhang, X., Gratzinger, D., Hong, W. J., … Majeti, R. (2016). A humanized bone marrow ossicle xenotransplantation model enables improved engraftment of healthy and leukemic human hematopoietic cells. *Nature Medicine*, *22*(7), 812–821. http://dx.doi.org/10.1038/nm.4103.
Rizo, A., Vellenga, E., de Haan, G., & Schuringa, J. J. (2006). Signaling pathways in self-renewing hematopoietic and leukemic stem cells: Do all stem cells need a niche? *Human Molecular Genetics*, *15*(Spec No. 2), R210–R219. http://dx.doi.org/10.1093/hmg/ddl175.
Rosen, C. J., Ackert-Bicknell, C., Rodriguez, J. P., & Pino, A. M. (2009). Marrow fat and the bone microenvironment: Developmental, functional, and pathological implications. *Critical Reviews in Eukaryotic Gene Expression*, *19*(2), 109–124.
Rouault-Pierre, K., Lopez-Onieva, L., Foster, K., Anjos-Afonso, F., Lamrissi-Garcia, I., Serrano-Sanchez, M., … Bonnet, D. (2013). HIF-2alpha protects human hematopoietic stem/progenitors and acute myeloid leukemic cells from apoptosis induced by endoplasmic reticulum stress. *Cell Stem Cell*, *13*(5), 549–563. http://dx.doi.org/10.1016/j.stem.2013.08.011.
Sacchetti, B., Funari, A., Michienzi, S., Di Cesare, S., Piersanti, S., Saggio, I., … Bianco, P. (2007). Self-renewing osteoprogenitors in bone marrow sinusoids can organize a hematopoietic microenvironment. *Cell*, *131*(2), 324–336. http://dx.doi.org/10.1016/j.cell.2007.08.025.
Samsonraj, R. M., Rai, B., Sathiyanathan, P., Puan, K. J., Rotzschke, O., Hui, J. H., … Cool, S. M. (2015). Establishing criteria for human mesenchymal stem cell potency. *Stem Cells*, *33*(6), 1878–1891. http://dx.doi.org/10.1002/stem.1982.
Santamaria, C., Muntion, S., Roson, B., Blanco, B., Lopez-Villar, O., Carrancio, S., … Del Canizo, M. C. (2012). Impaired expression of DICER, DROSHA, SBDS and some microRNAs in mesenchymal stromal cells from myelodysplastic syndrome patients. *Haematologica*, *97*(8), 1218–1224. http://dx.doi.org/10.3324/haematol.2011.054437.
Scheller, E. L., & Rosen, C. J. (2014). What's the matter with MAT? Marrow adipose tissue, metabolism, and skeletal health. *Annals of the New York Academy of Sciences*, *1311*, 14–30. http://dx.doi.org/10.1111/nyas.12327.
Schepers, K., Campbell, T. B., & Passegue, E. (2015). Normal and leukemic stem cell niches: Insights and therapeutic opportunities. *Cell Stem Cell*, *16*(3), 254–267. http://dx.doi.org/10.1016/j.stem.2015.02.014.

Schepers, K., Pietras, E. M., Reynaud, D., Flach, J., Binnewies, M., Garg, T., … Passegue, E. (2013). Myeloproliferative neoplasia remodels the endosteal bone marrow niche into a self-reinforcing leukemic niche. *Cell Stem Cell*, *13*(3), 285–299. http://dx.doi.org/10.1016/j.stem.2013.06.009.

Schneider, R. K., Ziegler, S., Leisten, I., Ferreira, M. S., Schumacher, A., Rath, B., … Ziegler, P. (2014). Activated fibronectin-secretory phenotype of mesenchymal stromal cells in pre-fibrotic myeloproliferative neoplasms. *Journal of Hematology & Oncology*, 7, 92. http://dx.doi.org/10.1186/s13045-014-0092-2.

Schofield, R. (1978). The relationship between the spleen colony-forming cell and the haemopoietic stem cell. *Blood Cells*, *4*(1–2), 7–25.

Silberstein, L., Goncalves, K. A., Kharchenko, P. V., Turcotte, R., Kfoury, Y., Mercier, F., … Scadden, D. T. (2016). Proximity-based differential single-cell analysis of the niche to identify stem/progenitor cell regulators. *Cell Stem Cell*, *19*, 530–543. http://dx.doi.org/10.1016/j.stem.2016.07.004.

Simsek, T., Kocabas, F., Zheng, J., Deberardinis, R. J., Mahmoud, A. I., Olson, E. N., … Sadek, H. A. (2010). The distinct metabolic profile of hematopoietic stem cells reflects their location in a hypoxic niche. *Cell Stem Cell*, 7(3), 380–390. http://dx.doi.org/10.1016/j.stem.2010.07.011.

Song, L. X., Guo, J., He, Q., Yang, L. P., Gu, S. C., Zhang, X., … Chang, C. K. (2012). Bone marrow mesenchymal stem cells in myelodysplastic syndromes: Cytogenetic characterization. *Acta Haematologica*, *128*(3), 170–177. http://dx.doi.org/10.1159/000339427.

Sorrentino, A., Ferracin, M., Castelli, G., Biffoni, M., Tomaselli, G., Baiocchi, M., … Valtieri, M. (2008). Isolation and characterization of CD146 + multipotent mesenchymal stromal cells. *Experimental Hematology*, *36*(8), 1035–1046. http://dx.doi.org/10.1016/j.exphem.2008.03.004.

Spencer, J. A., Ferraro, F., Roussakis, E., Klein, A., Wu, J., Runnels, J. M., … Lin, C. P. (2014). Direct measurement of local oxygen concentration in the bone marrow of live animals. *Nature*, *508*(7495), 269–273. http://dx.doi.org/10.1038/nature13034.

Takubo, K., Goda, N., Yamada, W., Iriuchishima, H., Ikeda, E., Kubota, Y., … Suda, T. (2010). Regulation of the HIF-1alpha level is essential for hematopoietic stem cells. *Cell Stem Cell*, 7(3), 391–402. http://dx.doi.org/10.1016/j.stem.2010.06.020.

Thery, C., Zitvogel, L., & Amigorena, S. (2002). Exosomes: Composition, biogenesis and function. *Nature Reviews. Immunology*, *2*(8), 569–579. http://dx.doi.org/10.1038/nri855.

Uy, G. L., Rettig, M. P., Motabi, I. H., McFarland, K., Trinkaus, K. M., Hladnik, L. M., … DiPersio, J. F. (2012). A phase 1/2 study of chemosensitization with the CXCR4 antagonist plerixafor in relapsed or refractory acute myeloid leukemia. *Blood*, *119*(17), 3917–3924. http://dx.doi.org/10.1182/blood-2011-10-383406.

Verstovsek, S., Manshouri, T., Pilling, D., Bueso-Ramos, C. E., Newberry, K. J., Prijic, S., … Estrov, Z. (2016). Role of neoplastic monocyte-derived fibrocytes in primary myelofibrosis. *The Journal of Experimental Medicine*, *213*(9), 1723–1740. http://dx.doi.org/10.1084/jem.20160283.

Visnjic, D., Kalajzic, Z., Rowe, D. W., Katavic, V., Lorenzo, J., & Aguila, H. L. (2004). Hematopoiesis is severely altered in mice with an induced osteoblast deficiency. *Blood*, *103*(9), 3258–3264. http://dx.doi.org/10.1182/blood-2003-11-4011.

von der Heide, E. K., Neumann, M., Vosberg, S., James, A. R., Schroeder, M. P., Tanchez, J. O., … Baldus, C. D. (2016). Molecular alterations in bone marrow mesenchymal stromal cells derived from acute myeloid leukemia patients. *Leukemia*. http://dx.doi.org/10.1038/leu.2016.324. [Epub ahead of print].

Vukovic, M., Guitart, A. V., Sepulveda, C., Villacreces, A., O'Duibhir, E., Panagopoulou, T. I., … Kranc, K. R. (2015). Hif-1alpha and Hif-2alpha synergize to suppress AML development but are dispensable for disease maintenance. *The Journal of Experimental Medicine*, *212*(13), 2223–2234. http://dx.doi.org/10.1084/jem.20150452.

Walkley, C. R., Olsen, G. H., Dworkin, S., Fabb, S. A., Swann, J., McArthur, G. A., … Purton, L. E. (2007). A microenvironment-induced myeloproliferative syndrome caused by retinoic acid receptor gamma deficiency. *Cell*, *129*(6), 1097–1110. http://dx.doi.org/10.1016/j.cell.2007.05.014.

Walkley, C. R., Shea, J. M., Sims, N. A., Purton, L. E., & Orkin, S. H. (2007). Rb regulates interactions between hematopoietic stem cells and their bone marrow microenvironment. *Cell*, *129*(6), 1081–1095. http://dx.doi.org/10.1016/j.cell.2007.03.055.

Wei, J., Wunderlich, M., Fox, C., Alvarez, S., Cigudosa, J. C., Wilhelm, J. S., … Mulloy, J. C. (2008). Microenvironment determines lineage fate in a human model of MLL-AF9 leukemia. *Cancer Cell*, *13*(6), 483–495. http://dx.doi.org/10.1016/j.ccr.2008.04.020.

Winkler, I. G., Sims, N. A., Pettit, A. R., Barbier, V., Nowlan, B., Helwani, F., … Levesque, J. P. (2010). Bone marrow macrophages maintain hematopoietic stem cell (HSC) niches and their depletion mobilizes HSCs. *Blood*, *116*(23), 4815–4828. http://dx.doi.org/10.1182/blood-2009-11-253534.

Wohrer, S., Rabitsch, W., Shehata, M., Kondo, R., Esterbauer, H., Streubel, B., … Valent, P. (2007). Mesenchymal stem cells in patients with chronic myelogenous leukaemia or bi-phenotypic Ph + acute leukaemia are not related to the leukaemic clone. *Anticancer Research*, *27*(6B), 3837–3841.

Xie, Y., Yin, T., Wiegraebe, W., He, X. C., Miller, D., Stark, D., … Li, L. (2009). Detection of functional haematopoietic stem cell niche using real-time imaging. *Nature*, *457*(7225), 97–101. http://dx.doi.org/10.1038/nature07639.

Yamazaki, S., Ema, H., Karlsson, G., Yamaguchi, T., Miyoshi, H., Shioda, S., … Nakauchi, H. (2011). Nonmyelinating Schwann cells maintain hematopoietic stem cell hibernation in the bone marrow niche. *Cell*, *147*(5), 1146–1158. http://dx.doi.org/10.1016/j.cell.2011.09.053.

Yamazaki, S., Iwama, A., Takayanagi, S., Eto, K., Ema, H., & Nakauchi, H. (2009). TGF-beta as a candidate bone marrow niche signal to induce hematopoietic stem cell hibernation. *Blood*, *113*(6), 1250–1256. http://dx.doi.org/10.1182/blood-2008-04-146480.

Ye, H., Adane, B., Khan, N., Sullivan, T., Minhajuddin, M., Gasparetto, M., … Jordan, C. T. (2016). Leukemic stem cells evade chemotherapy by metabolic adaptation to an adipose tissue niche. *Cell Stem Cell*, *19*(1), 23–37. http://dx.doi.org/10.1016/j.stem.2016.06.001.

Yue, R., Zhou, B. O., Shimada, I. S., Zhao, Z., & Morrison, S. J. (2016). Leptin receptor promotes adipogenesis and reduces osteogenesis by regulating mesenchymal stromal cells in adult bone marrow. *Cell Stem Cell*, *18*(6), 782–796. http://dx.doi.org/10.1016/j.stem.2016.02.015.

Zambetti, N. A., Ping, Z., Chen, S., Kenswil, K. J., Mylona, M. A., Sanders, M. A., … Raaijmakers, M. H. (2016). Mesenchymal inflammation drives genotoxic stress in hematopoietic stem cells and predicts disease evolution in human Pre-leukemia. *Cell Stem Cell*, *19*, 613–627. http://dx.doi.org/10.1016/j.stem.2016.08.021.

Zeng, Z., Shi, Y. X., Samudio, I. J., Wang, R. Y., Ling, X., Frolova, O., … Konopleva, M. (2009). Targeting the leukemia microenvironment by CXCR4 inhibition overcomes resistance to kinase inhibitors and chemotherapy in AML. *Blood*, *113*(24), 6215–6224. http://dx.doi.org/10.1182/blood-2008-05-158311.

Zhao, M., Perry, J. M., Marshall, H., Venkatraman, A., Qian, P., He, X. C., … Li, L. (2014). Megakaryocytes maintain homeostatic quiescence and promote post-injury regeneration of hematopoietic stem cells. *Nature Medicine*, *20*(11), 1321–1326. http://dx.doi.org/10.1038/nm.3706.

CHAPTER SEVEN

Targeting the Bone Marrow Niche in Hematological Malignancies

D. Verma, D.S. Krause[1]

Georg-Speyer-Haus, Institute for Tumor Biology and Experimental Therapy, Frankfurt, Germany
[1]Corresponding author: e-mail address: krause@gsh.uni-frankfurt.de

Contents

1. INTRODUCTION

The bone marrow microenvironment (BMM) represents a very complex system comprising various different cell types such as osteoblastic cells, osteoclasts, osteocytes, endothelial cells, fibroblasts, mesenchymal stem cells, fibroblasts, macrophages, CxCl12-abundant reticular cells, sympathetic nerve cells, and other cells, but also features noncellular elements like the extracellular matrix, oxygen tension, cytokines, and mechanical forces. These factors regulate and influence different functions characterizing hematopoietic stem cells (HSC), in particular localization, proliferation, quiescence, differentiation, and self-renewal.

Hematological malignancies, which can arise from various cells of origin from within the different hematopoietic lineages, usually due to hematopoietic cell-intrinsic lesions (Krause & Van Etten, 2007), are frequently located in the

Advances in Stem Cells and their Niches, Volume 1
ISSN 2468-5097
http://dx.doi.org/10.1016/bs.asn.2016.12.004

BMM. Cancer stem cells in leukemia, i.e., leukemic stem cells, the malignant counterpart to HSC, for example, are thought to reside in the BMM, where they are regulated similar to HSC and are protected from chemo- and tyrosine kinase inhibitor therapy (Ishikawa et al., 2007; Krause & Scadden, 2015; Morrison & Scadden, 2014; Zhang et al., 2013).

In addition, it has even become apparent that alterations in the BMM can lead to abnormalities in the hematopoietic system. Osteoprogenitor-specific ablation of *dicer 1*, which codes for an enzyme required for mRNA processing, for example, was associated with the development of a myelodysplasia-like syndrome with few mice showing signs of acute myeloid leukemia (AML) (Raajimakers et al., 2010). Further, upregulation of β-catenin in the osteoprogenitor lineage gave rise to AML (Kode et al., 2014) and germline knockout of retinoic acid receptor γ (Walkley et al., 2007), and osteocyte-specific ablation of the Gsα subunit of the G-protein-coupled receptor (Fulzele et al., 2013) led to a myeloproliferation. In humans, donor-derived leukemia after allogeneic HSC transplantation is a rare phenomenon, but it has been described (Sala-Torra et al., 2006), suggesting that the BMM may act as bad "soil" leading to bad "seed."

2. TARGETING THE BMM FOR THE TREATMENT OF HEMATOLOGICAL MALIGNANCIES

Evidence is accumulating demonstrating a cross talk between the BMM and malignant leukemic cells, which seems essential for the progression of the hematological malignancy. Most current therapies for hematological malignancies, especially leukemias and lymphomas and to a lesser extent plasma cell myeloma, are directly aimed at the malignant hematopoietic cells. However, due to various mechanisms of resistance, persistent or resistant LSC can lead to disease relapse and progression and a multifaceted "mode of attack" by incorporating strategies such as targeting of the BMM in addition to cancer cell-specific therapies holds great promise. Some of these strategies are discussed here (Fig. 1).

2.1 Targeting Leukemia Cell Chemotaxis to the BMM

2.1.1 The CXCR4/SDF1-α Axis

The CXCR4/SDF1-α axis is involved in regulating the retention of HSC and LSC in the BMM (Krause & Scadden, 2015; Morrison & Scadden, 2014). Various studies have reported that the impairment or inhibition of this axis resulted in the reduction of tumor load in various leukemias and, subsequently, delayed leukemia progression.

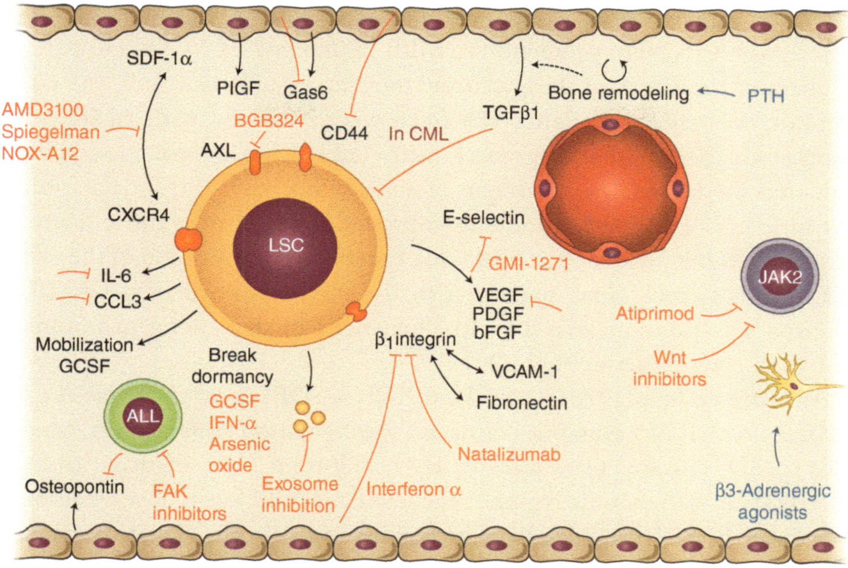

Fig. 1 Schematic representation of the bone marrow microenvironment, its means of interaction with leukemic stem cells and strategies to target these interactions for therapeutic purposes. *Ang2*, angiopoietin2; *bFGF*, basic fibroblast growth factor; *BMP*, bone morphogenetic protein; *BMPR*, bone morphogenetic protein receptor; *CCL3*, chemokine (C–C motif) ligand 3; *Gas6*, growth-arrest-specific-gene 6; *GCSF*, granulocyte colony-stimulating factor; *IFN-α*, interferon-α; *IL-6*, interleukin-6; *LIF*, leukemia-inhibitory factor; *PDGF*, platelet-derived growth factor; *PlGF*, placental growth factor; *SCF*, stem cell factor; *SDF-1α*, stroma cell-derived factor 1α; *TGFβ1*, transforming growth factor β1; *TPO*, thrombopoietin; *VCAM-1*, vascular cell adhesion molecule; *VEGF (A)*, vascular endothelial growth factor (A).

AMD3100 (plerixafor) is a clinically used CXCR4 antagonist used for mobilization of HSC. Impairment of the CXCR4/SDF1-α axis allows HSC and their progenitors to egress into the peripheral blood, and the following studies are built on the concept that mobilization of LSC from a chemoprotective niche may lead to improved eradication of LSC.

Treatment of acute promyelocytic leukemia (APL) in a cathepsin G-PML-RARα knock-in murine model with AMD3100 increased the mobilization of APL blasts into peripheral blood and spleen. This mobilization of APL blasts led to improved response to classical chemotherapy, which consisted of cytarabine (ara-C) and daunorubicin resulting in the reduction of tumor burden and prolonged overall survival of APL mice (Nervi et al., 2009). Infant patients with acute lymphoblastic leukemia (ALL) due to rearrangements of the mixed lineage leukemia (MLL) gene, generally, have a good prognosis, but approximately 50% of patients suffer from a relapse. Using a patient-derived xenograft murine model of ALL, it became apparent that CXCR4/SDF1-α-mediated signaling via the stroma provided chemoprotection to MLL-rearranged ALL cells. However, the chemoprotection was abrogated and the efficacy of lestaurtinib, an FLT3 inhibitor, was enhanced, when the mice were cotreated with AMD3100 (Sison et al., 2013). In an aggressive retroviral murine transduction/transplantation model of CML treatment of mice with AMD3100 and dasatinib, a tyrosine kinase inhibitor, modestly improved the control of the leukemia, but increased the incidence of death due to leukemic infiltration of the central nervous system. This suggested that the combination of a CXCR4 inhibitor plus dasatinib may be most beneficial in the setting of minimal residual disease (Agarwal et al., 2012).

An alternate strategy to impair the CXCR4/SDF1-α axis is by the use of NOX-A12, an RNA oligonucleotide in L-configuration (Spiegelmer), which binds to SDF1-α and neutralizes it. By coculturing bone marrow stromal cells (BMSC) with chronic lymphocytic leukemia (CLL) cells, it was demonstrated that BMSC provide chemoresistance to CLL cells. Treatment with NOX-A12 abrogated this chemoresistance and rendered the CLL cells more sensitive to bendamustine and fludarabine treatment (Hoellenriegel et al., 2014).

Based on the evidence demonstrating the prominence of the CXCR4/SDF1-α axis for the interaction of the BMM and leukemic cells a trial using MDX-1338, an antibody to CXCR4, for the treatment of relapsed AML, diffuse large B-cell lymphoma, CLL, and follicular lymphoma has been completed (clinicalTrials.gov identifier: NCT01120457), and a trial testing the chemosensitization prior to HSC transplantation by the use of a CXCR4 small-molecule antagonist plus chemotherapy in patients with acute leukemia (clinicalTrials.gov identifier: NCT02605460) is currently recruiting.

2.2 Targeting Leukemia Cell Adhesion to the BMM

2.2.1 CD44

CD44, a cell-surface glycoprotein, known to bind to hyaluronan, osteopontin, and E-selectin, and known to be involved in cell–cell interactions, migration, and cell adhesion, has been shown to be essential for engraftment of leukemia-initiating cells in the BMM in AML (Jin, Hope, Zhai, Smadja-Joffe, & Dick, 2006) and CML (Krause, Lazarides, von Andrian, & Van Etten, 2006). Additionally, treatment of immunosuppressed mice transplanted with CD34+ blast crisis CML cells with the tyrosine kinase inhibitor dasatinib plus a monoclonal antibody to CD44 reduced the frequency of human progenitor cells in the murine BMM and reduced the self-renewal capacity of LSC (Holm et al., 2013). Further, targeting of CD44 by nanoparticle-mediated silencing (Gul-Uludağ et al., 2014) or treatment with an anti-CD44 (and anti-CD49d) antibody (Erb, Megaptche, Gu, Buchler, & Zoller, 2014) led to induction of apoptosis and reduced adhesion of AML cells to mesenchymal stem cells, as well as dislodgement of the lymphoid cell lines EL4 and Jurkat from the BMM and spleen, which was associated with increased sensitivity to chemotherapy (Singh, Erb, & Zöller, 2013), respectively. No clinical trial on the use of anti-CD44-targeted therapy in the hematological malignancies is currently ongoing.

2.2.2 The Vascular Cell Adhesion Molecule-1/Very Late Antigen-4 Axis and N-Cadherin

N-cadherin and the vascular cell adhesion molecule (VCAM)-1/very late antigen (VLA)-4 axis also mediate chemoresistance in leukemic cells by modulating adhesion of leukemic cells to mesenchymal stem cells in the BMM. Activation of the VCAM1/VLA-4 signaling axis by coculture of leukemia cells with stromal cells led to global transcription changes in stromal cells including activation of nuclear factor (NF) κb signaling, which resulted in stromal cell-dependent chemoresistance in leukemia cells. Therapeutic inhibition of NFκb led to improvement of leukemia outcome (Jacamo et al., 2014).

Coculture of CML stem and progenitor cells with mesenchymal stromal cells led to inhibition of apoptosis and maintenance of colony-forming ability and engraftment potential despite exposure to a tyrosine kinase inhibitor, which was found to be due to activation of the N-cadherin receptor on leukemia cells. Activation of N-cadherin in CML cells by Wnt, which is secreted by stromal cells, led to increased interaction between N-cadherin and β-catenin, enhancement of the nuclear translocation of β-catenin, transcription of β-catenin target genes, and, hence, chemoresistance in CML cells. Taken together, these finding suggest that N-cadherin/Wnt-β-catenin signaling and the VCAM/VLA-4 axis may be interesting therapeutic targets, which may be exploited for the treatment of leukemia in conjunction with tyrosine kinase inhibitors or chemotherapy (Zhang et al., 2013).

2.2.3 Focal Adhesion Kinase-Mediated Pathways

Focal adhesion kinase (FAK) is a focal adhesion-associated protein, which is phosphorylated when integrins on the cell surface are bound, for instance, by extracellular matrix proteins. FAK is involved in cellular adhesion and spreading processes. FAK is overexpressed in BCR-ABL1 + B-cell acute lymphoblastic leukemia (B-ALL), which simultaneously harbor mutations in the lymphoid transcription factor, Ikaros (*IKZF1* gene). These mutations in *IKZF1* are usually dominant negative mutations, can be found in over 80% of BCR-ABL1 + B-ALL, and are associated with increased cell adhesion to the stroma of the BMM and poor outcome in humans and mice. Inhibition of FAK by VS-4718 and VS-6063 effectively blocked B-cell acute lymphoblastic leukemia (B-ALL) cell adhesion to stromal cells, induced their apoptosis, and was shown to be beneficial for survival of mice with B-ALL (Churchman et al., 2016; Joshi et al., 2014). A clinical trial on the treatment of AML and B-ALL with VS-4718 had been initiated, but was, subsequently, withdrawn (clinicalTrial.gov identifier: NCT02215629). Treatment of APL, T-ALL, B-ALL, CML, and AML cell lines in vitro with VS-4718 had an antiproliferative effect and treatment with VS-4718 in a murine model of AML due to injection of the MV-4-11 cell line led to delayed progression of disease (Tam et al., 2014).

2.2.4 Bruton Tyrosine Kinase-Mediated Pathways

CLL is the most common adult leukemia and is an incurable malignancy of uncontrollably growing mature B-lymphocytes. Similar to other hematological malignancies, the tumor microenvironment has been shown to influence the course of CLL. In particular, the BMM has been reported to promote growth, survival, and chemoresistance in CLL cells (Buchner et al., 2010; Burger, 2010; Burger, Ghia, Rosenwald, & Caligaris-Cappio, 2009; Herishanu et al., 2011; Pleyer, Egle, Hartmann, & Greil, 2009; Smit et al., 2007; Stevenson, Krysov, Davies, Steele, & Packham, 2011; Zenz, Mertens, Kuppers, Dohner, & Stilgenbauer, 2010). B-cell antigen receptor (BCR) signaling is important for the pathogenesis of most of the B-cell malignancies, including CLL (Stevenson et al., 2011), and Bruton tyrosine kinase (BTK) is a key component of the BCR signaling complex. Targeting BTK is a rational strategy for treating B-cell malignancies, including CLL. An irreversible small-molecule BTK inhibitor, PCI-32765 (Honigberg et al., 2010), demonstrated promising clinical outcomes in phase I and phase II studies for the treatment of B-cell non-Hodgkin lymphomas and CLL (Burger et al., 2010; Byrd et al., 2011; Fowler et al., 2010). Mechanistically, it was demonstrated that PCI-32765 inhibited BCR-controlled signaling and integrin $\alpha4\beta1$-mediated adhesion of CLL cells to fibronectin, an important extracellular matrix protein of the BMM, and to VCAM-1. CXCL12-, CXCL13-, and CCL19-induced signaling, adhesion, and migration of primary CLL cells were also inhibited by BTK inhibition (de Rooij et al., 2012).

2.3 Targeting the Interaction Between the Vasculature and Leukemia Cells

2.3.1 E-selectin

Components of the vasculature in the BMM are also involved in regulating HSC (Morrison & Scadden, 2014), and evidence is emerging that these components may also be involved in the regulation of LSC. Endothelial cells represent an element of the vascular niche in the BMM and express E-selectin, an adhesion molecule, which has been reported to regulate HSC quiescence, self-renewal, and chemoresistance (Winkler et al., 2012). Deficiency of E-selectin in mice or treatment of mice with the E-selectin inhibitor, GMI-1271, prompted the quiescence of HSC and enhanced HSC survival after chemotherapy or irradiation induced injury (Winkler et al., 2012). Similarly, E-selectin is also a crucial component in the leukemic BMM influencing LSC in CML and AML. In a murine retroviral model of MLL-AF9-induced AML E-selectin expression in the vasculature was upregulated, leading to increased adhesion of AML blasts to the vascular niche. Treatment with GMI-1271 resulted in disrupted adhesion of AML cells to E-selectin and mobilization of AML cells into peripheral blood. Also, E-selectin-deficient mice transplanted with MLL-AF9+ AML LSC exhibited increased sensitivity toward chemotherapy with ara-c compared to wild-type mice, suggesting that E-selectin plays a role in the chemoresistance of LSC (Winkler et al., 2014). These data confirmed previous findings in xenograft models that AML LSC are frequently localized near endosteal vascular endothelium (Ishikawa et al., 2007).

In CML, deficiency of E-selectin in mice being transplanted with BCR-ABL1-transduced bone marrow led to impaired engraftment of CML leukemia-initiating cells (LIC) (Krause, Lazarides, Lewis, von Andrian, & Van Etten, 2014). Treatment of mice with CML with the E-selectin inhibitor GMI-1271 plus imatinib resulted in the reduction of the leukemia load, increased survival of CML mice, and increased cycling of CML LSC (Aggoune, Magnani, Van Etten, & Krause, 2014). These data suggest that E-selectin plays an important role in the vascular niche in myeloid leukemia and that targeting E-selectin in these leukemias may be beneficial (Krause et al., 2014). Indeed, a clinical trial on the treatment of AML by the use of GMI-1271 is currently recruiting patients (clinicalTrials.gov identifier: NCT02306291).

2.4 Targeting Tumor-Induced Neoangiogenesis

Increased microvascular density, although more common in solid tumors, was also reported to be true for various hematological malignancies, such as leukemia, plasma cell myeloma, and myelodysplastic syndromes (Schmidt & Carmeliet, 2011). This was found to be associated with significantly increased expression of proangiogenic factors like vascular endothelial growth factor (VEGF), platelet-derived growth factor (PDGF), and basic fibroblast growth factor (bFGF), which are paracrine factors secreted by leukemia and other cells (Corrado et al., 2012; Schmidt et al., 2011). As the degree of vascularization of the BMM in a hematological malignancy may influence disease progression, targeting angiogenesis may prove to be a successful strategy.

BCR-ABL1+ and *JAK2-V617F* MPN are characterized by a significant increase in microvascular density. Increased expression of VEGF and increased microvascular density have been correlated with an elevated *JAK2-V617F* mutant allele burden in myeloproliferative neoplasia (MPN) (Medinger et al., 2009). Targeting neoangiogenesis in such patients could be a new strategy to improve treatment outcomes in these patients. A number of studies have shown delayed disease progression in various hematological malignancies after antiangiogenesis therapy (Table 1) (Medinger & Mross, 2010; Medinger & Passweg, 2014). In a murine xenograft model of T-cell ALL cotreatment with bevacizumab, a monoclonal antibody to VEGF-A, and doxorubicin prolonged the survival of the diseased mice (Wang et al., 2011). Similarly, Anti-VEGFR2 therapy proved to be beneficial for the treatment of APL and erythroleukemia in murine xenograft models (Dias et al., 2001; He et al., 2003; Zhu et al., 2003). Alternatively, a novel agent disrupting the vasculature, OXi4503 (combretastatin A1-diphosphate), has been shown to impair the tumor vasculature in AML, leading to leukemia cell apoptosis, but the effect was most prominent in conjunction with an anti-VEGF treatment like bevacizumab (Madlambayan et al., 2010).

2.5 Targeting the Interaction Between the Bone and Leukemia Cells

2.5.1 Bone Remodeling

Targeting of the osteolineage "compartment" in the BMM is another powerful means to target LSC, as suppression of LSC, positive for the oncogene *BCR-ABL1*, in CML could be achieved by induction of bone remodeling by parathyroid hormone (PTH), the most potent regulator of bone. PTH-induced bone remodeling resulted in the release of transforming growth factor-β1 (TGF-β1) from the bone matrix. TGF-β1 led to the suppression of LSC in CML, significantly prolonged the survival of diseased mice with an osteoblast-specific constitutive activation of the receptors for PTH and PTH-related peptide, and led to 15-fold reduction of LSC in wild-type mice treated with PTH. Interestingly, however, no such effect was observed in AML due to the oncogene *MLL-AF9*. This was at least partly due to lower expression of the receptor for TGF-β1 on MLL-AF9+

Table 1 Antiangiogenic Therapies in the Hematological Malignancies

Drug	Target	Disease Entities	References
Receptor tyrosine kinase inhibitors			
PTK787/ZK 222584 (Vatalanib®)	VEGFR1-3, PDGFRβ, c-Kit	AML, PMF, MDS, CML, DLBCL, MM	Barbarroja et al. (2009), Barbarroja et al. (2010), Giles et al. (2007), Lin et al. (2002), Roboz et al. (2006), and Wood et al. (2000)
SU5416 (Semaxanib)	VEGFR1-2, c-Kit, FLT3	AML, MDS, MM, MPN	Fiedler et al. (2003), Giles et al. (2003), and Mesters et al. (2001)
PKC-412 (Midostaurin)	VEGFR2, PKC, PDGFR, FLT3, c-Kit	AML	Gallogly and Lazarus (2016) and Starr (2016)
Cediranib (Recentin®)	VEGFR1-3, PDGFRβ, c-Kit	AML, MDS, CLL	Fiedler et al. (2010) and Wedge et al. (2005)
Anti-VEGF strategies			
Bevacizumab (Avastin®)	VEGF-A	AML, MDS, MPN, CLL, NHL, MM, CML	Karp et al. (2004) and Zahiragic et al. (2007)
Immunomodulatory drugs			
Thalidomide	bFGF, VEGF, IL-6	AML, MDS, MPN, CLL, NHL, MM	Barlogie et al. (2001), Kumar and Rajkumar (2006), Rajkumar et al. (2006), Raza et al. (2008), Singhal et al. (1999), Steins et al. (2002), Thomas et al. (2003), and Thomas et al. (2006)
Lenalidomide (Revlimid®)	bFGF, VEGF, IL-6	AML, MDS, CLL, NHL	Dimopoulos et al. (2007), Kotla et al. (2009), Kumar and Rajkumar (2006), Quintas-Cardama et al. (2009), Rajkumar and Blood (2006), Tefferi et al. (2006), and Weber et al. (2007)

AML, acute myeloid leukemia; *bFGF*, basic fibroblast growth factor; *CLL*, chronic lymphocytic leukemia; *CML*, chronic myeloid leukemia; *DLBCL*, diffuse large B-cell lymphoma; *IL-6*, interleukin-6; *MDS*, myelodysplastic syndrome; *MM*, multiple myeloma; *MPN*, myeloproliferative neoplasm; *NHL*, non-Hodgkin lymphoma; *PDGFR*, platelet-derived growth factor receptor; *VEGF*, vascular endothelial growth factor; *VEGFR*, vascular endothelial growth factor receptor.

Adapted from Medinger, M., & Mross, K. (2010). Clinical trials with anti-angiogenic agents in hematological malignancies. *Journal of Angiogenesis Research*, *2*, 10; Medinger, M., & Passweg, J. (2014). Angiogenesis in myeloproliferative neoplasms, new markers and future directions. *Memo*, *7*, 206–210.

cells and suggested differential effects of the niche on different myeloid leukemias (Krause et al., 2013). Based on this finding, a clinical trial will be initiated in Germany at five different partner sites, in which CML patients receiving a tyrosine kinase inhibitor for at least 6 months will be cotreated with PTH, in order to test, if cotreatment with PTH decreases *BCR-ABL1* transcript levels, deepens the molecular remission, and improves patent outcome (Krause et al., 2013).

2.5.2 Osteopontin

Osteopontin is an extracellular matrix protein found in the BMM, where it is produced by osteoblastic and mesenchymal stem cells. The number of hematopoietic stem and progenitor cells (HSPC) in an osteopontin-deficient BMM was increased, suggesting that osteopontin negatively regulates HSC (Stier et al., 2005). Osteopontin, however, also maintains LSC in B-ALL, which secrete osteopontin themselves, in a dormant state via supporting leukemic cell adhesion to its own structure. Abrogation of the osteopontin-signaling axis led to an increase of proliferation of the leukemic cells and an increased tumor burden. In conjunction with chemotherapy inhibition of osteopontin successfully reduced minimal residual disease (Boyerinas et al., 2013), suggesting that inhibition of osteopontin may represent a promising treatment strategy.

2.6 Targeting Disease-Specific Interactions With the BMM

2.6.1 The Sympathetic Nervous System

Myeloproliferative neoplasms are a group of diseases characterized by mutations in HSC, leading to excess growth of certain cells. Several mutations have been discovered in MPN, with one of them being the V617F mutation in *Janus kinase 2 (JAK2)*, which is found in >95% of cases of polycythemia vera and approximately 55% of essential thrombocythemia (Baxter et al., 2005; James et al., 2005; Kralovics et al., 2005; Levine et al., 2005). Mice expressing the human *JAK2(V617F)* mutation in HSC developed MPN and were reported to have a reduced and defective sympathetic nerve fiber system and a reduced number of nestin+ mesenchymal stem cells (Arranz et al., 2014), which are innervated by sympathetic nerve fibers and are responsible for the regulation of different important functions of normal HSC (Mendez-Ferrer, Lucas, Battista, & Frenette, 2008; Mendez-Ferrer et al., 2010). The impairment of the BMM, which was due to interleukin (IL)-1β secreted by the MPN cells, was shown to be essential for the pathogenesis of MPN. Upon treatment with neuroprotective or sympathomimetic drugs (β3-adrenergic agonists), the detrimental effect on sympathetic nerve fibers regulating nestin+ MSC was "rescued" and the progression of MPN was blocked (Arranz et al., 2014). Based on these findings, a clinical trial was initiated for the treatment of JAK2+ MPN by the use of a β3-sympathicomimetic drug, mirabegron (clinicalTrial.gov identifier: NCT02311569).

2.6.2 The Gas6–Axl Axis

Growth arrest-specific gene 6 (*Gas6*) was discovered as a gene upregulated during the growth arrest of fibroblast cells (Schneider, King, & Philipson, 1988). Gas6 acts as a ligand for three different receptors, namely Tyro3, Axl, and Mer, with differential affinities toward these receptors (Lemke & Rothlin, 2008). Among all these receptors, the Axl-Gas6 axis is known for its importance in solid tumors such as breast, lung, and pancreatic cancer, where it has been correlated with poor prognosis, increased invasion, and chemoresistance (Wu et al., 2014), but also for AML. AML cells instruct bone marrow stroma cells to upregulate the expression of Gas6, and further, the Axl-Gas6 paracrine axis provides proliferation benefits and chemoprotection to AML cells. Therapeutic inhibition of Axl by BGB324 has yielded promising results and could serve as a useful strategy for the treatment of FLT3 mutated or FLT3 wild-type AML cells (Ben-Batalla et al., 2013; Loges et al., 2010). Complementarily, an independent study reported shorter disease-free and overall survival in Gas6+ AML patients and suggested Gas6 expression level as a prognostic marker in cytogenetically normal AML patients (Whitman et al., 2014). In summary, these data suggest that the Axl-Gas6 axis may represent an important therapeutic target for the treatment of AML in future.

2.7 Targeting Secreted Cytokines

Another treatment strategy would be to target cytokines, which are secreted by tumor or stromal cells and are involved in the perpetuation and progression of the malignancy, or their respective receptors. The secretion of cytokines by stromal cells may be instructed by the leukemic cells or the cytokines may be derived from the leukemic cells themselves (Krause & Scadden, 2015).

2.7.1 CCL3

MPN modifies the BMM into a self-reinforcing leukemic niche, thereby impairing normal hematopoiesis. It has been shown that MPN leads to an increased number and to altered function of osteoblastic cells, which is at least partly mediated by CCL3 release from the leukemic cells. This was shown to lead to a decreased ability of osteoblastic cells to sustain normal

HSC (Schepers et al., 2013). CCL3 contributes to the maintenance of CML LSC (Baba et al., 2013). Similarly, impaired osteoblastic cell number and function were also observed in a murine model of AML, which was thought to be CCL3 dependent (Frisch et al., 2012). Bone disease in plasma cell myeloma may also be mediated by CCL3 (Frisch et al., 2012; Vallet et al., 2011). Therefore, in conjunction with chemotherapy, inhibition of CCL3 may also prove to be a feasible strategy for targeting the malignant BMM in MPN, AML, and plasma cell myeloma.

2.7.2 Placental Growth Factor

Placental growth factor (PIGF) represents another example for a cytokine secreted by BMSC, and it is known for playing a prominent role in the pathogenesis of BCR-ABL1-induced CML. Here, the CML cells induce upregulation of PIGF from BMSC, which aggravates the disease by stimulating bone marrow angiogenesis and promoting CML cell proliferation and metabolism. Loss of PIGF or inhibition of PIGF by the monoclonal antibody mAb5D11D4 resulted in the reduction of tumor load and significant prolongation of survival of mice with CML (Schmidt et al., 2011).

2.7.3 Other Cytokines

Upon leukemia development, normal HSC are thought to be outcompeted by disruption of the normal HSPC niche and sequestration of normal hematopoiesis. Stem cell factor, secreted by B-ALL cells, has previously been shown to be a causative factor, leading to the compromise and loss of normal HSPC and leukemic spread within the bone marrow cavity (Colmone et al., 2008). More recently, it was shown that HSC exposure to CML cells impaired certain stem cell functions of HSC, leading to reduction of long-term reconstitution potential, increased proliferation and differentiation, and an acquisition of gene expression, which most closely resembled the CML cells themselves. This effect was found to be due to IL-6, an inflammatory cytokine, which promoted leukemogenesis and imposed the properties of the malignant cells on the untransformed or normal HSPC. In fact, treatment of leukemic mice with an IL-6 neutralizing antibody rescued the changes in normal HSC induced by the leukemic cells and effectively controlled the disease. The protective role of anti-IL-6 antibody treatment also benefited human CD34+ cells when exposed to CML cells. These data suggest that the inclusion of anti IL-6 therapy for the treatment of CML patients may be beneficial (Welner et al., 2015).

Similarly, the BMM in MPN patients with mutations in *JAK2* or *MPL* is characterized by an inflammatory gene expression pattern, which is due to the secretion of cytokines by malignant and normal cells, correlating with disease aggressivity. Inhibition of the JAK kinase, for instance by ruxolitinib, reduced cytokine secretion and disease severity likely via inhibiting Stat3-dependent signaling of aberrantly secreted inflammatory cytokines (Kleppe et al., 2015). In this respect, a clinical trial on the treatment of patients with myelofibrosis has been initiated (clinicalTrial.gov identifier: NCT02784496).

2.7.4 Exosomes

Exosomes are microvesicles, which are secreted by tumor cells, mesenchymal stromal cells, endothelial cells, and many other cell types. They can bidirectionally transfer mRNA or microRNA, drug efflux transporters, and other proteins to neighboring cells, thereby modifying the microenvironment (Krause & Scadden, 2015). In imatinib-resistant CML, for example, carboxyamidotriazole-orotate may be beneficial in inhibiting exosome-stimulated angiogenesis (Corrado et al., 2012), but further studies are needed to test the feasibility and validity of targeting exosomes in hematological malignancies.

2.7.5 Galectin-3

Galectin-3 is a member of the lectin family and plays an important role in cell–cell adhesion and cell–matrix interactions. The expression of galectin-3 was induced in CML cell lines upon coculture on stromal cells. In CML and AML, it has been shown to mediate chemoresistance via reduction of apoptosis and in vivo overexpression of galectin-3 leads to enhanced lodgement in the niche (Yamamoto-Sugitani et al., 2011). No clinical trials on the use of anti-galectin-3 modalities in the treatment of hematological malignancies have currently been published.

2.8 Altering LSC Dormancy in the LSC Niche

As discussed earlier, the major concern of existing therapies against leukemia and other hematological malignancies is the frequent inefficiency at eradicating LSC, which are responsible for relapse of the disease. LSC may be quite similar to their normal counterparts, namely HSC, and, therefore, may equally be maintained in a dormant stage, as they are regulated by their interaction with the BMM (Essers & Trumpp, 2010). Maintenance of a dormant stage in LSC by the BMM likely provides LSC with protection against chemotherapy-induced apoptosis. "Awakening" from this dormant stage is thought to be responsible for disease relapse during or, especially, after withdrawal of therapy (Essers & Trumpp, 2010).

Specifically, it has been demonstrated that most of the AML cells close to the endosteal region are noncycling and that AML cells are protected from chemotherapy-induced apoptosis by the endosteal bone marrow niche (Ishikawa et al., 2007). Likewise, CD34+ cells isolated from CML patients contained noncycling or quiescent cells, which in vitro showed resistance toward imatinib treatment (Graham et al., 2002).

Several studies have demonstrated the capacity of certain agents, which successfully can break the dormancy of HSC and LSC within their respective microenvironments, thereby rendering them more susceptible toward chemotherapy. These agents include cytokines like granulocyte-colony stimulating factor (G-CSF), interferon α (IFNα), and the chemical compound arsenic trioxide, which targets the nuclear factor promyelocytic leukemia protein (PML) for proteosomal degradation (Essers & Trumpp, 2010).

A study showed that priming of human CD34+ CML cells with G-CSF rendered them more sensitive to treatment with imatinib mesylate (Drummond et al., 2009). In NSG mice transplanted with human AML cells, G-CSF treatment was successful at breaking the dormancy of leukemic cells residing close to the endosteal niche, leading to successful targeting of these cells by chemotherapy (Saito et al., 2010).

Until now there has been no evidence showing the direct effect of IFNα with regard to breaking the dormancy of LSC as it is true for normal HSC (Essers et al., 2009), though IFNα has been and is being used successfully for the treatment of CML and other MPN. Normally, CML patients have to undergo lifelong therapy with imatinib, even in the event of a complete molecular remission, as discontinuation of imatinib results in disease relapse in 60% of cases (Mahon et al., 2010). However, in a different French trial a small number of CML patients did not relapse after discontinuation of imatinib. Strikingly, the great majority of these patients had been on IFNα therapy prior to being switched to imatinib, suggesting that IFNα may possibly have "awakened" dormant LSC in these patients, leading to improved eradication by imatinib (Essers & Trumpp, 2010; Rousselot et al., 2007), though this hypothesis needs to be tested in further studies.

3. CONCLUSIONS

In summary, many potential adjunct treatment strategies exist to target different components in the BMM in order to augment existing therapies in the form of chemotherapy or tyrosine kinase inhibitor therapy in hematological malignancies. Combining multimodal treatment efforts similar to highly active antiretroviral therapy for the treatment of human immunodeficiency virus, in order to target the interactions between a hematological malignancy and its microenvironment, seems to be the logical consequence in view of the rapidly accumulating understanding of the normal and malignant bone marrow microenvironment. Uniting therapeutical forces to combat hematological malignancies and their still far from perfect outcomes is now within our reach.

REFERENCES

Agarwal, A., Fleischman, A. G., Petersen, C. L., MacKenzie, R., Luty, S., Loriaux, M., et al. (2012). Effects of plerixafor in combination with BCR-ABL kinase inhibition in a murine model of CML. *Blood*, *120*(13), 2658–2668.

Aggoune, D., Magnani, J. L., Van Etten, R. A., & Krause, D. S. (2014). The vascular niche is involved in regulating leukemic stem cells in murine chronic myelogenous leukemia. *Blood*, *124*(21), 516.

Arranz, L., Sanchez-Aguilera, A., Martin-Perez, D., Isern, J., Langa, X., Tzankov, A., et al. (2014). Neuropathy of haematopoietic stem cell niche is essential for myeloproliferative neoplasms. *Nature*, *512*(7512), 78–81. http://dx.doi.org/10.1038/nature13383.

Baba, T., Naka, K., Morishita, S., Komatsu, N., Hirao, A., & Mukaida, N. (2013). MIP-1alpha/CCL3-mediated maintenance of leukemia-initiating cells in the initiation process of chronic myeloid leukemia. *The Journal of Experimental Medicine*, *210*(12), 2661–2673.

Barbarroja, N., Torres, L. A., Luque, M. J., Carretero, R. M., Valverde-Estepa, A., Lopez-Sanchez, L. M., et al. (2009). Additive effect of PTK787/ZK 222584, a potent inhibitor of VEGFR phosphorylation, with Idarubicin in the treatment of acute myeloid leukemia. *Experimental Hematology*, *37*(6), 679–691.

Barbarroja, N., Torres, L. A., Rodriguez-Ariza, A., Valverde-Estepa, A., Lopez-Sanchez, L. M., Ruiz-Limon, P., et al. (2010). AEE788 is a vascular endothelial growth factor receptor tyrosine kinase inhibitor with antiproliferative and proapoptotic effects in acute myeloid leukemia. *Experimental Hematology*, *38*(8), 641–652.

Barlogie, B., Desikan, R., Eddlemon, P., Spencer, T., Zeldis, J., Munshi, N., et al. (2001). Extended survival in advanced and refractory multiple myeloma after single-agent thalidomide: Identification of prognostic factors in a phase 2 study of 169 patients. *Blood*, *98*(2), 492–494.

Baxter, E. J., Scott, L. M., Campbell, P. J., East, C., Fourouclas, N., Swanton, S., et al. (2005). Acquired mutation of the tyrosine kinase JAK2 in human myeloproliferative disorders. *Lancet*, *365*(9464), 1054–1061.

Ben-Batalla, I., Schultze, A., Wroblewski, M., Erdmann, R., Heuser, M., Waizenegger, J. S., et al. (2013). Axl, a prognostic and therapeutic target in acute myeloid leukemia mediates paracrine crosstalk of leukemia cells with bone marrow stroma. *Blood*, *122*(14), 2443–2452.

Boyerinas, B., Zafrir, M., Yesilkanal, A. E., Price, T. T., Hyjek, E. M., & Sipkins, D. A. (2013). Adhesion to osteopontin in the bone marrow niche regulates lymphoblastic leukemia cell dormancy. *Blood*, *121*(24), 4821–4831.

Buchner, M., Baer, C., Prinz, G., Dierks, C., Burger, M., Zenz, T., et al. (2010). Spleen tyrosine kinase inhibition prevents chemokine- and integrin-mediated stromal protective effects in chronic lymphocytic leukemia. *Blood*, *115*(22), 4497–4506.

Burger, J. A. (2010). Chemokines and chemokine receptors in chronic lymphocytic leukemia (CLL): From understanding the basics towards therapeutic targeting. *Seminars in Cancer Biology*, *20*(6), 424–430.

Burger, J. A., Ghia, P., Rosenwald, A., & Caligaris-Cappio, F. (2009). The microenvironment in mature B-cell malignancies: A target for new treatment strategies. *Blood*, *114*(16), 3367–3375.

Burger, J. A., O'Brien, S., Fowler, N., Advani, R., Sharman, J. P., Furman, R. R., et al. (2010). The Bruton's tyrosine kinase inhibitor, PCI-32765, is well tolerated and demonstrates promising clinical activity in chronic lymphocytic leukemia (CLL) and small lymphocytic lymphoma (SLL): An update on ongoing phase 1 studies. *Blood*, *116*(21), 32.

Byrd, J. C., Blum, K. A., Burger, J. A., Coutre, S. E., Sharman, J. P., Furman, R. R., et al. (2011). Activity and tolerability of the Bruton's tyrosine kinase (Btk) inhibitor PCI-32765 in patients with chronic lymphocytic leukemia/small lymphocytic lymphoma (CLL/SLL): Interim results of a phase Ib/II study. *Journal of Clinical Oncology*, *29*(15), abstr. 6508.

Churchman, M. L., Evans, K., Richmond, J., Robbins, A., Jones, L., Shapiro, I. M., et al. (2016). Synergism of FAK and tyrosine kinase inhibition in Ph+ B-ALL. *JCI Insight*, *1*(4), pii: e86082.

Colmone, A., Amorim, M., Pontier, A. L., Wang, S., Jablonski, E., & Sipkins, D. A. (2008). Leukemic cells create bone marrow niches that disrupt the behavior of normal hematopoietic progenitor cells. *Science*, *322*(5909), 1861–1865.

Corrado, C., Flugy, A. M., Taverna, S., Raimondo, S., Guggino, G., Karmali, R., et al. (2012). Carboxyamidotriazole-orotate inhibits the growth of imatinib-resistant chronic myeloid leukaemia cells and modulates exosomes-stimulated angiogenesis. *PloS One*, 7(8), e42310.

de Rooij, M. F., Kuil, A., Geest, C. R., Eldering, E., Chang, B. Y., Buggy, J. J., et al. (2012). The clinically active BTK inhibitor PCI-32765 targets B-cell receptor- and chemokine-controlled adhesion and migration in chronic lymphocytic leukemia. *Blood*, *119*(11), 2590–2594.

Dias, S., Hattori, K., Heissig, B., Zhu, Z., Wu, Y., Witte, L., et al. (2001). Inhibition of both paracrine and autocrine VEGF/ VEGFR-2 signaling pathways is essential to induce long-term remission of xenotransplanted human leukemias. *Proceedings of the National Academy of Sciences of the United States of America*, *98*(19), 10857–10862.

Dimopoulos, M., Spencer, A., Attal, M., Prince, H. M., Harousseau, J. L., Dmoszynska, A., et al. (2007). Lenalidomide plus dexamethasone for relapsed or refractory multiple myeloma. *The New England Journal of Medicine*, *357*(21), 2123–2132.

Drummond, M. W., Heaney, N., Kaeda, J., Nicolini, F. E., Clark, R. E., Wilson, G., et al. (2009). A pilot study of continuous imatinib vs pulsed imatinib with or without G-CSF in CML patients who have achieved a complete cytogenetic response. *Leukemia*, *23*(6), 1199–1201.

Erb, U., Megaptche, A. P., Gu, X., Buchler, M. W., & Zoller, M. (2014). CD44 standard and CD44v10 isoform expression on leukemia cells distinctly influences niche embedding of hematopoietic stem cells. *Journal of Hematology & Oncology*, 7, 29.

Essers, M. A., Offner, S., Blanco-Bose, W. E., Waibler, Z., Kalinke, U., Duchosal, M. A., et al. (2009). IFNalpha activates dormant haematopoietic stem cells in vivo. *Nature*, *458*(7240), 904–908.

Essers, M. A., & Trumpp, A. (2010). Targeting leukemic stem cells by breaking their dormancy. *Molecular Oncology*, *4*(5), 443–450.

Fiedler, W., Mesters, R., Heuser, M., Ehninger, G., Berdel, W. E., Zirrgiebel, U., et al. (2010). An open-label, phase I study of cediranib (RECENTIN) in patients with acute myeloid leukemia. *Leukemia Research*, *34*(2), 196–202.

Fiedler, W., Mesters, R., Tinnefeld, H., Loges, S., Staib, P., Duhrsen, U., et al. (2003). A phase 2 clinical study of SU5416 in patients with refractory acute myeloid leukemia. *Blood*, *102*(8), 2763–2767.

Fowler, N., Sharman, J. P., Smith, S. M., Boyd, T., Grant, B., Kolibaba, K. S., et al. (2010). The Btk inhibitor, PCI-32765, induces durable responses with minimal toxicity in patients with relapsed/refractory B-cell malignancies: Results from a phase I study. *Blood*, *116*(21), 425.

Frisch, B. J., Ashton, J. M., Xing, L., Becker, M. W., Jordan, C. T., & Calvi, L. M. (2012). Functional inhibition of osteoblastic cells in an in vivo mouse model of myeloid leukemia. *Blood*, *119*(2), 540–550.

Fulzele, K., Krause, D. S., Barry, K., Lotinun, S., Baron, R., Bonewald, L., et al. (2013). Myelopoiesis is regulated by osteocytes through Gsalpha-dependent signaling. *Blood*, *121*(6), 930–939.

Gallogly, M. M., & Lazarus, H. M. (2016). Midostaurin: An emerging treatment for acute myeloid leukemia patients. *Journal of Blood Medicine*, *7*, 73–83.

Giles, F. J., List, A. F., Carroll, M., Cortes, J. E., Valickas, J., Chen, B. L., et al. (2007). PTK787/ZK 222584, a small molecule tyrosine kinase receptor inhibitor of vascular endothelial growth factor (VEGF), has modest activity in myelofibrosis with myeloid metaplasia. *Leukemia Research*, *31*(7), 891–897.

Giles, F. J., Stopeck, A. T., Silverman, L. R., Lancet, J. E., Cooper, M. A., Hannah, A. L., et al. (2003). SU5416, a small molecule tyrosine kinase receptor inhibitor, has biologic activity in patients with refractory acute myeloid leukemia or myelodysplastic syndromes. *Blood*, *102*(3), 795–801.

Graham, S. M., Jorgensen, H. G., Allan, E., Pearson, C., Alcorn, M. J., Richmond, L., et al. (2002). Primitive, quiescent, Philadelphia-positive stem cells from patients with chronic myeloid leukemia are insensitive to STI571 in vitro. *Blood*, *99*(1), 319–325.

Gul-Uludağ, H., Valencia-Serna, J., Kucharski, C., Marquez-Curtis, L. A., Jiang, X., Larratt, L., et al. (2014). Polymeric nanoparticle-mediated silencing of CD44 receptor in CD34(+) acute myeloid leukemia cells. *Leukemia Research*, *38*(11), 1299–1308.

He, R., Liu, B., Yang, C., Yang, R. C., Tobelem, G., & Han, Z. C. (2003). Inhibition of K562 leukemia angiogenesis and growth by expression of antisense vascular endothelial growth factor (VEGF) sequence. *Cancer Gene Therapy*, *10*(12), 879–886.

Herishanu, Y., Perez-Galan, P., Liu, D., Biancotto, A., Pittaluga, S., Vire, B., et al. (2011). The lymph node microenvironment promotes B-cell receptor signaling, NF-kappaB activation, and tumor proliferation in chronic lymphocytic leukemia. *Blood*, *117*(2), 563–574.

Hoellenriegel, J., Zboralski, D., Maasch, C., Rosin, N. Y., Wierda, W. G., Keating, M. J., et al. (2014). The Spiegelmer NOX-A12, a novel CXCL12 inhibitor, interferes with chronic lymphocytic leukemia cell motility and causes chemosensitization. *Blood*, *123*(7), 1032–1039.

Holm, F., Mason, C. N., Runza, V., Weigand, S., Sadarangani, A., & Jamieson, C. H. M. (2013). CD44 monoclonal antibody-enhanced clearance of chronic myeloid leukemia stem cells from the malignant niche. *Blood*, *122*, 858.

Honigberg, L. A., Smith, A. M., Sirisawad, M., Verner, E., Loury, D., Chang, B., et al. (2010). The Bruton tyrosine kinase inhibitor PCI-32765 blocks B-cell activation and is efficacious in models of autoimmune disease and B-cell malignancy. *Proceedings of the National Academy of Sciences of the United States of America*, *107*(29), 13075–13080.

Ishikawa, F., Yoshida, S., Saito, Y., Hijikata, A., Kitamura, H., Tanaka, S., et al. (2007). Chemotherapy-resistant human AML stem cells home to and engraft within the bone-marrow endosteal region. *Nature Biotechnology*, *25*(11), 1315–1321.

Jacamo, R., Chen, Y., Wang, Z., Ma, W., Zhang, M., Spaeth, E. L., et al. (2014). Reciprocal leukemia-stroma VCAM-1/VLA-4-dependent activation of NF-kappaB mediates chemoresistance. *Blood*, *123*(17), 2691–2702.

James, C., Ugo, V., Le Couedic, J. P., Staerk, J., Delhommeau, F., Lacout, C., et al. (2005). A unique clonal JAK2 mutation leading to constitutive signalling causes polycythaemia vera. *Nature*, *434*(7037), 1144–1148.

Jin, L., Hope, K. J., Zhai, Q., Smadja-Joffe, F., & Dick, J. E. (2006). Targeting of CD44 eradicates human acute myeloid leukemic stem cells. *Nature Medicine, 12*(10), 1167.

Joshi, I., Jena, N., Yoshida, T., Paraskevopoulou, L., Zhang, Z. H., Liu, Z. Y., et al. (2014). Focal adhesion kinase inhibitors reverse the stromal adhesion phenotype of ikaros-mutant B-ALL, induce apoptosis, and synergize with ABL1 tyrosine kinase inhibitors: A new paradigm for pathogenesis and therapy of high-risk B-ALL. *Blood, 124*(21), 285.

Karp, J. E., Gojo, I., Pili, R., Gocke, C. D., Greer, J., Guo, C., et al. (2004). Targeting vascular endothelial growth factor for relapsed and refractory adult acute myelogenous leukemias: Therapy with sequential 1-beta-D-arabinofuranosylcytosine, mitoxantrone, and bevacizumab. *Clinical Cancer Research, 10*(11), 3577–3585.

Kleppe, M., Kwak, M., Koppikar, P., Riester, M., Keller, M., Bastian, L., et al. (2015). JAK-STAT pathway activation in malignant and nonmalignant cells contributes to MPN pathogenesis and therapeutic response. *Cancer Discovery, 5*(3), 316–331.

Kode, A., Manavalan, J. S., Mosialou, I., Bhagat, G., Rathinam, C. V., Luo, N., et al. (2014). Leukaemogenesis induced by an activating beta-catenin mutation in osteoblasts. *Nature, 506*(7487), 240–244.

Kotla, V., Goel, S., Nischal, S., Heuck, C., Vivek, K., Das, B., et al. (2009). Mechanism of action of lenalidomide in hematological malignancies. *Journal of Hematology & Oncology, 2*, 36.

Kralovics, R., Passamonti, F., Buser, A. S., Teo, S. S., Tiedt, R., Passweg, J. R., et al. (2005). A gain-of-function mutation of JAK2 in myeloproliferative disorders. *The New England Journal of Medicine, 352*(17), 1779–1790.

Krause, D. S., Fulzele, K., Catic, A., Sun, C. C., Dombkowski, D., Hurley, M. P., et al. (2013). Differential regulation of myeloid leukemias by the bone marrow microenvironment. *Nature Medicine, 19*(11), 1513–1517.

Krause, D. S., Lazarides, K., Lewis, J. B., von Andrian, U. H., & Van Etten, R. A. (2014). Selectins and their ligands are required for homing and engraftment of BCR-ABL1+ leukemic stem cells in the bone marrow niche. *Blood, 123*(9), 1361–1371.

Krause, D. S., Lazarides, K., von Andrian, U. H., & Van Etten, R. A. (2006). Requirement for CD44 in homing and engraftment of BCR-ABL-expressing leukemic stem cells. *Nature Medicine, 12*(10), 1175–1180.

Krause, D. S., & Scadden, D. T. (2015). A hostel for the hostile: The bone marrow niche in hematologic neoplasms. *Haematologica, 100*(11), 1376–1387.

Krause, D. S., & Van Etten, R. A. (2007). Right on target: Eradicating leukemic stem cells. *Trends in Molecular Medicine, 13*(11), 470–481.

Kumar, S., & Rajkumar, S. V. (2006). Thalidomide and lenalidomide in the treatment of multiple myeloma. *European Journal of Cancer, 42*(11), 1612–1622.

Lemke, G., & Rothlin, C. V. (2008). Immunobiology of the TAM receptors. *Nature Reviews. Immunology, 8*(5), 327–336.

Levine, R. L., Wadleigh, M., Cools, J., Ebert, B. L., Wernig, G., Huntly, B. J., et al. (2005). Activating mutation in the tyrosine kinase JAK2 in polycythemia vera, essential thrombocythemia, and myeloid metaplasia with myelofibrosis. *Cancer Cell, 7*(4), 387–397.

Lin, B., Podar, K., Gupta, D., Tai, Y. T., Li, S., Weller, E., et al. (2002). The vascular endothelial growth factor receptor tyrosine kinase inhibitor PTK787/ZK222584 inhibits growth and migration of multiple myeloma cells in the bone marrow microenvironment. *Cancer Research, 62*(17), 5019–5026.

Loges, S., Schmidt, T., Tjwa, M., van Geyte, K., Lievens, D., Lutgens, E., et al. (2010). Malignant cells fuel tumor growth by educating infiltrating leukocytes to produce the mitogen Gas6. *Blood, 115*(11), 2264–2273.

Madlambayan, G. J., Meacham, A. M., Hosaka, K., Mir, S., Jorgensen, M., Scott, E. W., et al. (2010). Leukemia regression by vascular disruption and antiangiogenic therapy. *Blood, 116*(9), 1539–1547.

Mahon, F. X., Rea, D., Guilhot, J., Guilhot, F., Huguet, F., Nicolini, F., et al. (2010). Discontinuation of imatinib in patients with chronic myeloid leukaemia who have maintained complete molecular remission for at least 2 years: The prospective, multicentre stop imatinib (STIM) trial. *The Lancet Oncology*, *11*(11), 1029–1035.

Medinger, M., & Mross, K. (2010). Clinical trials with anti-angiogenic agents in hematological malignancies. *Journal of Angiogenesis Research*, *2*, 10.

Medinger, M., & Passweg, J. (2014). Angiogenesis in myeloproliferative neoplasms, new markers and future directions. *Memo*, *7*, 206–210.

Medinger, M., Skoda, R., Gratwohl, A., Theocharides, A., Buser, A., Heim, D., et al. (2009). Angiogenesis and vascular endothelial growth factor-/receptor expression in myeloproliferative neoplasms: Correlation with clinical parameters and JAK2-V617F mutational status. *British Journal of Haematology*, *146*(2), 150–157.

Mendez-Ferrer, S., Lucas, D., Battista, M., & Frenette, P. S. (2008). Haematopoietic stem cell release is regulated by circadian oscillations. *Nature*, *452*(7186), 442–447.

Mendez-Ferrer, S., Michurina, T. V., Ferraro, F., Mazloom, A. R., Macarthur, B. D., Lira, S. A., et al. (2010). Mesenchymal and haematopoietic stem cells form a unique bone marrow niche. *Nature*, *466*(7308), 829–834.

Mesters, R. M., Padro, T., Bieker, R., Steins, M., Kreuter, M., Goner, M., et al. (2001). Stable remission after administration of the receptor tyrosine kinase inhibitor SU5416 in a patient with refractory acute myeloid leukemia. *Blood*, *98*(1), 241–243.

Morrison, S. J., & Scadden, D. T. (2014). The bone marrow niche for haematopoietic stem cells. *Nature*, *505*(7483), 327–334.

Nervi, B., Ramirez, P., Rettig, M. P., Uy, G. L., Holt, M. S., Ritchey, J. K., et al. (2009). Chemosensitization of acute myeloid leukemia (AML) following mobilization by the CXCR4 antagonist AMD3100. *Blood*, *113*(24), 6206–6214.

Pleyer, L., Egle, A., Hartmann, T. N., & Greil, R. (2009). Molecular and cellular mechanisms of CLL: Novel therapeutic approaches. *Nature Reviews. Clinical Oncology*, *6*(7), 405–418.

Quintas-Cardama, A., Kantarjian, H. M., Manshouri, T., Thomas, D., Cortes, J., Ravandi, F., et al. (2009). Lenalidomide plus prednisone results in durable clinical, histopathologic, and molecular responses in patients with myelofibrosis. *Journal of Clinical Oncology*, *27*(28), 4760–4766.

Raajimakers, M. H., Mukherjee, S., Guo, S., Zhang, S., Kobayashi, T., Schoonmaker, J. A., et al. (2010). Bone progenitor dysfunction induces myelodysplasia and secondary leukaemia. *Nature*, *464*(7290), 852–857.

Rajkumar, S. V., & Blood, E. (2006). Lenalidomide and venous thrombosis in multiple myeloma. *The New England Journal of Medicine*, *354*(19), 2079–2080.

Rajkumar, S. V., Blood, E., Vesole, D., Fonseca, R., Greipp, P. R., & Eastern Cooperative Oncology, G. (2006). Phase III clinical trial of thalidomide plus dexamethasone compared with dexamethasone alone in newly diagnosed multiple myeloma: A clinical trial coordinated by the Eastern Cooperative Oncology Group. *Journal of Clinical Oncology*, *24*(3), 431–436.

Raza, A., Mehdi, M., Mumtaz, M., Ali, F., Lascher, S., & Galili, N. (2008). Combination of 5-azacytidine and thalidomide for the treatment of myelodysplastic syndromes and acute myeloid leukemia. *Cancer*, *113*(7), 1596–1604.

Roboz, G. J., Giles, F. J., List, A. F., Cortes, J. E., Carlin, R., Kowalski, M., et al. (2006). Phase 1 study of PTK787/ZK 222584, a small molecule tyrosine kinase receptor inhibitor, for the treatment of acute myeloid leukemia and myelodysplastic syndrome. *Leukemia*, *20*(6), 952–957.

Rousselot, P., Huguet, F., Rea, D., Legros, L., Cayuela, J. M., Maarek, O., et al. (2007). Imatinib mesylate discontinuation in patients with chronic myelogenous leukemia in complete molecular remission for more than 2 years. *Blood*, *109*(1), 58–60.

Saito, Y., Uchida, N., Tanaka, S., Suzuki, N., Tomizawa-Murasawa, M., Sone, A., et al. (2010). Induction of cell cycle entry eliminates human leukemia stem cells in s a mouse model of AML. *Nature Biotechnology*, *28*(3), 275–280.

Sala-Torra, O., Hanna, C., Loken, M. R., Flowers, M. E., Maris, M., Ladne, P. A., et al. (2006). Evidence of donor-derived hematologic malignancies after hematopoietic stem cell transplantation. *Biology of Blood and Marrow Transplantation*, *12*(5), 511–517.

Schepers, K., Pietras, E. M., Reynaud, D., Flach, J., Binnewies, M., Garg, T., et al. (2013). Myeloproliferative neoplasia remodels the endosteal bone marrow niche into a self-reinforcing leukemic niche. *Cell Stem Cell*, *13*(3), 285–299.

Schmidt, T., & Carmeliet, P. (2011). Angiogenesis: A target in solid tumors, also in leukemia? *Hematology-American Society Hematology Education Program*, *2011*, 1–8.

Schmidt, T., Kharabi Masouleh, B., Loges, S., Cauwenberghs, S., Fraisl, P., Maes, C., et al. (2011). Loss or inhibition of stromal-derived PIGF prolongs survival of mice with imatinib-resistant Bcr-Abl1(+) leukemia. *Cancer Cell*, *19*(6), 740–753.

Schneider, C., King, R. M., & Philipson, L. (1988). Genes specifically expressed at growth arrest of mammalian cells. *Cell*, *54*(6), 787–793.

Singh, V., Erb, U., & Zöller, M. (2013). Cooperativity of CD44 and CD49d in leukemia cell homing, migration, and survival offers a means for therapeutic attack. *Journal of Immunology*, *191*(10), 5304–5316.

Singhal, S., Mehta, J., Desikan, R., Ayers, D., Roberson, P., Eddlemon, P., et al. (1999). Antitumor activity of thalidomide in refractory multiple myeloma. *The New England Journal of Medicine*, *341*(21), 1565–1571.

Sison, E. A. R., Rau, R. E., McIntyre, E., Li, L., Small, D., & Brown, P. (2013). MLL-rearranged acute lymphoblastic leukaemia stem cell interactions with bone marrow stroma promote survival and therapeutic resistance that can be overcome with CXCR4 antagonism. *British Journal of Haematology*, *160*(6), 785–797.

Smit, L. A., Hallaert, D. Y., Spijker, R., de Goeij, B., Jaspers, A., Kater, A. P., et al. (2007). Differential Noxa/Mcl-1 balance in peripheral versus lymph node chronic lymphocytic leukemia cells correlates with survival capacity. *Blood*, *109*(4), 1660–1668.

Starr, P. (2016). Midostaurin the first targeted therapy to improve survival in AML: Potentially practice-changing. *American Health and Drug Benefits*, *9*(Special issue), 1–21.

Steins, M. B., Padro, T., Bieker, R., Ruiz, S., Kropff, M., Kienast, J., et al. (2002). Efficacy and safety of thalidomide in patients with acute myeloid leukemia. *Blood*, *99*(3), 834–839.

Stevenson, F. K., Krysov, S., Davies, A. J., Steele, A. J., & Packham, G. (2011). B-cell receptor signaling in chronic lymphocytic leukemia. *Blood*, *118*(16), 4313–4320.

Stier, S., Ko, Y., Forkert, R., Lutz, C., Neuhaus, T., Grunewad, E., et al. (2005). Osteopontin is a hematopoietic stem cell niche component that negatively regulates stem cell pool size. *The Journal of Experimental Medicine*, *201*, 1781.

Tam, W. F., Ring, J. E., Trombino, A. F., Weaver, D. T., Pachter, J. A., Padval, M., et al. (2014). VS-4718, a potent focal adhesion kinase (FAK) inhibitor, exhibits anticancer activity in leukemia models in vitro and in vivo. *Blood*, *124*(21), 982.

Tefferi, A., Cortes, J., Verstovsek, S., Mesa, R. A., Thomas, D., Lasho, T. L., et al. (2006). Lenalidomide therapy in myelofibrosis with myeloid metaplasia. *Blood*, *108*(4), 1158–1164.

Thomas, D. A., Estey, E., Giles, F. J., Faderl, S., Cortes, J., Keating, M., et al. (2003). Single agent thalidomide in patients with relapsed or refractory acute myeloid leukaemia. *British Journal of Haematology*, *123*(3), 436–441.

Thomas, D. A., Giles, F. J., Albitar, M., Cortes, J. E., Verstovsek, S., Faderl, S., et al. (2006). Thalidomide therapy for myelofibrosis with myeloid metaplasia. *Cancer*, *106*(9), 1974–1984.

Vallet, S., Pozzi, S., Patel, K., Vaghela, N., Fulciniti, M. T., Veiby, P., et al. (2011). A novel role for CCL3 (MIP-1alpha) in myeloma-induced bone disease via osteocalcin downregulation and inhibition of osteoblast function. *Leukemia*, *25*(7), 1174–1181.

Walkley, C. R., Olsen, G. H., Dworkin, S., Fabb, S. A., Swann, J., McArthur, G. A., et al. (2007). A microenvironment-induced myeloproliferative syndrome caused by retinoic acid receptor gamma deficiency. *Cell*, *129*(6), 1097–1110.

Wang, L., Shi, W. Y., Yang, F., Tang, W., Gapihan, G., Varna, M., et al. (2011). Bevacizumab potentiates chemotherapeutic effect on T-leukemia/lymphoma cells by direct action on tumor endothelial cells. *Haematologica*, *96*(6), 927–931.

Weber, D. M., Chen, C., Niesvizky, R., Wang, M., Belch, A., Stadtmauer, E. A., et al. (2007). Lenalidomide plus dexamethasone for relapsed multiple myeloma in North America. *The New England Journal of Medicine*, *357*(21), 2133–2142.

Wedge, S. R., Kendrew, J., Hennequin, L. F., Valentine, P. J., Barry, S. T., Brave, S. R., et al. (2005). AZD2171: A highly potent, orally bioavailable, vascular endothelial growth factor receptor-2 tyrosine kinase inhibitor for the treatment of cancer. *Cancer Research*, *65*(10), 4389–4400.

Welner, R. S., Amabile, G., Bararia, D., Czibere, A., Yang, H., Zhang, H., et al. (2015). Treatment of chronic myelogenous leukemia by blocking cytokine alterations found in normal stem and progenitor cells. *Cancer Cell*, *27*(5), 671–681.

Whitman, S. P., Kohlschmidt, J., Maharry, K., Volinia, S., Mrozek, K., Nicolet, D., et al. (2014). GAS6 expression identifies high-risk adult AML patients: Potential implications for therapy. *Leukemia*, *28*(6), 1252–1258.

Winkler, I. G., Barbier, V., Nowlan, B., Jacobsen, R. N., Forristal, C. E., Patton, J. T., et al. (2012). Vascular niche E-selectin regulates hematopoietic stem cell dormancy, self renewal and chemoresistance. *Nature Medicine*, *18*(11), 1651–1657.

Winkler, I. G., Barbier, V., Pattabiraman, D. R., Gonda, T. J., Magnani, J. L., & Levesque, J. P. (2014). Vascular niche E-selectin protects acute myeloid leukaemia stem cells from chemotherapy. *Blood*, *124*(21), 620.

Wood, J. M., Bold, G., Buchdunger, E., Cozens, R., Ferrari, S., Frei, J., et al. (2000). PTK787/ZK 222584, a novel and potent inhibitor of vascular endothelial growth factor receptor tyrosine kinases, impairs vascular endothelial growth factor-induced responses and tumor growth after oral administration. *Cancer Research*, *60*(8), 2178–2189.

Wu, X., Liu, X., Koul, S., Lee, C. Y., Zhang, Z., & Halmos, B. (2014). AXL kinase as a novel target for cancer therapy. *Oncotarget*, *5*(20), 9546–9563.

Yamamoto-Sugitani, M., Kuroda, J., Ashihara, E., Nagoshi, H., Kobayashi, T., Matsumoto, Y., et al. (2011). Galectin-3 (Gal-3) induced by leukemia microenvironment promotes drug resistance and bone marrow lodgement in chronic myelogenous leukemia. *Proceedings of the National Academy of Sciences of the United States of America*, *108*(42), 17468–17473.

Zahiragic, L., Schliemann, C., Bieker, R., Thoennissen, N. H., Burow, K., Kramer, C., et al. (2007). Bevacizumab reduces VEGF expression in patients with relapsed and refractory acute myeloid leukemia without clinical antileukemic activity. *Leukemia*, *21*(6), 1310–1312.

Zenz, T., Mertens, D., Kuppers, R., Dohner, H., & Stilgenbauer, S. (2010). From pathogenesis to treatment of chronic lymphocytic leukaemia. *Nature Reviews. Cancer*, *10*(1), 37–50.

Zhang, B., Li, M., McDonald, T., Holyoake, T. L., Moon, R. T., Campana, D., et al. (2013). Microenvironmental protection of CML stem and progenitor cells from tyrosine kinase inhibitors through N-cadherin and Wnt-beta-catenin signaling. *Blood*, *121*(10), 1824–1838.

Zhu, Z., Hattori, K., Zhang, H., Jimenez, X., Ludwig, D. L., Dias, S., et al. (2003). Inhibition of human leukemia in an animal model with human antibodies directed against vascular endothelial growth factor receptor 2. Correlation between antibody affinity and biological activity. *Leukemia*, *17*(3), 604–611.

FSC
www.fsc.org
MIX
Papier | Fördert gute Waldnutzung
FSC® C083411

Kolibri 360 GmbH
Ferdinand-Jühlke-Straße 7
99095 Erfurt, Deutschland
produktsicherheit@kolibri360.de

43 Advances in Biochemical Engineering/ Biotechnology

Managing Editor: A. Fiechter

Applied Molecular Genetics

Guest Editor: J. Reiser

With contributions by
C. L. Cooney, R. L. Dabora, S.-O. Enfors,
V. Glumoff, H. Hellebust, H. Heslot,
M. Kälin, K. Köhler, U. Ochsner, S. B. Primrose,
J. Reiser, L. Strandberg, A. Veide

With 19 Figures and 12 Tables

Springer-Verlag Berlin Heidelberg GmbH

ISBN 978-3-662-15033-7 ISBN 978-3-540-47151-6 (eBook)
DOI 10.1007/978-3-540-47151-6

Originally published by Springer-Verlag Berlin Heidelberg New York in 1990
Softcover reprint of the hardcover 1st edition 1990
Library of Congress Catalog Coard Number 72-152360
Printed of Germany

Typesetting: Th. Müntzer, Bad Langensalza;

2152/3020-543210 — Printed on acid-free paper

Editorial

This volume of Advances in Biochemical Engineering/Biotechnology as well as the following one are dedicated to Professor Armin Fiechter on the occasion of his 65th birthday. Contributions were solicited from Armin's colleagues but were limited to subjects which related to the biotechnology scope of this series.

One might judge a successful scientist by several criteria and Armin fullfills them all. One criterion of success is one's record of scientific achievement. Armin's work on yeast physiology, process development and instrumentation has been very fruitful and the necessity of adopting new technology or applying concepts from other disciplines has never hindered progress in his laboratory. One very important measure of achievement and the one which is likely to be the most important in both professional and human terms, is the training of successful students. Armin is one of the most eminent promoters of biotechnology in Switzerland and abroad and he was instrumental in establishing the Institute for Biotechnology at the ETH in 1982 of which he is still director. By pioneering the combination of fundamental and applied aspects he has greatly influenced the advancement of biotechnology in teaching and research in Switzerland.

At the age of 65, the pace of Armin's activities shows no sign of slackening. At the latest count there are some 20 doctoral students working under his supervision. He is currently the editor-in-chief of the Journal of Biotechnology and of this series, both of which were co-founded by him and have gained momentum under his influence.

This special issue is dedicated to Armin Fiechter with admiration.

Zürich, July 1990 Jakob Reiser

Armin Fiechter

Table of Contents

Controlling Bacteriophage Infections in Industrial Bioprocesses

S. B. Primrose
Life Sciences Division, Amersham International plc, White Lion Road, Amersham, Bucks HP7 9LL, UK

Bacteriophage infections of microbial processes are relatively common but can be economically devasting. If the designers and operators of bioreactors have an understanding of the biology and ecology of bacteriophages the risks and hazards of bacteriophage infections can be minimised. If an infection should occur it is important to recognize the fact as soon as possible. The symptoms of a phage infection and procedures for cleaning-up infected bioprocess plants are described.

1 Introduction

Bacteriophages are viruses which infect and lyse bacteria and as such can, and do, wreak havoc on industrial bioprocesses. Despite this there is no body of scientific or technical literature on the subject. Biochemical engineers and bioprocess technologists are seldom taught the basics of phage biology. Consequently, those charged with the operation of bioprocesses do not recognize the early signs of a phage infection, do not know how to carry out a post-infection clean-up, nor how to minimize or prevent future attacks. A phage attack can be catastrophic and result in lack of production for periods ranging from a few days to many months. Fortunately for the industry, yeast and fungal cultures do not suffer

Advances in Biochemical Engineering/
Biotechnology, Vol. 43
Managing Editor: A. Fiechter

from viral attacks. Fungal viruses are known but their infective properties are such that they do not cause widespread culture lysis.

There is little published information on the importance of phages in industrial bioprocesses, with only the dairy industry freely admitting to the problem [1–4]. Other processes known to be seriously affected are the bioproduction of acetone-butanol [5, 6] and monosodium glutamate [7] and processes utilizing actinomycetes [8, 9], and *Escherichia coli* and its close relatives. The microbial processing of milk products probably is the most important industrial process affected by phages.

2 The Biology of Bacteriophages

An understanding of the biology of bacteriophages is fundamental if infection is to be prevented or controlled. There are many specialized texts on the subject but only a few [10–12] are appropriate for the non-specialist.

2.1 Physical Properties of Bacteriophages

Bacteriophage particles exhibit considerable diversity in size and shape (Fig. 1). The vast majority of phages have a head-tail morphology, an architectural principle not found in other groups of viruses and which may be a reflection on the way bacterial viruses infect susceptible cells. Many different structural variations have been noted, e.g. contractile versus non-contractile tails, presence or absence of base-plates, collars, etc, but this morphological diversity is irrelevant as far as infection of industrial processes is concerned. On analysis, practically all phage

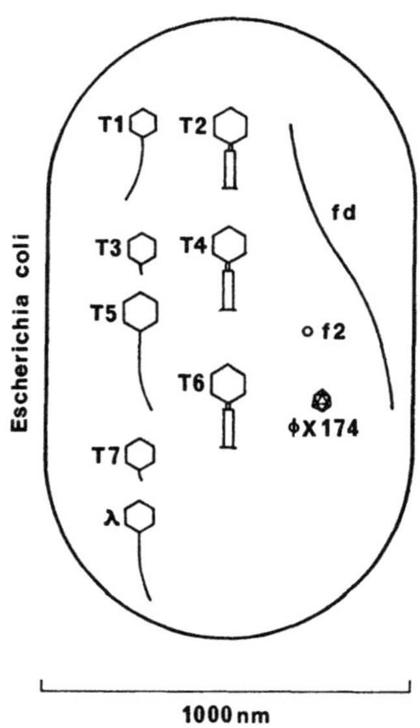

Fig. 1. Relative size of a bacterium (*Escherichia coli*) and an assortment of bacteriophages which infect it

particles consist of protein and nucleic acid only although some do have a lipid coat. The nucleic acid can be either DNA or RNA, but not both, and in tailed phages is located in the head.

The overall length of a phage is in the general range of 50–200 millimicrons and the head averages 50–90 mu in width. Thus phages are small enough to pass through most bacteriological filters which have pore sizes in excess of 0.2 microns.

The phage particle is essentially a survival mechanism designed to protect the phage genome from the rigours of the environment to which it is exposed when it destroys one host and before it infects another. It is clearly efficient otherwise phages would long ago have become extinct. Most phages will survive in cell lysates for long periods (months to years) at normal environmental temperatures provided they do not dry out. Many phages also are resistant to drying. Although freezing and thawing can reduce viral titres, survival is still very high. Bacteriophages are about as sensitive to heat as the majority of non-sporing bacteria. At 100 °C they are inactivated almost instantly. Between 65 °C and 85 °C inactivation rates can be determined conveniently under laboratory conditions. Below 65 °C some phages are inactivated very rapidly but most hardly at all. The medium in which phages are heated has a great influence on the rate of inactivation. Inactivation is most rapid in pure water; the addition of salts, especially calcium or magnesium, and proteins reduces the rate of inactivation considerably.

Many of the classical disinfectants active against bacteria also inactivate bacteriophages and can be used in plant clean-up. Thus compounds which react with proteins or nucleic acids should destroy phages. However, some of the newer surface-active agents may have little effect on the viability of phages. Under certain conditions phage particles may not exist as monodisperse units but as aggregates. The application of disinfectants to phage aggregates may not result in complete inactivation.

Some well-studied bacteriophages, e.g. coliphage T1, have acquired notoriety because of the ease with which they spread through laboratory complexes. What property or properties of a phage confer upon it the ability to be readily disseminated is not known. Apparent ease of spread may be related to drying. Alternatively it could be related to the ability to form aggregates and the impact aggregate size has on gravitational removal from aerosols. Unfortunately, those phages which most easily infect a bioprocess plant are those which spread the easiest, making their elimination more difficult. Many of them also are very heat resistant!

2.2 The Process of Infection

The process of infection begins when a bacteriophage particle undergoes a chance collision with a host cell. If the phage possesses an adsorption site that is chemically complementary to a specific receptor on the bacterial cell surface then irreversible adsorption occurs. The receptors for most phages are located on a cell wall although for some phages cellular appendages such a flagella or pili can act as adsorption sites. Following adsorption, the viral nucleic acid enters the cell but for most phages the mechanism whereby this occurs is not known.

Once inside the cell, the phage nucleic acid undergoes replication. As the number of phage nucleic acid molecules builds up inside the cell, viral genes are transcribed and translated and structural proteins are synthethized. Eventually the nucleic acid and protein assemble into complete phage particles. Coincident with phage assembly is the synthesis of a virally-specified lysozyme. This enzyme attacks the peptidoglycan layer of bacterial cell walls. The wall is progressively weakened until it is ruptured by the internal osmotic pressure of the cell, and the progeny phages are liberated into the environment along with the other contents of the cell.

The average number of new phage particles liberated by each infected cell is known as the *burst size*. The value of the burst size varies from 30–300 and the time from infection to lysis ranges from 20–100 minutes. The exact value of each depends on the particular host-phage system and the environmental conditions. The particles liberated by lysis can in turn infect other cells in the population with a repetition of the same cycle. Consequently, even if only a single infectious phage particle is introduced into a culture, practically the entire bacterial population may be destroyed in a few hours.

The last cycle of infection is particularly dramatic. When as many as 1–5% of the cells in the culture are lysing there are no obvious physical or physiological signs of infection. However, the phage particles released can infect all the remaining cells and shortly thereafter the entire culture becomes glass-clear. In aerated cultures excessive foaming also may occur as a result of release of host cell proteins into the growth medium. When the entire culture finally lyses the number of phage particles present is very high. Thus in a 10000 L culture whose density was 10^{10} cells ml^{-1} prior to lysis by a phage with a burst size of 100, the total number of phage particles will be 10^{19}. As will be seen later, the magnitude of this phage count makes clean-up very difficult.

2.3 The Genetics of Bacteriophage Resistance

The rate of spontaneous mutation at any genetic locus is of the order of 10^{-5}–10^{-8} per cell per generation. Thus as a culture increases in density it becomes increasingly heterogenous. If a bacterium is susceptible to a particular phage then high cell density bacterial cultures will contain a proportion of phage-resistant mutant cells. These mutant cells are easily isolated and can be shown to have an inheritable inability to adsorb the phage as a result of an alteration in the structure of their cell walls. There is a common misconception that such mutants provide a defence against future infections with the same phage. Nothing could be further from the truth. Just as bacteria mutate, so do phage. The rate of mutation of phage genes is similar to that of bacterial genes. If the phage population exceeds 10^8 particles, as it usually does (see above), host-range mutants will be present which can infect the cells resistant to wild-type phage. Further bacterial mutants resistant to the host-range phage mutant can be isolated but new phage mutants which infect it also can be isolated [13] (Table 1). Because one always is dealing with numbers of bacteria and phage in excess of 10^{10} it is virtually impossible to stop the alternate

Table 1. Infectivity of host-range (hr) mutants of a bacteriophage for wild-type and various resistant mutants (R1, R2) of the host bacterium

Phage	Bacteria		
	Wild-type	R1	R2
Wild-type	+	—	—
hr1	+	+	—
hr2	+	+	+

selection of host and phage mutants. An example of this in an industrial bioprocess has been reported [14].

Even where suitable phage-resistant mutants can be isolated following infection this is not usually a viable option. Most of the bacterial strains used in bioprocesses are selected for high yield of product. The requisite high yields are obtained only under precise culture conditions and any genetic alteration to the cell can reduce the yield. This is particularly true with phage-resistant mutants which result from cell wall alterations, many of which can affect cell permeability.

Although bacteriophages have been isolated for almost all known bacteria there is wide variation in the susceptibility of different strains of any one bacterial species. For example, phages infecting *E. coli* are extremely easy to isolate but a few strains of *E. coli* have been identified for which it has proved impossible to isolate phages [15]. A number of factors in combination, e.g. presence of multiple host-restriction and modification systems plus abnormal wall structure, probably contribute to the extreme phage resistance of certain strains [16].

In the case of the lactic streptococci, phage-resistance can be plasmid-encoded [17] and resistance can result from inhibition of phage adsorption, restriction of DNA and inhibition of phage maturation. In most instances the mechanism of resistance has been determined for only a few phage isolates: it remains to be seen if any of these resistance mechanisms will operate against a wide range of lactic bacteriophages or in unrelated bacterial species.

2.4 Bacteriophage Ecology

Bacteriophages, like all viruses, are obligate intracellular parasites. In order to multiply they need not only susceptible host cells but cells which are actively growing. Thus the numbers of a particular phage in the environment can be correlated with the numbers of its host cells. Not surprisingly, as the external environment warms up in spring and summer the numbers of bacterial cells and their associated phages increase [18]. As the external environment cools in autumn and winter the numbers of phage decline. However, many of these phage are not inactivated, at least not rapidly. Rather they adsorb to clay minerals [19] in soil and water as a result of seasonal changes in pH.

The physical location of a bioprocess plant can greatly influence the probability of phage infections occurring. Thus plants where soil microorganisms such as actinomycetes (antibiotics) or *Brevibacterium* (mono-sodium glutamate) are cultured on a large scale will be more susceptible to phage infections if located in the middle of an agricultural region. With the advent of genetic engineering, bioprocesses utilizing *E. coli* increasingly are being used. Such bioprocesses should not be conducted close to sewage disposal plants. Coliphages are present in high numbers in sewage and the aerosols produced by activated sludge plants can result in phage particles being dispersed many miles downwind. Even plants in agricultural areas are not safe if the local farmers spread sewage sludge or farmyard waste on their land.

At a more local level, plant operators should ensure that pools of the production organism do not occur below sampling ports or in waste channels. Such locations are breeding grounds for unwanted phages.

3 Recognizing a Bacteriophage Infection

As indicated earlier, if a phage is multiplying in a bacterial culture then the number of infected cells can increase from a few percent to nearly 100% in as little as 30 minutes. Even the best-trained operator is unlikely to notice any changes in the bioprocess operating parameters when less than 5% of the cells are infected. However, during the last cycle of infection the cells metabolic rate will decrease sharply. This is most easily detected as a sharp increase in the dissolved oxygen tension. If mass spectrometry is being used to analyse exit gases then a rapid decrease in CO_2 concentration also should be noted. If the pH of the growth medium is being controlled then a decrease in the rate of alkali addition might be noted [9].

As the cells begin to lyse the turbidity of the culture will decline. On line monitoring of culture density is unusual but if plant operators suspect a phage infection they should monitor turbidity every few minutes (the reason will become clear later). Because lysing cells liberate a lot of protein the amount of foam will increase rapidly as will the consumption of anti-foam [9].

As with any contaminant, bacteriophages result in clearly observable differences in the temporal profiles of the various culture parameters being monitored. The difference between phages and other contaminants is the rapidity with which they cause very large deviations from normality.

4 Cleaning Up a Bacteriophage Infected Plant

As soon as a phage infection is suspected the bioreactor should be shut down immediately. As noted earlier, in a 10000 L vessel there can be as many as 10^{19} phage particles although this will depend on the cell density when lysis occurs. Continued aeration of the culture medium will only disseminate large numbers of phage particles into the environment of the bioprocess plant making clean-up far more difficult. This dispersal will be facilitated if excess foaming of lysed cells has overwhelmed the anti-foam system. Spread of phage particles will be facilitated if the exit gas lines from a number of bioreactors are manifolded. Although it is fundamentally bad practice to design a bioprocess plant with the air intakes and air exits in close proximity, this author has seen numerous examples. In such situations the entire aeration system, e.g. compressor, filters, etc, also will be contaminated.

The first task is to develop an assay for the phage particles. This is very easy to do and should take only one day [20]. The second task is to heat sterilize the contents of the bioreactor. The exact time-temperature profile will depend on the design of the bioprocess plant but clearly the hotter and the longer the treatment, the better. Before discarding the reactor contents, the newly developed phage assay procedure should be used to ensure the absence of viable phage.

Because bacteriophages can pass through most, if not all, bacteria-proof filters, any pipework connected to the fermenter, however remotely, needs to be extensively steamed. This includes air inlet lines both upstream and downstream of the compressor. Much more rigour in sterilization of pipework and other plant is needed at this stage than would be the case if a bacterial contaminant had predominated in the culture. Where possible liquid samples or surface swabs should be taken and assayed for phage.

The next step is to decontaminate the exterior surfaces of all pipes and plant. The ease with which this can be done will depend on the size and design of the plant. Where possible the entire plant should be washed down with water, which is as hot as is practicable. Again, samples of washings should be checked with the phage assay procedure to determine the extent of contamination and the efficacy of its reduction, particularly if open drains exist in which pools of liquid might collect and serve as reservoirs for infection. Nor should cooling ponds be ignored.

5 Prevention of Bacteriophage Attack

With an understanding of the biology of bacteriophages it is possible to design, build and operate a bioprocess facility in which the probability of bacteriophage infection will be very low. The extent to which this can be done will depend entirely on whether a new plant and process is to be operated or whether an existing plant and process is to continue.

5.1 Facility Location

If a new facility is being built consideration should be given to its location. If enteric organisms are to be grown the plant should be sited well away from sources of coliphages. If soil organisms are to be grown then the facility should not be too close to cultivated land. The local weather also can have an effect. The normal British climate with its regular rainfall means that airborne phages will be "grounded". The long dry summers in the US Midwest are conducive to the spread of soil bacteriophages in dust.

5.2 Plant Design

As with other contaminants, bacteriophages usually gain access to the culture medium via the air handling system. Whereas bacteria and othe microbial contaminants can be excluded by careful selection and maintenance of an inlet air filter this is not necessarily the case for bacteriophages. The size of bacteriophages is such that they will pass through membranes whose pore size is small enough to remove bacteria. Many of the air filters in common use are not absolute filters but rely on the tortuous path of gas molecules through a packed bed of fibres to remove contaminating microbes. The performance of such fibrous filters depends on air velocity and particle size [21]. At the high flow rates used in most large-scale microbial cultures inertial forces are crucial to particle removal. Unfortunately bacteriophages are too small to have a significant moment of inertia and are not removed [22].

The choice of air compressor can facilitate sterilisation. If compression is done adiabatically the subsequent rise in temperature in the compressed air *may* be sufficient to inactivate any phages present. The efficiency of this dry heat sterilization can be improved by passing the air through a lagged retention chamber. Bartholomew et al. [7] have presented data on the exposure time/temperature combinations necessary to reduce the probability of phage penetration to less than 10^{-15}. Alternatively the compressed air can be flash heated to at least 90–95 °C by steam injection and then cooled to remove condensate before filtration. Although not an easy solution from a design point of view it does afford the best protection.

The air inlet should be well-separated from the air outlet, to prevent circulation of any contaminants [19]. Raising the air inlet as high as possible also may be beneficial. A series of coarse filters should be fitted to remove dust particles and other debris and, if possible, a hydrophobic filter to remove phage being transmitted in water droplets. Some commercially-available filters, e.g. Pall Emflon, may be particularly effective at removing bacteriophages. Maintenance of these filters must be rigorous.

The other design feature of importance relates to waste disposal. There should be no pools of liquid containing live culture anywhere in the vicinity of the bioprocess plant. All sampling points should be fitted with tun dishes and any spillage piped to a kill tank. There should be no open drains, no drains with horizontal sections where culture medium containing live organisms can collect, and all drains should flow to a kill tank. Finally, downstream processing operations should be designed such that there is no waste containing any live organisms.

5.3 Strain Selection

Where a microbial process has been in operation for some time there is little that can be done to select a naturally phage-resistant strain. Where a new process is being developed careful tought should be given to the choice of starting organism. For example, *E. coli* mutants which have lost most of the "dispensable" layers of the cell wall are likely to be less susceptible to phage infection than wild-type strains since most of the potential phage receptors have been eliminated.

If a genetically-engineered organism is to be cultured then thought should be given to the deliberate inclusion in the genetic construct of one or more host-restriction and modification (HRM) systems. HRM is a mechanism whereby foreign nucleic acids which gain entry to a cell can be destroyed. Most HRM systems reduce the infectivity of bacteriophage DNA by a factor of 10^2 to 10^4 and so can provide a defence against bacteriophage attack. The genes for many different HRM systems have been cloned and can be inserted into other bacteria almost at will.

Another advantage of genetically-engineered organisms is that they offer the potential for strain rotation. In the dairy industry, the recurring problem of phage infection can be minimised by constantly changing the bacterial strains used to ones with different phage sensitivities [1, 2]. This cannot be done with most of the other traditional bioprocesses where there is a single long lineage of improved strains, all derived from the same parent. However, if the key genetic information is plasmid borne, as in genetically-engineered organisms, it is possible to move the plasmid to host strains with different phage sensitivities. Although similar yields of product cannot be guaranteed, strain rotation in this way can minimise the downtime following a serious phage attack.

5.4 Operating Practice

There are three simple procedures which should be introduced in all microbial processes which are susceptible to phage attack. The first of these is to develop an emergency control procedure which can be implemented in the event of a phage infection. The key aspects of this procedure were detailed in Sect. 4.

Second is the development of a suitable phage assay procedure. Once this procedure is established, samples for phage assay should be withdrawn regularly from all culture vessels when the density of cells within them is such that there is visible turbidity. Hopefully, no phage will ever be detected. If it is, the emergency control procedures can be implemented immediately. Environmental samples also should be examined for the presence of bacteriophages capable of infecting the production strain. These samples should be taken only from aqueous environments, e.g. cooling ponds, semi-permanent puddles in the grounds, waste-disposal systems, etc. A build-up of phages in the environment should be taken as an indication that a plant clean-up is required if infection of the bioreactors is to be avoided.

Finally, plant operators should be taught to recognize the key symptoms of a phage infection (Sect. 3) and to take immediate remedial action.

6 References

1. Daly, C (1983) Antonie van Leeuwenhoek 49: 297
2. Lawrence, RC, Thomas, TD (1979) The Fermentation of Milk with Lactic Acid Bacteria. In: Microbial Technology: Current State, Future Prospects (eds. Bull, AT, Ellwood, DC, Ratledge, C), P. 187, Cambridge, Cambridge University Press

3. Klaenhammer, TR (1984) Advances in Applied Microbiology 30, 1
4. Vedamuthu, ER, Washam, C (1983) Cheese. In: Biotechnology, Volume 5, Food and Food Production with microorganisms. (eds. Rehm, H-J, Reed, G), p 231, Basel, Verlag Chemie
5. McCoy, E, McDaniel, LE, Sylvester, JC (1944) J. Bacteriol. 47, 433
6. Ogata, S, Hongo, M (1979) Advances in Applied Microbiology 25, 241
7. Barthomomew, WH, Engstrom, DE, Goodman, NS, O'Toole, AL, Shelton, JL and Tannen, LP (1974) Biotechnology and Bioengineering 16, 1005
8. Casida, LE Jr (1968) Industrial Microbiology. New York, John Wiley
9. Hongo, M, Oki, T, Ogata, S (1973) Phage contamination and control. In: The Microbial Production of Amino Acids (eds. Yamada K, Kinoshita S, Tsunoda T and Aida K) p67. New York, John Wiley
10. Douglas, J (1975) Bacteriophages. London, Chapman and Hall
11. Dimmock, NJ, Primrose SB (1987) Introduction to Modern Virology. Oxford, Blackwell Scientific Publications
12. Smith, KM, Ritchie, DA (1980) Introduction to Virology. London, Chapman and Hall
13. Baylor, MB, Hurst, DD, Allen, SL, Bertani, ET (1967) Genetics 42, 104
14. Oki, T, Matsui, T, Ozaki, A (1967) Agric. biol. Chem. 31, 861
15. Primrose, SB unpublished observations
16. Primrose, SB, Seeley, ND, Logan, KB, Nicholson, JW (1982) Applied and Environmental Microbiology 43, 694
17. Daly, T, Fitzgerald, G (1987) Mechanisms of bacteriophage insensitivity in the lactic streptococci. In: Streptococcol Genetics (eds. Ferretti JJ and Curtiss III, R) p. 259 Washington, American Society for Microbiology
18. Primrose, SB, Seeley, ND, Logan, KB (1981) The Recovery of Viruses from Water: Methods and Applications. In Viruses and Waste Water Treatment (eds. Goddard, M, Butler, M) p. 211, Oxford Pergamon Press
19. Stotzky, G, Schiffenbauer, M, Lipson, SM, Yu, BH (1981) Surface Interactions between Viruses and Clay Minerals and Microbes: Mechanisms and Implications. In: Viruses and Waste Water Treatment (eds. Goddard, M, Butler, M) p. 199, Oxford, Pergamon Press
20. Primrose, SB, Seeley, ND, Logan, KB (1982) Methods for the Study of Virus Ecology. In: Experimental Microbial Ecology (eds. Burns, RG, Slater, JH) p. 66, Oxford, Blackwell Scientific Publications
21. Humphrey, AE (1960) Advances in Applied Microbiology 2, 301
22. Sadoff, HL, Almlof, JW (1956) Industrial and Engineering Chemistry 48, 2199

Intracellular Lytic Enzyme Systems and Their Use for Disruption of *Escherichia coli*

R. L. Dabora[1] and C. L. Cooney[2,3]

This article focusses on lytic enzyme systems available in *E. coli* and their potential use for cellular disruption. In the systems described here the genetic information for lysis would be carried within the microbial host, either integrated or naturally occurring on chromosomal DNA, or on extra-chromosomal elements such as plasmids. Each microbe would carry complete information for endogenous enzymatic lysis, and lysis would occur in a controlled manner after being triggered by an external factor such as temperature or inducer addition. The lytic systems explored in this review include the autolytic enzymes, colicin lytic enzymes, and bacteriophage lytic enzymes from phage phiX174, T4, lambda, MS2 and Qβ. Many of the colicin lytic enzymes and all of the bacteriophage lytic enzymes described here have been cloned, and in some instances examined as cellular disruption methods. None of the *E. coli* autolytic enzymes have been cloned, but information pertinent for use as a disruption method is described.

1 Merck and Co., Inc., P.O. Box 7, Elkton, VA 22827;
2 Department of Chemical Engineering and Biotechnology Process Engineering Center, Massachusetts Institute of Technology, Cambridge, MA 02139;
3 To whom correspondence should be addressed

Advances in Biochemical Engineering/
Biotechnology, Vol. 43
Managing Editor: A. Fiechter

1 Introduction

Many problems of downstream processing are created by the biological system and the methods of culturing. Many of these same problems often can be resolved through genetic manipulation and/or biochemical control. One example is the recovery of intracellular proteins from a bacterial cell, namely *Escherichia coli*. The usual approach to the initial release of protein is the use of mechanical or chemical methods which exposes proteins to shear and creates a polydispersed mixture of cell debris. An alternative approach is the design of a process utilizing the natural autolytic capabilities of the cell by placing them under external control by an environmental factor such as temperature or chemical induction. Alternatively, lytic enzymes from bacteriophage or colicin systems could be used in a similar manner. In both cases, lysis of the bacterial cell occurs from lytic enzymes produced within the cell, affording gentle, controlled, and efficient cellular disruption.

Although the use of lytic enzymes as an alternative method of cell disruption has been suggested, there are few studies examining the potential of these systems [1]. The use of these enzymes for large-scale disruption in plasmid or other cloned systems has not been fully explored, in part because the lytic genes have not been identified or characterized. Also, special growth and media conditions, as well as lysis triggering may be required for efficient lysis. This review focusses on several lytic systems available for use in *E. coli*. While use of these systems may not be a replacement for existing cell disruption methods, they offer an alternative to, or may be used in conjunction with, current practices. Analysis of lytic systems may aid in the design or improvement of an alternative, integrative, and efficient cellular disruption process.

2 Methods of Microbial Cell Disruption

A number of different methods of microbial cell disruption are in use today; these are described in several reviews on the theory and practice of both small and large-scale disruption processes [1, 2, 3, 4]. In general, these processes can be divided into two classes: physical and chemical. For large-scale processes the primary physical methods include liquid shear (French press, homogenization), agitation with abrasives (such as glass beads), and solid shear. The primary large-scale chemical methods include osmotic shock, detergents, alkali, and enzymatic methods such as lysozyme addition. Liquid shear is the most commonly reported large-scale microbial cell disruption method [3].

Constraints placed on the disruption method include the choice of microorganism, the equipment available, and the product being recovered. Since continued emphasis is being placed on integrative recovery and purification processes [2], the procedure choice may not be one which releases the product in highest yield, but may be one which preferentially releases the desired product over other proteins present, thus, effecting purification.

The use of enzymes and particularly cloned lytic enzymes as alternative disruption methods has been considered only in an exploratory manner. There are only a few references to the use of recombinant DNA for potential improvement or replacement of current disruption unit operations [5, 6, 7]. Some of the potential enzymatic methods for cell disruption processes which could be adapted to large-scale, but have not been fully developed, include the use of: autolytic enzymes, colicin lytic enzymes, and bacteriophage lytic enzymes.

Much of the review of the above processes will focus on *E. coli*. The motivation for studying the potential of enzymatic alternatives in this microorganism is the following: the genetics and recombinant DNA techniques of *E. coli* are well characterized and defined [8], several lytic enzyme systems have been identified in this microorganism although the potential for disruption processes has not been fully developed [1], it is a good model system for intracellular protein recovery, and many recombinant-derived proteins are currently produced in *E. coli* [9]. Development of alternative cell disruption methods in this microorganism may therefore be of industrial importance.

3 Autolytic Enzymes

3.1 Structure of the *E. coli* Cell Envelope

The cell envelope of *E. coli* and of all Gram negative bacteria is made up of several layers: an inner or cytoplasmic membrane, cell wall, periplasmic space, and an outer membrane [10]. The cell wall is particularly important for maintaining the shape and integrity of the cell. Degradation or destruction of the wall by an intra or extracellular method will usually result in cell lysis [11].

The *E. coli* cell wall is known to consist of a peptidoglycan layer containing chains of alternating *N*-acetylglucosamine and *N*-acetylmuramic acid residues in β-1,4 linkages. These long glucan chains are crosslinked by short peptides which are attached to the carboxyl groups of *N*-acetylmuramic acid [10]. The cell wall is thin and is not highly cross-linked (only 50% of the peptide chains are crosslinked [12]). For this reason only limited degradation of the wall should be necessary for cell lysis [13].

3.2 Types of Autolysins

A number of enzymes present in *E. coli* which can degrade the peptidoglycan layer have been identified. These are the murein hydrolases or "autolytic enzymes". Autolytic enzymes which have been identified in *E. coli* include: muramidase, glucosaminidase, amidase, endopeptidase, transglycosylase, and carboxypeptidase [14, 15, 16].

3.3 Use of Autolytic Enzymes in Cell Disruption

The use of autolysis as a method of cell disruption is not common, particularly in *E. coli.* Several microorganisms such as *Bacillis, Clostridia, Staphylococcus aureus,* and yeast are more readily autolysed. Yeast autolysis in particular is practiced on an industrial scale; yeast autolysates are used as: a source of protein [17], a source of amino acids [18], or a food flavoring component [19]. The primary reason for using autolysis in yeast is that the cell wall is very thick and rigid due to the beta-glucan component of the cell wall. Yeast contain *endo*-β-(1,3)-glucanases as well as nonspecific *exo*-β-glucanases which are able to hydrolyze the glucan layer [17]; these enzymes can be induced by aging, increasing temperature, or adding organic compounds such as butanol, ethyl acetate or toluene [20].

There is one example of the use of ampicillin-induced autolysis in *E. coli* for the recovery of protein products [21]. Human leukocyte interferon was recovered after autolysis of the culture by the addition of 75 mg/L ampicillin and precipitation of the product by lowering the pH to 1.9.

3.4 Role of the Autolysins

A variety of different roles for the autolytic enzymes have been proposed. One is their role in growth, particularly in cell wall synthesis, where it is believed the cell wall must be broken to allow for further cell wall biosynthesis. Evidence from this system in *Streptococcus faecalis* suggests that this is clearly true for this microorganism [22]. However, it would be expected from this hypothesis that autolysis-deficient mutants would not be able to grow and divide normally. In fact, autolytic mutants have been described in which growth and division occur normally [23], making it difficult to clearly identify cell wall synthesis as the primary role of the autolysins.

Another role suggested for the autolytic enzymes has been in cell division and motility. In several bacterial strains, decrease or loss of autolytic activity results in a failure of cell separation and/or loss of motility [22]. In organisms such as *B. subtilis* the decrease or loss of autolytic activity can result in long chain formation and loss of flagella while in *Staphylococcus*, large clumps of cells are detected.

Autolytic mutants which do not exhibit altered growth or division have been isolated. For this reason it has been proposed that the autolytic enzymes are separate from the enzymes involved in cell wall synthesis and growth and that the autolytic enzymes are responsible only for the pathological response to β-lactam antibiotics or other autolytic triggering systems [24]. There remains an incomplete understanding of the involvement of the autolytic enzymes in various cell processes.

3.5 Triggering of Autolysis in *E. coli*

There are three ways to trigger autolysis in *E. coli*. These include: subjecting the cells to osmotic shock, inhibiting peptidoglycan synthesis, and infection with phage phiX174 or induction of the cloned phiX174 lysis gene.

Different methods have been used to bring about autolysis by osmotic shock; these include: the addition of 1.0 mM EDTA, distilled water, 0.5 M sodium acetate, or distilled water followed quickly by 0.5 M sodium acetate [25]. In all cases there seems to be a correlation between the extent of autolysis and the extent of peptidoglycan degradation. Within 2 h of autolysis onset, there was a 20–60% decrease in optical density and 40% of the cell wall label was released into the supernatant [25].

Several methods have been used to inhibit cell wall synthesis and hence trigger autolysis. One method is with antibiotics. Those which have been reported to be effective are D-cycloserine, moenomycin and β-lactams (specifically cephaloridine, ampicillin, and penicillin G). With the use of β-lactam antibiotics, autolysis is known to proceed after penetration of the antibiotic through the outer membrane and binding of the antibiotic to penicillin binding proteins (for review see Ref. [16]). A second method of inhibition of cell wall synthesis is by depriving the cell of a cell wall component such as diaminopimelate (DAP) or glucosamine in a strain which is auxotrophic for that component. Third, synthesis can be inhibited by transferring a strain which is temperature sensitive in some cell wall biosynthetic step to the nonpermissive temperature.

The third way to trigger autolysis in *E. coli* is by infection with the phage phiX174 or by inducing the cloned phage lysis gene *E* [26], although the role of the host cell autolytic enzymes in this lytic system has not been elucidated completely. While a functional autolytic system appears to be required for lysis by cloned phiX174, the autolysis triggering pathway may be different than that induced by β-lactam antibiotics [27]. Lysis by bacteriophage phiX174 gene *E* is discussed in further detail in a later section.

3.6 Characteristics of Autolysis

The rate and extent of autolysis is dependent upon growth conditions as well as the induction method used. The highest rates of autolysis are seen in exponential phase cultures using EDTA, cephaloridine, or moenomycin as the triggering agents [28]. The initial killing and lysis rates were comparable with all three systems [23]. A 99% decrease in viability was observed within 60 min of induction of exponentially growing cultures. Experiments with chemostats indicate that longer generation times are associated with increasing tolerance to autolysis induced by β-lactam antibiotics as measured by both lysis and loss of cell viability [29]. The rate of killing of *E. coli* by β-lactam antibiotics is strictly proportional to the rate of bacterial growth.

All three methods of inducing autolysis have common features. First, all require exponentially growing cells to bring about lysis. Second, autolysis can be inhibited by the addition of 10 mM Mg^{+2} to the culture medium [25, 26, 28, 30].

One exception to the growth requirement for autolysis induction is in *relA* mutants [31]. When placed under non-growth conditions (for example, by amino acid starvation), most autolysin inducers are not capable of triggering autolysis. However, several compounds (Nocardicin A, MT141, CGP14233, and imipenem) were capable of triggering autolysis in *relA* mutants which had been starved up to 30 min. Analysis of the cell wall products revealed that the autolysins of non-growing versus growing cultures were very similar.

One difference between the autolysin triggering methods is the extent of peptidoglycan degradation after autolysis induction. Autolysis by osmotic shock is accompanied by extensive peptidoglycan degradation. It was found that 65–70% of the peptidoglycan was degraded in EDTA-induced cells [32], 45% in sodium phosphate treated cells [28], while only 20–35% was degraded in cephaloridine or moenomycin treated cells [33]. In cells infected with phage phiX174, very little peptidoglycan degradation occurs [34]. The results indicate that extensive peptidoglycan degradation is not a prerequisite for efficient cell lysis [32].

The primary problem with using the autolysins as a method for cell disruption is the need for exponentially growing cells. Efficient cell lysis may not be achievable with high cell densities. In addition, the triggering method may require resuspending cultures in large volumes of aqueous solutions or the undesirable addition of antibiotics. Special strains such as DAP^- mutants could be used, but induction of autolysis would require addition of DAP for growth initially, and the removal of the component for lysis induction. Since the autolytic enzymes act upon the cell wall it may be possible to clone one or more enzyme and induce lysis through induction of a specific autolysin. This requires the elucidation of the genes and/or enzymes most useful for this purpose.

4 Colicin Lytic Enzymes

The colicins are a class of antibiotic proteins some of which are produced in *E. coli*. There are different classes of colicins and they are characterized by different modes of action [35]. Colicins E1, A, Ia, Ib, and K form ion-permeable channels in the cytoplasmic membrane causing secondary effects such as decrease in active transport, loss of motility, reduction in ATP levels, and the arrest of protein and nucleic acid synthesis (for reviews see Ref. [36, 37]). Colicin E3 and Cloacin DF13 inhibit protein synthesis by ribosome inactivation while Colicin E2 causes degradation of bacterial DNA through a nonspecific endonuclease (reviews in Refs. [36 and 38]). Similarities which have been found between the three different mechanisms include: the production and release of colicin upon induction with mitomycin C or UV light, the immunity of the colicin producing cells to the colicin they produce, and sensitivity to other colicins (see review in Ref. [39]). The reversibility of killing action by trypsin suggests that the primary site of action of the colicins is the bacterial membrane [35]. *E. coli* mutants have been isolated which prevent colicin production. One class of mutants lacks a recepter on the outer membrane; the other class shows a tolerance to the lethal effects of colicins and is characterized by changes in the outer membrane [37].

The lethal and lytic effects which occur during colicin release have been attributed to the presence of a lethality gene within the colicin operon. The *kil* gene of ColE1 [40], the *H* gene of Cloacin DF13 [41, 42], *celB* gene of ColE2 [43], *hic/celC* genes of ColE3 [44, 45], and *cal* from ColA [46] have all been identified. They all show some homology in their DNA sequences [44, 47, 48, 49]. The site of action of the lethality genes seems to be the bacterial membrane; no activity against the cell wall has been reported.

Several of the lysis genes have been cloned under controllable promotors. The *celB* gene of ColE2, when placed under *lac* promoter control, was sufficient for lysis upon induction [50]. The gene seemed to be very lethal even in strains with *lacI*^{q} background. The culture had to be cultivated at 30 °C (a growth temperature of 37 °C was lethal); temperature shift to 42 °C resulted in lysis of the culture as detected by halos surrounding colonies on X-gal (5-bromo-4-chloro-3-indolyl-β-D-galactosidase) plates. When grown in liquid culture, turbidity decreased within one to two hours after induction. The release of intracellular β-galactosidase was slow; only 10–20% of the total activity was released after 3 h of induction. In contrast, Colicin E2 comprised 60–70% of the total cell protein released into the culture [43]. The authors suggest that the preferential release of colicin and other cell proteins (such as alkaline phosphatase) over intracellular β-galactosidase may be the result of altered membrane permeability after *celB* induction [43]. Although the reason for selectivity of release is not known, it is possible that size, hydrophobicity, compartmentalization, or charge of the protein is an important characteristic. Additional evidence for altered membrane permeability is given by the increased sensitivity of induced cultures to exogenously added lysozyme [43]. Although lysozyme is normally excluded by the outer membrane, mitomycin C-induced cultures showed increased sensitivity to lysozyme action as measured by a decrease in turbidity.

Other characteristics of the culture upon induction of the *celB* gene include the activation of *E. coli* chromosomal phospholipase, and elimination of turbidity decrease (but not colicin production or release) in the presence of 10 to 20 mM Mg^{2+} [43]. Colicin release and the decrease in culture turbidity upon *celB* induction therefore may not be coupled. No host cell genes seem to be involved in the lysis mechanism as established by the lack of *E. coli* mutants recovered which are resistant to *celB* gene product action [50].

Induction of the cloned *kil* gene from the ColE1 operon under *lac* promoter control shows some physiological similarities to that of *celB* [51]. Strains containing plasmids with the cloned *kil* gene also could not be grown at temperatures greater than 30 °C. Within thirty minutes of induction, culture turbidity and viability decreased (greater than 95%), there was a decrease in protein synthesis, and a decreased ability to accumulate α-methylglucoside (a measure of active transport). In addition, the decrease in turbidity was eliminated in medium containing 10 or 20 mM Mg^{2+}. Several classes of *E. coli* mutants were isolated with increased resistance to *kil* function, suggesting the involvement of host cell proteins with the lytic effects of the *kil* gene product. These mutants also showed altered sensitivity to various detergents and antibiotics. The mutants were normal for phage lambda and T4 production, suggesting different lysis mechanisms for the two systems [51].

The *hic* gene from Colicin E3 has been cloned and characterized under control of its own promoter. The N-terminal region of the *hic* gene product is hydrophobic and may therefore associate with the membrane. A twenty-fold reduction in viability occurs within ninety minutes after mitomycin C induction of the *hic* gene. Release of cellular proteins appears to be nonspecific [44].

Other colicin lytic genes which have been identified are gene *H* of CloDF13 [41] and the *lys* gene of ColE3 [45]. The gene product of gene *H* (6 kDa) was found to associate with the membrane. This protein contains a hydrophobic region in its C-terminus. High concentrations of gene product *H* in mitomycin-induced cells bring about cell lysis [42]. Gene *H* has been shown to be necessary for the both the lytic and killing effects upon induction.

Autolytic mutants defective in EDTA or β-lactam induced lysis were examined for the effect of the ColA lysis gene. It was found that lysis and production of colicin A occurred normally in these mutants suggesting alternate lysis pathway systems [52].

One interesting practical use of the colicin lysis genes is in the construction of an excretion vector in *E. coli* [53]. Using this vector the cloned penicillinase from *Bacillus* sp. was excreted along with the periplasmic enzyme alkaline phosphatase. Up to 97% of the total penicillinase activity was excreted into the culture medium, while no β-galactosidase activity was detected in the medium [54]. Excretion was believed to occur due to weak expression (activation) of the *kil* gene on the plasmid. No cell lysis was observed. In a separate study, human growth hormone attached to a *Bacillus* signal sequence along with weak expression of the *kil* gene caused excretion of growth hormone through the outer membrane. The signal sequence allows passage of the protein through the inner membrane while weak expression of *kil* causes permeability of the outer membrane without lysis [55]. Using this excretion vector 11.2 mg/L (55%) of the human growth hormone was recovered in the culture medium after 24 h of cultivation. Forty-two percent of the total remained in the periplasm. The excreted protein appeared to be processed correctly and showed similar activity to authentic human growth hormone.

Problems which may be encountered with use of the colicin lytic genes include high lethality and instability. Strains containing cloned genes under *lac* promoter control show instability at high growth temperatures (over 30 °C) in addition to loss of viability even in the absence of induction. At the present time, the mode of action of many of these enzymes has not been elucidated, making control of their action difficult.

Promising results using this system include the differential release of proteins. In addition, the increased sensitivity to lysozyme makes use of these genes in conjunction with other chemical or physical means of disruption a viable alternative. Additional information including the study of factors controlling differential release may yield valuable information regarding both the mechanism of action and possibility of use of the colicin lytic enzymes as a cell disruption method.

5 Bacteriophage Lytic Enzymes

The first suggestion for the use of bacteriophage in a cellular disruption process in *E. coli* utilized lysis from without by the bacteriophage T2 [56]. Two *E. coli* enzymes, L-lysine decarboxylase and L-arginine decarboxylase, were recovered in high yields with high specific activities using this phage. The phage lysis method was found to be more effective for the recovery of these particular proteins when compared with several other cell disruption methods including autolysis, lysozyme treatment, and grinding methods. Other phage lytic systems which have been studied are described below.

5.1 Bacteriophage Lambda

Phage lambda is known to bring about rapid lysis of an infected host cell. It has been determined that three phage gene products are necessary for cell lysis in lambda-infected *E. coli*: S, R, and Rz [57]. Gene products R and Rz are believed to degrade the host cell wall while gene product S appears to be responsible for altering the cytoplasmic membrane, thus allowing gene products R and Rz to reach the cell wall layer [58]. The R gene product has been identified as a transglycosylase [59] while the Rz product may be an endopeptidase [57]. Gene product S has been localized primarily to the inner membrane upon production [60].

The lysis region containing the *S*, *R*, and *Rz* genes has been cloned and put under control of the *lac* promoter [61]. Induction of the promoter by the inducer isopropyl-β-D-thiogalactoside (IPTG) was sufficient to bring about rapid lysis within forty minutes after induction. It has been determined that both S and R gene products are essential for lysis via the cloned lysis genes; gene product Rz is only necessary for lysis in medium containing high Mg^{2+} concentrations [62]. All information for lysis of bacteriophage lambda is contained within its lysis region; no *E. coli* genes have been found which influence lysis once the decision for lysis has been established [63].

The S gene protein product appears to be involved with the transport of gene products R and Rz to the periplasm. It has been found in S^- mutants that large quantities of bacteriolytic (endolysin) activity accumulate within the cell [61]. Both R and Rz gene products appear to have no effect on lysis when constrained to the cytoplasm.

The cloned lambda lysis region, defective in gene product S production ($S^- R^+ Rz^+$), has been investigated as a method for small-scale cell disruption [6]. Induction of lysis genes *R* and *Rz* followed by freeze-thawing, results in a similar lysis efficiency (measured by β-galactosidase release), as using a French press. It is suggested that freeze-thaw treatment in these cultures causes similar alterations to the membrane as S gene product function would, allowing effective action of gene products R and Rz on the cell wall. Both logarithmic phase and stationary phase cultures showed similar efficient release. In fact, no additional β-galactosidase activity was recovered when induction and freeze thaw was followed by French press or toluene-SDS treatment. Although this freeze-thaw method may not be conducive to large-scale processes, the addition of toluene-SDS after induction may be useful for efficient lysis using cultures of high cell density. The successful recovery of enzyme activity after induction of stationary phase cultures is particularly encouraging. It is not known if induced cultures are weakened and therefore more easily disrupted by French press or other mechanical methods.

Premature lysis after induction of cloned genes ($S^+R^+Rz^+$) can be brought about by the addition of cyanide (10 mM), chloramphenicol (40 μg/ml), or chloroform (2%) [61]. Lysis occurs almost immediately after the addition of these compounds at 15 or 28 min after induction; lysis normally occurs within 40 min. In cultures containing plasmids with defective *S* function ($S^-R^+Rz^+$) only chloroform is able to bring about the premature lysis effect. The addition of these compounds has the potential for shortening lysis times during cell disruption by the phage lambda lysis genes.

In addition to lysis, viability of the host cell upon induction is *S* gene-dependent. In fact viability decreases within only 8 min after induction, and before the onset of lysis [62]. Viability has been shown to decrease in $R^-Rz^-S^+$ clones indicating that lysis is not required for a decrease in viability [61]. The percentage of the population surviving lysis induction is dependent upon the inducer concentration. At high concentrations of inducer (1×10^{-3} M IPTG) only 3.6% of the population had survived 60 min after induction [62].

In addition to using cloned lambda lysis genes as a method for cell disruption, Auerbach and Rosenberg [5], have patented the use of *E. coli* strains containing defective temperature sensitive lambda lysogens as a method for cell disruption. These strains contain a lambda prophage which lacks genes necessary for replication or structural protein assembly; functional phage therefore cannot be produced. The lambda lysis genes of the prophage are under temperature sensitive control through the use of the lambda *pL* promoter and the cI_{857} repressor. The strains grow normally at permissive temperatures (less than 38 °C) and lysis is induced in mid-log phase by temperature shift to 42–44 °C. Alternatively the temperature shift can be very short (5 min) followed by cooling of the culture to 2 °C–38 °C. Lysis still occurs at these lower temperatures, making recovering of labile proteins possible. Recovery of product can be enhanced by concentrating cells before induction, changing the osmotic strength of the medium after induction, or by using mechanical methods such as agitation in conjunction with temperature induced lysis. The information for lysis is contained on an external plasmid, or integrated into the *E. coli* chromosome.

Using the temperature sensitive lambda lysogens, 95% of cloned *E. coli* LT-B antigen was recovered in yields of 8.5% of total cellular protein in a 10 L fermenter 6 h after induction by temperature shift to 42 °C [5]. During the process, the viscosity increased within 2–4 h after induction indicating lysis of the culture, and then decreased within 4–6 h due to the action of nucleases. The decreased viscosity allowed ultrafiltration of the sample to remove both cellular debris and cells which had not lysed.

Examples provided in this patent as well as other examples of the use of cloned lambda lysis genes, suggest that there is potential for this method as an effective and scaleable method for cell disruption. Efficient intracellular protein release can be carried out using the lysis region cloned on a plasmid or integrated into the chromosome. Integration of a defective lambda prophage allows the stable maintenance of lysis information in the absence of selection, without production of phage, while the use of temperature is an effective, controlled, and facile method for the induction of lysis.

5.2 Bacteriophage T4

Several similarities have been found between the lytic genes of bacteriophage T4 and lambda. Lysis by phage T4 requires the action of two gene products: e and t. Gene *e* encodes for a lysozyme which has been identified as a muramidase [64], while gene *t* seems to have a similar function to the lambda *S* gene; it is required for lysis, but does not appear to have lysozyme activity. Phage are formed by t^- mutants, but lysis of the *E. coli* host does not occur except by addition of chloroform [65]. Gene *t* is required for the cessation of cellular metabolism which occurs during lysis [66] and is believed to degrade or alter the cytoplasmic membrane thus allowing gene product e to reach the periplasm and gain access to the cell wall [65].

The T4 lysozyme gene *e* only, has been cloned under the control of the *tacII* promoter [67, 68]. It was found that the cells could tolerate the production of lysozyme and could produce levels up to 2% of cellular protein [68]. Cells were grown to high cell densities before induction since there was a selctive disadvantage to clones which produced the lysozyme product. These experiments further confirmed that other gene products are necessary for T4 lysis. The cloned lysozyme also has been used in a cellular disruption procedure in which protein products are recovered after induction of gene *e* followed by freeze-thawing [67]. The release of protein product using this system was 8 mg/L.

Although phage T4 and lambda may have similar lytic mechanisms, little is available on the use of T4 lytic enzymes in cell disruption. The similarities between lambda S^- and T4 t^- mutants or clones suggest lysis mechanisms may be identical in the two phage systems. However, to our knowledge no reports are available on the study and/or use of cloned bacteriophage T4 genes *e* and *t* together in a microbial cell disruption process.

E. coli mutants which affect lysis by phage T4 have been isolated [69]. These mutants are impaired in the synthesis of peptidoglycan and have been shown to allow lysis and release of phage in e^- T4 phage. The mutants also possess increased sensitivity to antibiotics, detergents, and some colicins. Furthermore, they map to the *thr-ara-leu* region of the *E. coli* chromosome, which encodes a number of genes involved in cell division [70]. *E. coli* host functions therefore appear to affect lysis by phage T4.

5.3 Bacteriophage phiX174

Phage phiX174 is a small, single-stranded, DNA bacteriophage which is known to bring about lysis of infected cells within 20–25 min after infection [71]. The phiX174 gene *E* alone has been shown to be necessary for host cell lysis [72, 73, 74]. It is not known how gene product E functions, but it does not appear to have any lysozyme-like activity [34]. Gene product E may have a membrane associated function which is predicted by its amino acid sequence [75]. It has been suggested that gene product E "triggers" the autolytic system of the *E. coli* host cell [76].

The phiX174 gene *E* has been cloned and put under control of the *lac* promoter [73, 74]. Lysis of the host cell is induced by addition of the inducer IPTG. The time interval for completion of lysis after induction varied for different host cells. Young and Young [73], found that lysis occurs within 35–40 min or 55–60 min after induction while Henrich et al., [74] and Henrich and Plapp [7], found lysis occurs within 150–180 min. In the latter case it was determined by microscopic analysis that 85% of the cells had been disrupted.

A comparison between sonication and the cloned phiX174 lysis gene system for small-scale disruption was made [7]. The efficiency of intracellular protein release was determined by measuring levels of release of β-galactosidase. The phiX174 disruption method exceeded the efficiency of sonication by two-fold at cell densities below 5×10^{10} cells/mL; it was less efficient at higher cell densities. However, quantitative comparisons of the two methods may be difficult due to differences in handling of samples in the two methods; the sonicated samples were centrifuged 45 min at 30,000 × g while the induced samples were centrifuged 5 min at 10,000 × g. The different centrifugation times could make a difference in amounts of measurable soluble β-galactosidase. In a separate study, it was found that partially induced cultures released β-galactosidase more readily by sonication than uninduced cultures suggesting that cells become structurally weak before lysis [77].

Although much information has been gathered on the physiological characteristics of gene *E* induced lysis, the mechanism of action of this gene has not been completely elucidated. The gene *E* protein has a molecular weight of 10,000 daltons [78] and consists of 91 amino acids [75]. The nucleotide sequence suggests that gene product E is hydrophobic and interacts with the cell membrane [75].

Inner and outer membrane fractions of phiX174-infected *E. coli* cells were found to contain alterations in their protein profiles. The inner membrane fraction consisted of reduced levels of proteins as well as new proteins while the outer membrane showed decreases in several major proteins [79]. No attempt was made to identify these proteins. In a separate study, gene product E was found in the cytoplasmic membrane fraction after infection with phage phiX174 or induction by the cloned gene *E*. The cytoplasmic membrane of induced minicells also contained gene product E [80]. Altman et al. [60] found that gene product E was primarily associated with the inner membrane although small amounts appeared in the outer membrane fraction.

C-terminal deletions of gene *E* were constructed under the control of the *lac* promoter [81]. All were unable to bring about lysis of the host cell, including one in which only 17 of the 91 codons was deleted. When these deletions were fused to the *lacZ* gene, four out of five deletions had their lytic ability restored, although lysis was a little delayed. By measuring the amount of β-galactosidase produced with these fusions it was estimated that less than 1000 molecules per cell were needed for cell lysis and probably less than 100 were needed for the decrease in viability seen upon induction. Furthermore, it has been estimated that 100 to 300 molecules of gene product E per cell are synthesized in phiX174-infected cells [78].

It was also hypothesized from this work that gene product E needs to oligomerize in order to bring about its lytic effect. Beta-galactosidase normally forms a tetramer and the *lacZ* fusion may restore the normal C-terminal oligomerization function of gene product E. In a separate study deletions of the C-terminal region of gene *E* were fused to the chloramphenicol acetyl transferase (CAT) gene. This fusion also was able to restore the lytic function of the deleted gene, supporting the oligomerization theory, since CAT also forms a tetramer. A *trpE* fusion did not work, however [82].

The conclusion made from these studies [81, 82], as well as from one other [83], was that while the oligomerization function is located in the C-terminal domain, the lytic activity of gene product E is localized to the N-terminal region of the protein. Buckley and Hayashi [82] have shown that the lytic activity is localized to the N-terminal 29 amino acids, the region which is believed to span the inner membrane.

No lytic activity has been found in mature phage or infected cells [34]. By analysis of the cell wall after phiX174 infection, rapid degradation of the peptidoglycan was observed. It was determined that two enzymes had acted upon it: an endopeptidase and an endoglucosidase [76]. However, the mechanism of action of the phage did not appear to be due to inhibition of cell wall biosynthesis since the activity of the translocase, a major cell wall biosynthetic enzyme, was normal [76].

In an effort to understand the mechanism of lysis by phage phiX174 a number of studies on external or internal environment factors affecting the lytic process have been undertaken. Some of the factors affecting gene *E* induced lysis include: autolytic system of the host cell, and requirement for an energized proton-motive-force [84].

Lysis of *E. coli* in stationary phase does not occur upon infection with phiX174 [26]. However, if cultures infected during stationary phase are diluted in fresh medium, lysis occurs even in the presence of chloramphenicol or rifampicin (both are inhibitors of protein synthesis).

Addition of Mg^{2+} has been shown to inhibit lysis by both the cloned gene *E* and whole phage at concentrations of 0.05 M and 0.2 M respectively [30, 73]. In addition, spheroplasts are formed in the presence of magnesium [73]. There is no induction of lysis in stationary phase or in medium containing a poor carbon source [30]. At pH below 6.0 lysis is suppressed in the cloned gene *E* system; if the pH is adjusted to 6.8, lysis occurs. Lysis also is suppressed at pH higher than 8.0. All of the above factors, Mg^{2+} concentration, pH, and growth phase of the culture, have been shown to influence the autolytic system of *E. coli* in a similar manner. Further evidence for the involvement of the autolytic system comes from the abnormal shapes which occur in *E. coli* with low level expression of gene *E* from a plasmid [73]. These clones were insensitive to the induction of lysis, but a high percentage of abnormally shaped cells were observed upon induction, suggesting interference with division or elongation processes of the host cell.

To determine more accurately the influence of the autolytic system of the host cell, studies on the effect of autolysis deficient mutants on lysis by gene *E* were carried out. Plasmids containing gene *E* under *lac* promoter control were introduced into two different temperature sensitive autolysis mutants [27, 85]. Both mutants exhibit normal responses to autolysis by β-lactam antibiotics or cycloserine at 30 °C, and show inhibition of autolysis at 42 °C. In mutant VC30 [85], induction of gene *E* by IPTG at the nonpermissive temperature (42 °C) resulted in no lysis. However, if these cultures were downshifted to 30 °C (the permissive temperature), lysis occurred. In mutant VC44 [27], lysis occurred at 42 °C after induction of gene *E*. However, if the culture was preincubated at 42 °C before induction of gene *E*, there was no lysis of the culture. The results from these studies indicate that a functional autolytic system is necessary for gene *E* mediated lysis, but that the lytic triggering pathways for gene *E* and β-lactam antibiotics are different.

Lysis has been shown to begin 5–8 min after induction of the cloned gene *E* [86]. Chloramphenicol and rifampicin were added at various times after induction, and found to prevent lysis if added 30 sec after induction. There was no effect of these compounds if added more than 1.5 min after induction. This indicates that gene *E* needs 3.5–6.5 min to bring about its lytic effects. It also was determined that a proton-motive-force dependent step occurs 3 to 5 min before lysis [86]; this may coincide with insertion of gene product E into the cytoplasmic membrane.

Further evidence for the involvement of the membrane in gene *E* induced lysis comes from a study of *E. coli* host mutants. It was found that the rate of lysis by gene *E* was increased in *fadR* mutants (this mutation is involved in the regulation of fatty acid degradation; in rich medium the mutants show an increase in autolytic activity). Lysis was found to be reduced in a double mutant, *fabB*, *fadE*, which has reduced cell membrane fluidity when grown in the appropriate medium. Mutations in *envC*, a mutation which causes a chain-forming morphology along with increased sensitivity to crystal violet and some detergents, also caused a decrease in the rate of lysis by cloned gene *E* [87]. The state or fluidity of the *E. coli* membrane is apparently important for gene *E*-induced lysis.

As more information is gathered on the mechanism of phiX174 induced lysis, it may be possible to utilize *E. coli* host mutants to enhance the rate or extent of lysis. In addition, the elucidation of the mechanism will aid in the design of media, culture, and lysis conditions necessary for optimal lysis efficiency.

5.4 Bacteriophage MS2

Bacteriophage MS2 is a single-stranded RNA bacteriophage containing a single lysis gene sufficient for lysis. Lysis gene *L*, when put under control of either the λ*pL* promoter [88] or the *lac* promoter [89], causes lysis of the host cell within 40 to 60 min after induction. The lysis gene has been shown to overlap two other genes in a +1 frameshift and translation of the gene is coupled to that of the coat protein [88]. This study was carried out primarily to study basic lysis control mechanisms; it has not been examined for use in cell disruption per se.

The mechanism of lysis induction in MS2 has not been fully unconvered. It has been suggested that lysis proceeds through the triggering of autolytic enzymes even though there is no increase in the degradation of labelled cell wall upon induction of gene *L*. Lysis by gene *L* does not occur at pH 5.0, which is similar to penicillin-induced autolysis [90]. Upon induction of lysis, an increased permeability of the cells is noted, as determined by the ability of the substrate *ortho*-nitrophenyl-β-D-galactoside (ONPG) to permeate the cell pellet for detection of measurable β-galactosidase.

Evidence for the involvement of the membrane in MS2-induced lysis comes from a study of *E. coli* mutants deficient in biosynthesis of membrane-derived oligosaccharides (MDO) [91]. These mutants show a large separation between the cytoplasmic and outer membranes when examined under electron microscopy. It is quite possible that the MDOs maintain the spacing between the different layers of the cell envelope [92]. In MDO^- mutants, the activity of the lysis gene is altered; upon induction of the gene, lysis of the host cell does not occur as measured by culture turbidity. In addition, it was discovered that the lysis protein could not be found in the membrane or soluble fraction. Further analysis of the membrane indicated that the lysis protein may be degraded within this fraction [91]. The cloned lysis gene *E* of phage phiX174 was also unable to cause complete lysis in the MDO^- strain; there was a slight decrease in turbidity upon induction, followed by an increase in cell density.

Work by Goessens et al. [93], suggests that the lytic activity of gene *L* is in the C-terminal portion of the protein. A synthetic peptide containing the terminal 25 amino acids from this protein was constructed. It was found that this peptide dissipates the proton motive force possibly by causing the formation of hydrophilic pores in the membrane.

5.5 Bacteriophage Q*β*

In bacteriophage Qβ, the maturation protein A2 has been shown to be the sole gene product necessary for lysis [94]. This lysis gene has been cloned under *lac* promoter control and lysis of the culture occurs within 30 min of induction. Although the mechanism of lysis has not yet been fully uncovered, it is thought that lysis is regulated by intracellular concentrations of protein A2, requiring a threshold level before onset of lysis. Not surprisingly, the A2 protein shows no homology with either the MS2 lysis gene product L or the phiX174 lysis gene product E [94].

6 "Designer" Lytic Proteins

As more information on the structure and function of lytic enzymes becomes available there are more possibilities for the creative construction of hybrid or altered lytic proteins with changes in specificity or efficiency.

Harkness and Lubitz [95], have constructed a hybrid bacteriophage lytic gene comprised of the N-terminal region of gene *E* from phiX174 and the C-terminal region of gene *L* from MS2 (these regions are considered to contain the lytic regions of the respective lysis genes). The hybrid protein was able to bring about lysis of the host cell upon induction. In fact, the time of onset of lysis after induction was 25 min, which is between the induction times of 8 and 40 min for cloned phiX174 lysis gene *E* and MS2 lysis gene *L*, respectively. The rate of lysis was increased relative to the other lysis genes. The authors attribute the increase in rate to the membrane-spanning region of the hybrid protein which should be present in both C and N-terminal regions of the protein. This gives evidence that altered rates of efficiency of lysis are achievable by "designer" lytic proteins.

Although differences in specificity of lytic enzymes have not been shown through hybrid proteins, there is evidence that it is achievable. Kato et al. [55] found that weak expression of the ColE1 *kil* gene caused extracellular excretion in *E. coli* of human growth hormone containing a *Bacillus* signal sequence. Pugsley and Schwartz [43] found that expression of the ColE2 operon resulted in release of 10% of the total cell protein within 5 h after induction. It was found that 60–70% of the protein released was colicin E2 and its immunity protein; β-galactosidase and RNA polymerase could not be detected. The authors suggest that certain cytoplasmic proteins may be too large or have improper charges or hydrophobicities, preventing them from being released. As more knowledge is gained about these systems, the design of lytic proteins with altered specificities as well as efficiencies may be possible.

7 Directions and Potential for Future Research

Although a number of different lytic systems have been identified in *E. coli* and some preliminary studies have been made, the use of these systems as alternative methods for cell disruption has not been fully developed. The mechanism of action of these lytic enzymes are quite different, allowing possible flexibility of cellular disruption procedures to suit a particular purification process. Research areas which could be explored in these systems include an examination of the selectivity of protein release, effect of medium design or strain selection, cell density or growth phase requirements, and use in combination with other methods of cell disruption.

Differences in the selectivity of protein release may occur within a particular lytic process, as well as between different lytic systems. Molecular weight, hydrophobicity, or charge of the protein product may all influence release. For example, if small pores are formed in the membrane or cell wall after induction of lysis, it may be possible to release small molecular weight proteins preferentially over proteins with larger molecular weight. In addition, periplasmic proteins may be released preferentially over intracellular proteins. Changes in pH or ionic strength may further reveal preferential release of particular proteins due to charge effects. Similarly, one particular lytic mechanism (e.g. membrane pore formation vs cell wall degradation) may be chosen over another in order to achieve preferential release. Examination of the release of different proteins under varied conditions would determine the specificity of release, if any, using the different lytic systems.

The lysis system should be designed as an integrated system taking into consideration such factors as medium design and strain selection to achieve an overall process goal. For example, medium components have been shown to influence the rate and extent of lysis. The concentration of Mg^{+2} is known to affect lysis by a number of lytic enzymes, including the colicin lytic enzymes, phiX174 gene product E, and the autolytic enzymes. In addition, the presence of a particular chemical such as Mg^{+2} may influence the quality or size of the cellular debris and hence has potential for improvement in centrifugation efficiencies through the formation of large and more uniform particles. The quality of cell debris after lysis also will be a function of the particular lytic method chosen.

Information for lysis may be carried either on a plasmid or integrated within the host chromosome. The choice of strain is therefore an important factor in designing the lytic system since regulatory mechanisms must be present in the host strain in order to carry out lysis in a stable and controlled manner. The phenotypic characteristics of the strain chosen may also influence the ease of lysis or extent of structural instability of the host cell after induction. For example, the *fadR* mutation of the *E. coli* host strain causes an increase in the rate of lysis by phiX174 gene product E when grown under specific conditions. *E. coli* strain W7 which is missing an enzyme involved in cell wall synthesis may prove more susceptible to lysis by the phage phiX174 lysis gene *E*.

Several of the lytic systems have the disadvantage that growing or resuspended cultures are needed for lysis induction, resulting in lower product yield due to lowered cell densities. It is therefore essential to determine the importance of growth phase of the cell culture on the lysis efficiency. The phiX174 lytic enzyme, which is a membrane-affecting protein, has a requirement for logarithmic phase growth in order to achieve efficient lysis. The lytic mechanism of this phage is quite different from phage lambda and T4 which contain lysozyme-like enzymes as well as membrane-affecting proteins in their lytic systems. It is possible that cloned intracellular lysozyme systems such as these may allow induction of lysis at higher cell densities and/or during stationary phase. Results using S^- mutants of phage lambda suggest that stationary phase cultures, although with low cell densities, are efficiently lysed [6]. Alternatively, logarithmic phase cells could be concentrated by centrifugation or ultrafiltration prior to lysis induction, in order to prevent dilution of the protein product desired. These types of studies could be initiated in autolytic or phage phiX174 systems where logarithmic phase cells appear to be essential for efficient lysis. At present, the cloned colicin lytic enzymes may be too lethal to achieve high cell densities before lysis induction.

In addition to the problems of achieving high cell densities in large-scale processes, other factors which may pose specific problems with increased scale include increased viscosity, excessive foaming, release of proteases, and heat transfer for induction of lysis through temperature.

The use of temperature as a method of induction is attractive as an alternative to chemical induction, but requires that the protein product is not inactivated by the temperature increase. Short induction times followed by cooling may be sufficient to release lytic proteins which are active at lower temperatures, while preventing the inactivation of protein products.

Preliminary work on the phage phiX174 and phage lambda lytic systems suggest that the cloned lytic systems could be used in conjunction with other traditional methods of disruption. Induction of lysis for short time periods may weaken the cell wall structure, making the use of high pressure homogenization more attractive through reduced passes or pressures.

A more complete understanding of the mechanisms involved in these lytic systems will assist in the development of the enzymes involved as alternatives to current cell disruption processes. Basic research, as well as side by side comparison studies of the different methods are needed to identify and contrast particular applications of each method. Cloning of the enzymes will enhance the ability to understand mechanisms and factors influencing lysis characteristics.

8 References

1. Hughes DE, Wimpenny JWT, Lloyd D (1971) The disintegration of microorganisms. In: Norris JR, Ribbons DW (eds) Methods in microbiology. Academic, New York, p 1
2. Scawen MD, Atkinson A, Darbyshire J (1980) Large-scale enzyme purification. In: Grant RA (ed), Applied protein chemistry, Applied Science Publishers, London, p 281
3. Darbyshire J (1981) Large scale enzyme extraction and recovery. In: Wiseman A (ed) Topics in enzyme and fermentation biotechnology, Ellis Horwood, New York, p 147
4. Engler CR (1985) Disruption of microbial cells. In: Moo-Young M (ed-in-chief) Cooney CL, Humphrey AE (eds) Comprehensive biotechnology, vol 2, Pergamon, New York, p 305
5. Auerbach JI, Rosenberg M (1987) US Patent # 4637980
6. Crabtree S, Cronan JE (1984) J. Bact. 158: 354
7. Henrich B, Plapp R (1984) J. Biochem. Biophys. Meth. 10: 25
8. Glover DM (1984) Gene cloning. Chapman and Hall, New York
9. Milewski EA (1984) The NIH guidelines for research involving recombinant DNA molecules. In: Bollon AP (ed) Recombinant DNA products: insulin, interferon and growth hormone, CRC Press, Boca Raton, p 156
10. Park JT (1987) The murein sacculus. In: F.C. Neidhardt (ed) *Escherichia coli* and *Salmonella typhimurium*. American Society for Microbiology, vol 1, Washington DC, p 7
11. Coakley WT, Bater AJ, Lloyd D (1977) Adv. Micro. Phys. 16: 279
12. Gmeiner J (1980) J. Bact. 143: 510
13. Braun V, Gnirke H, Henning U, Rehn K (1973) J. Bact. 114: 1264
14. Pelzer von H (1963) Z. Naturf. 18B: 950
15. Holtje JV, Mirelman D, Sharon N, Schwarz U (1975) J. Bact. 124: 1067
16. Tomasz A (1983) Mode of action of β-lactam antibiotics — a microbiologist's view. In: Demain AL, Solomon NA (eds) Antibiotics containing the β-lactam structure I, Springer, Berlin Heidelberg New York (vol 67/I)

17. Phaff HJ (1977) Enzymatic yeast cell wall degradation. In: Feeney RJ, Whitaker JR (eds) Food proteins: improvement through chemical and enzymatic modification, American Chemical Society, Washington DC, p 244 (no 160)
18. Belikov VM, Latov VK, Tsyriapkin VA, Sergeev VA (1976) Microbiologichesky Promyshlennost [Microbiological Industry] 3:1
19. Akin C, Murphy RM (1981) US patent 4285976
20. Arnold WN (1981) Autolysis. In: Arnold WN (ed) Yeast cell envelopes: biochemistry, biophysics, and ultrastructure, vol 2, CRC Press, Boca Raton, p 129
21. Weissman C, Schein C (1982) Eur. Pat. Appl. EP # 61250
22. Rogers HJ (1979) The function of bacterial autolysins. In: Berkeley RCW, Gooday GW, Ellwood DC, (eds) Microbial polysaccharides and polysaccharases, Academic, New York
23. Leduc M, Kasra R, Singer H, Heijenoort J van (1984) FEMS Micro. Lett. 23: 137
24. Tomasz A (1984) Building and breaking of bonds in the cell wall of bacteria — the role for autolysins. In: Nombela C (ed) Microbial cell wall synthesis and autolysis, Elsevier, New York
25. Leduc M, Heijenoort J van (1980) J. Bact. 142: 52
26. Blasi U, Halfmann G, Lubitz W (1984) Induction of autolysis of *Escherichia coli*. In: C. Nombela (ed) Microbial cell wall synthesis and autolysis, Elsevier, New York
27. Halfmann G, Lubitz W (1986) J. Bact. 166: 683
28. Leduc M, Kasra R, Heijenoort J van (1982) J. Bact. 152: 26
29. Tuomanen E, Cozens R, Tosch W, Zak O, Tomasz A (1986) J. Gen. Micro. 132: 1297
30. Lubitz W, Halfmann G, Plapp R (1984) J. Gen. Micro. 130: 1079
31. Tuomanen E, Tomasz A (1986) J. Bact. 167: 1077
32. Leduc M, Frehel C, Heijenoort J van (1985) J. Bact. 161: 627
33. Kitano K, Tomasz A (1979) Antimicrob. Agents Chemother. 16: 838
34. Markert A, Zillig W (1965) Vir. 25: 88
35. Nomura M (1964) PNAS 52: 1514
36. Konisky J (1978) The bacteriocins. In: Ornston LN, Sokatch JR (eds) The bacteria, vol VI, Academic, New York, p 71
37. Luria SE, Suit JL (1982) Transmembrane channels produced by colicin molecules. In: Martinosi A (ed) Membranes and transport, vol 2, Plenum, New York, p 279
38. Cramer WA, Dankert JR, Uratani Y (1983) Biochim. Biophys. Acta 737: 173
39. Luria SE, Suit JL (1987) Colicins and Col plasmids. In: Neidhardt FC (ed) *Escherichia coli* and *Salmonella typhimurium*, vol 2, American Society for Microbiology, Washington DC, p 1615
40. Sabik JF, Suit JL, Luria SE (1983) J. Bact. 153: 1479
41. Hakkaart MJJ, Veltkamp E, Nijkamp HJJ (1981) Mol. Gen. Genet. 183: 318
42. Hakkaart MJJ, Veltkamp E, Nijkamp HJJ (1981) Gen. Genet. 183: 326
43. Pugsley AP, Schwarz M (1984) EMBO J. 3: 2393
44. Watson RJ, Lau PCK, Vernet T, Visentin LP (1984) Gene 29: 175
45. Jakes KS, Zinder ND (1984) J. Bact. 157: 582
46. Lloubes R, Baty D, Lazdunski C (1986) Nucl. Acids Res. 14: 2621
47. van den Elzen PJM, Walters HHB, Veltkamp E, Nijkamp HJJ (1983) Nucl. Acids Res. 11: 2465
48. Cole ST, Saint-Joanis B, Pugsley AP (1985) Mol. Gen. Genet. 198: 465
49. Chan PT, Ohmori H, Tomizawa J, Lebowitz J (1985) J. Biol. Chem. 260: 8925
50. Pugsley AP, Schwarz M (1983) J. Bact. 156: 109
51. Altieri J, Suit JL, Fan M-LJ, Luria SE (1986) J. Bact. 168: 648
52. Howard SP, Leduc M, van Heijenoort J, Lazdunski C (1987) FEMS Micro. Lett. 42: 147
53. Kobayashi T, Kato C, Kudo T, Horikoshi K (1986) J. Bact. 166: 728
54. Kudo T, Kato C, Horikoshi K (1983) J. Bact. 156: 949
55. Kato C, Kobayashi T, Kudo T, Furusato T, Murakami Y, Tanaka T, Baba H, Oishi T, Ohtsuka E, Ikehara M, Yanagida T, Kato H, Moriyama S, Horikoshi K (1987) Gene 54: 197
56. Sher IH, Mallette MF (1952) J. Biol. Chem. 200: 257
57. Young R, Way J, Way S, Yin J, Syvanen M (1979) J. Mol. Biol. 132: 307
58. Reader RW, Siminovitch L (1971) Vir. 43: 623
59. Bienkowska-Szewczyk K, Lipinska B, Taylor A (1981) Mol. Gen. Genet. 184: 111

60. Altman E, Young K, Garrett J, Altman RA, Young R (1985) J. Vir. 53: 1008
61. Garrett J, Fusselman R, Hise J, Chiou L, Smith-Grillo D, Schulz J, Young R (1981) Mol. Gen. Genet. 182: 326
62. Garrett JM, Young R (1982) J. Vir. 44: 886
63. Friedman DI, Olsen ER, Georgopoulos C, Tilly K, Herskowitz I, Banuett F (1984) Micro. Rev. 48: 299
64. Tsugita A, Inouye M (1968) J. Biol. Chem. 243: 391
65. Josslin R (1970) Vir 40: 719
66. Mukai F, Streisinger G, Miller B (1967) Vir. 33: 398
67. Wetzel RB (1985) Eur. Pat. Appl. EP # 155189
68. Perry LJ, Heyneker HL, Wetzel R (1985) Gene 38: 259
69. Raj CVS, Wu HC (1973) J. Bact. 1973: 656
70. Bachmann B (1987) Linkage map of *Escherichia coli* K-12, edition 7. In: Neidhardt FC (ed) *Escherichia coli* and *Salmonella typhimurium*, vol 2, American Society for Microbiology, Washington DC, p 807
71. Hutchinson CA, Sinsheimer RL (1963) J. Mol. Biol. 7: 206
72. Hutchinson CA, Sinsheimer RL (1966) J. Mol. Biol. 18: 429
73. Young KD, Young R (1982) J. Vir. 44: 993
74. Henrich B, Lubitz W, Plapp R (1982) Mol Gen. Genet. 185: 493
75. Barrell BG, Air GM, Hutchinson CA (1976) Nature 264: 34
76. Lubitz W, Plapp R (1980) Curr. Micro. 4: 301
77. Dabora RL (1989) Studies on the action of the cloned phiX174 lysis gene E, Thesis. Massachusetts Institute of Technology, Cambridge, Massachusetts
78. Pollock TJ, Tessman ES, Tessman I (1978) J. Vir. 28: 408
79. Lubitz W, Schmid R, Plapp R (1981) Curr. Micro. 5: 45
80. Blasi U, Geisen R, Lubitz W, Henrich B, Plapp R (1983) Localization of the bacteriophage phiX174 lysis gene product in the cell envelope of *Escherichia coli*. In: Hakenbeck T (ed) The target of penicillin, Walter de Gruyter, New York
81. Maratea D, Young K, Young R (1985) Gene 40: 39
82. Buckley KJ, Hayashi M (1986) Mol. Gen. Genet. 204: 120
83. Blasi U, Lubitz W (1985) J. Gen. Vir. 66: 1209
84. Blasi U, Harkness RE, Witte A, Halfmann G, Lubitz W (1986) Endogenous induction of bacterial lysis by cloned phiX174 gene E product. In: Seidl PH, Schleifer KH (eds) Biological properties of peptidoglycan, Walter de Gruyter, New York
85. Lubitz W, Harkness RE, Ishiguro E (1984) J. Bact. 159: 385
86. Witte A, Lubitz W, Bakker EP (1987) J. Bact. 169: 1750
87. Wadle D, Henrich B, Plapp R (1986) Curr. Micro. 14: 65
88. Kastelein RA, Remaut E, Fiers W, van Duin J (1982) Nature 295: 35
89. Coleman J, Inouye M, Atkins J (1983) J. Bact. 153: 1098
90. Holtje JV, van Duin J (1984) MS2-phage induced lysis of *E. coli* depends upon the activity of the bacterial autolysins. In: Nombela C (ed) Microbial cell wall synthesis and autolysis, Elsevier, New York
91. Holtje JV, Fiedler W, Rotering H, Walderich B, van Duin J (1988) J. Biol. Chem. 263: 3539
92. Kennedy EP (1982) PNAS 79: 1092
93. Goessens WHF, Driessen AJM, Wilschut J, van Duin J (1988) EMBO J. 7: 867
94. Winter RB, Gold L (1983) Cell 33: 877
95. Harkness RE, Lubitz W (1987) FEMS Micro. Lett. 48: 9

Impact of Genetic Engineering on Downstream Processing of Proteins Produced in *E. coli*

S.-O. Enfors, H. Hellebust, K. Köhler, L. Strandberg and A. Veide
Department of Biochemistry and Biotechnology, The Royal Institute of Technology, S-100 44 Stockholm, Sweden

Genetic engineering can be used to give a protein properties that are advantageous for downstream processing. Many heterologous proteins are degraded at high rates by proteases. Depending on which type of proteolytic degradation is encountered the strategy may be different: induction of inclusion bodies, change of the amino acid sequence in the sensitive site of the product, or protection by fusion of the product with other proteins. The number of unit operations needed to purify a protein may be reduced by addition of other polypeptides or amino acids to the product. Affinity chromatography, immobilized metal ion affinity chromatography, and extraction in aqueous two-phase systems are unit operations which can be made more versatile by the fusion technique.

1 List of Symbols and Abbreviations

Ala	Alanine
Arg	Arginine
Asn	Asparagine
Asp	Aspartic acid
DHFR	Dihydrofolate reductase
E. coli	*Escherichia coli*
Gly	Glycine
His	Histidine
IFN	Interferon
IGF	Human insulin-like growth factor
IgG	Immunoglobulin G
IL	Interleukin
kDa	1000 dalton
NTA	nitilotriacetate
ompT	outer membrane protein T
PEG	poly (ethylene) glycol
rDNA	recombinant DNA
Ser	Serine
SpA	Staphylococcal protein A
ZZ	artifical IgG-binding protein derived from staphylococcal protein A

Advances in Biochemical Engineering/
Biotechnology, Vol. 43
Managing Editor: A. Fiechter

2 Introduction

Genetic engineering is becoming increasingly important in biotechnology, not only for expressing new proteins in a cell but also for facilitating the downstream operations. By this technique a protein may be redesigned to acquire properties that are important for the preparation and/or use of the product as illustrated in Fig. 1. In this paper two important issues in the downstream processing of proteins that have been approached in this way, namely proteolytic attack on the product and purification of the product, will be discussed.

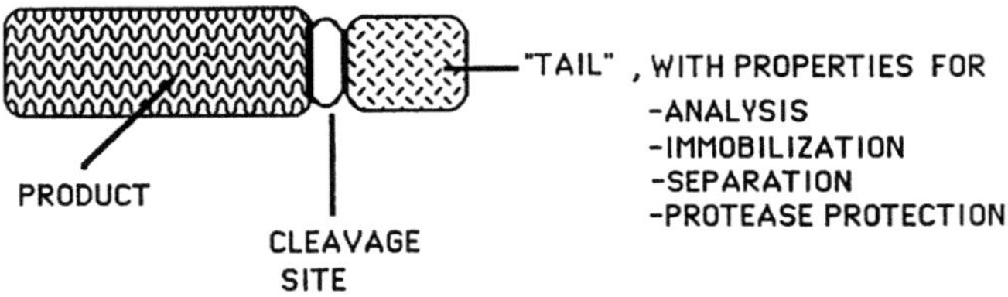

Fig. 1. Principle of a fusion protein with properties for facilitated downstream operations

3 Role of Proteases in the Cell

There are a number of proteases with different tasks in a cell. In *E. coli* some twenty proteases have been characterized (Fig. 2). For a review see references [1, 2]. The normal functions of these enzymes are very diverse and may include degradation of aberrant proteins, degradation or modification of proteins as a means of metabolic control, and splitting of the signal peptide in the process of transport through the membrane.

All kinds of proteins in a cell seem to be subject to a certain rate of proteolysis but homologous *E. coli* proteins are stable compared to those in animal cells [3]. In *E. coli* a few percent of the proteins are unstable with a half life less than

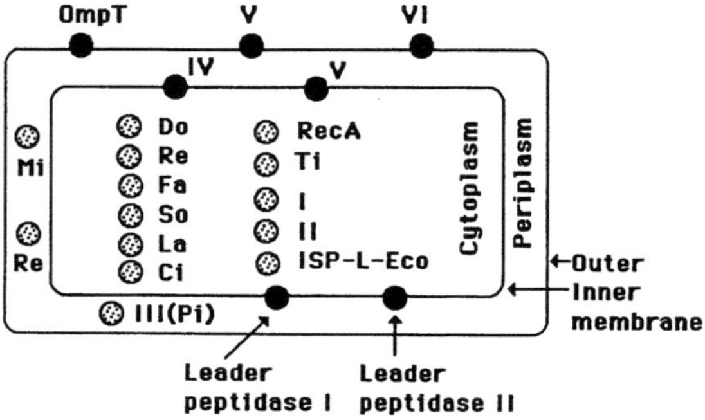

Fig. 2. Cellular localization of the best known *E. coli* proteases

the cell's generation time while the majority of the proteins have a half life which is longer. The unstable proteins probably belong to the group of proteins that are involved in metabolic control. Aberrant proteins can be obtained e.g. by application of amino acid analogs in the medium. One of the proteases of *E. coli* protease *La* which is the product of the lon gene, is well known to participate in the degradation of such aberrant proteins [4]. It seems as if recombinant proteins are sometimes processed as error proteins by the host proteases.

The mechanisms involved in control of proteolysis in *E. coli* are poorly understood. One mechanism of degradation of foreign proteins by the host proteases may involve improperly folded proteins and this will be discussed below.

4 Proteolysis During Production of rDNA Proteins in *E. coli*

IFN α-A was quickly degraded in *E. coli* when the cells were subjected to starvation of glucose or to oxygen limitation [5]. On the other hand, IL-2 and IFN-γ were not degraded under the same conditions. The latter phenomenon was shown to depend on the fact that these proteins, but not IFN α-A, were produced in the form of inclusion bodies in the cell. When soluble samples of the proteins were incubated with disrupted *E. coli* cells, all three species were proteolytically modified. When IFN α-2 was produced in *E. coli* as well as in *Methylophilus methylotrophus*, degradation of the product was temperature dependent and increased with temperature more than expected from the Arrhenius equation [6]. At temperatures below 29 °C the product was stable in both hosts without involvement of inclusion body formation, which indicated that the variation in stability at different temperatures was related to the properties of the protein rather than to the organism's proteases. Increased exposure of hydrophobic surfaces to proteases at higher temperature was assumed as one possible mechanism. Since production under control of the heat inducible P_R promotor is common, this phenomenon may be of more general interest for the production of rDNA proteins.

It is also likely that starvation and possibly also limitation with respect to some nutrients can increase the rate of proteolysis in the cell [7, 1]. Starvation with respect to carbon/energy sources also initiates synthesis of several unique proteins, some of which belong to the heat shock proteins [8]. Since large scale production of rDNA proteins in microorganisms generally has to be performed under some kind of nutrient limitation, the proteolysis may be influenced by the method applied for process control. These examples show that process conditions in the bioreactor may influence proteolysis in different ways.

5 Control of Proteolysis

The traditional methods of reducing the effects of proteases during protein processing are to apply low temperature and/or protease inhibitors. However, actions already taken during the biosynthesis stage such as the use of protease negative mutants in cases when such cells can be obtained may be important. Lately genetic engineering has also offered a tool for solving problems of proteolysis in production of rDNA proteins.

The strategy to solve the proteolysis problem must, however, depend on the type of proteolysis and the ultimate use of the product. Severe proteolysis, resulting in complete loss of product, gives of course a low yield but does not otherwise influence the final product quality. A more subtle modification of the product may give it other properties with respect to biological activity and immunogenicity [9] without changing the physicochemical properties of the protein that are utilized for its purification. That sort of proteolysis may be much more difficult to handle in downstream processing.

5.1 Protection by Inclusion Body Formation

Fusion of heterologous proteins to *E. coli* β-galactosidase or parts thereof has long been used to improve the expression of heterologous proteins [10, 11]. In most cases this improvement is likely to be caused by the fact that if intracellular proteins precipitate, then so called inclusion bodies, are frequently formed. It is likely that the mechanism of stabilization by inclusion body formation is steric hindrance for the protease to reach the majority of the product in the precipitate. Some types of these inclusion bodies can be renaturated in steps involving first complete unfolding in a strong denaturing solution such as 8 M urea, 6 M guanidinium hydrochloride or by a strong base such as 0.01 M sodium hydroxide followed by renaturation by dilution or dialysis [12].

The mechanism of inclusion body formation is not clear. Accumulated calf prochymosin in *E. coli* contained some intermolecular disulphide bridges [13]. However, other mechanisms were also assumed to be involved, and the reducing cytoplasmic environment is generally considered not to support disulfide formation. The observation that inclusion bodies are frequently associated with over-production, not only of rDNA proteins but also of homologous proteins, has resulted in the hypothesis that inclusion bodies are the result of precipitation caused by high intracellular concentrations exceeding the solubility of the protein [12]. However, this hypothesis is contradicted by the difficulty of renaturing inclusion bodies in comparison with the ease of renaturation of most precipitated proteins. Furthermore, in the case of SpA-β-galactosidase production in *E. coli*, the inclusion body formation is most intensive during the early stage after induction of the P_R-promotor, when the intracellular concentration of this product is lowest [14]. This molecule is an interesting model for studies of inclusion body formation since the conditions can be controlled at will to give variations in inclusion body formation [15]. Conditions which favour inclusion body formation of SpA-β-galactosidase fusion protein after induction of the lambda P_R-promotor by heat (39–42 °C) are high temperature, ammonia instead of amino acids as N-source and declining pH (no pH-control) [14, 15]. Inclusion bodies of SpA-β-galactosidase that were induced in this way were shown to be stable towards proteolysis by ompT that otherwise splits the protein [16]. Inclusion body formation in systems with other products and other promotors have also been shown to be temperature sensitive in a similar way [17, 18].

In a recent review, Mitraki and King [19] suggested that inclusion bodies are formed by intermolecular interactions between partially denatured intermediates taking place during the folding pathway of the protein. Such reactions are supposed to be very sensitive to the physico-chemical environment in the cytoplasm. The different cultivation conditions used in control of inclusion body formation of SpA-β-galactosidase in combination with the intense and selective protein synthesis might have provided such denaturing micro-environments in the cytoplasm.

5.2 Protection by Protein Engineering

There are at least two methods through which a protein can be stabilized towards proteolysis by changing the structure of the protein: Removal of the sensitive site and protective fusion. Removal of the sensitive site was sucessfully applied in stabilization of SpA-β-galactosidase (Fig. 3, fusion No. 4). This molecule was split by ompT during the recovery of the product from disintegrated cells. After this initial hydrolysis two proteins similar to SpA and β-galactosidase resulted [16]. The ompT target Lys-Arg in the linker region between the two molecules

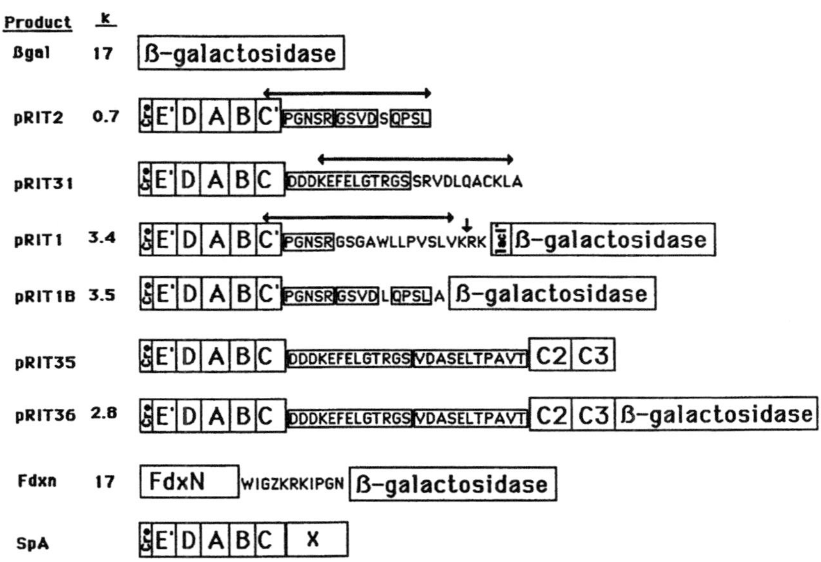

Fig. 3. Structures, proteolytically sensitive regions and partitioning coefficients of some β-galactosidase fusion proteins, and fusion components. The *arrows* indicate proteolytically sensitive regions [20]. Partition coefficients (*k*) refer to PEG 4000/potassium phosphate (Köhler, unpublished data). The fusion between ferredoxin and β-galactosidase was constructed by Dr. A. Pühler, University of Bielefeld

was removed and the new SpA-β-galactosidase molecule (Fig. 3, fusion No. 5) was resistant to proteolysis during downstream processing.

It was shown during studies of the fate of two of the main degradation products, that after the initial cleavage of SpA:β-galactosidase (Fig. 3, fusion No. 4), the β-galactosidase part was stable while the SpA part was further degraded [20]. However, the SpA molecule was not completely degraded but was cleaved in the C-region but not further up stream (Fig. 3, fusion No. 4). Thus, the proteases left the complete, and probably well folded, regions E, D, A and B of the SpA-molecule. This proteolytic activity in the truncated C′-region of SpA and in the linker region could not be observed until the β-galactosidase part of SpA-β-galactosidase from pRIT1 had been removed by ompT. Thus, β-galactosidase protected the sensitive sites in the C-region of the SpA-molecule.

This protection is not a unique property of β-galactosidase which is a large molecule with a molecular weight of 464 kDa since the two IgG-binding regions, C2 and C3 streptococcal protein G with a molecular weight of 15 kDa (Fig. 3, fusion No. 6) offered the same protection to the sensitive sites of protein A (Fig. 3, fusion No. 3).

It can be assumed that there are at least three main types of proteolysis of rDNA proteins in *E. coli*. Firstly, complete degradation of proteins, like the one mediated by protease La; Secondly, site specific cleavages, like those caused by ompT, which has a specificity for basic amino acids. This cleavage may open up the molecule in such a way that protease with predominant activity in improperly folded regions, as observed in the experiment shown Fig. 3, become active.

The protective properties of some, presumably well folded, protein structures have been utilized by Uhlèn and co-workers in the so-called dual fusion technique [21]. This involves fusion of two protein ligands, e.g. the SpA derived IgG-binding ligand ZZ and the albumen binding fraction of streptococcal protein G to the ends of a proteolytically sensitive molecule. It was shown that such fusions reduced the proteolysis of IGF-II to a considerable extent. The fact that a split in the product leads to loss of at least one of the two affinities (IgG or albumin) means that only un-cleaved products are obtained after serial affinity chromatography based on these ligands.

6 Facilitated Purification

Fusion of a protein or amino acids to a protein product has been used in many cases to give the fusion protein properties suitable for separation. Table 1 lists some of the ligands and separation techniques used. If a native protein is required or if the ligand used for separation is not acceptable during the utilization of the product then additional steps must be included to release the ligand from the product. Chemical as well as enzymatic hydrolysis can be used (Table 2). The advantage with the chemical methods are that they are simple and cheap, but they are less applicable in cases of large products since the probability that the sensitive sequence will be present also in the product increases. Enzymatic methods

Table 1. Some ligands and separation techniques used for facilitated downstream processing of proteins

Product	Ligand	Separation method	Reference
IGF-I	ZZ	IgG-affinity chromatography	[22, 23, 24]
IGF-II	ZZ+ albumin binding protein G fragment	IgG- and albumin affinity chromatography	[21]
urogastrone	poly-Arg	Ion exchange chromatography	[26]
mouse DHFR	poly-His	IMA-chromatography with Ni^{2+}	[25]
protein A	β-galacto-sidase	Aqueous two-phase extraction	[28, 29]

Table 2. Some methods used to split the ligand from the product after fusion for facilitated downstream processing. * indicates the hydrolysed bond and X refers to any amino acid

Method	Sensitive region	Reference
Hydroxylamine	X-Asn*Gly-X	[24]
Low pH	X-Asp*Pro-X	[31]
CNBr	X-Met*X	[32]
Factor Xa	X-Ile-Glu-Gly-Arg*X	[33]
Carboxy peptidase B	C-terminal basic amino acids	[26]
Carboxyl peptidase A	C-terminal aromatic amino acids	[25]

on the other hand may be more selective but are also more expensive and add an additional protein separation step unless the enzyme may be used in an immobilized state.

6.1 Chromatographic Techniques

Several methods have been used to give a protein specific affinity through fusion to facilitate its purification by affinity chromatography (Table 1). One of the first methods applied in commercial production is the protein A-derived IgG-binding ligand ZZ developed by Uhlén and co-workers [22, 23]. It has been used for the purification of a number of substances. The ligand ZZ was derived from the B region of SpA (Fig. 3) which is one of 5 IgG-binding regions in SpA. To make the ZZ-ligand resistant to hydroxylamine cleavage, Gly29 was replaced by Ala. To optimize the affinity of the ligand multiple copies of the modified B-region were tested [23] and a tandem arrangement of the modified B-region was finally used in the so called ZZ-ligand. Insertion of an Asn next to the native Gly at the N-terminus of IGF-I before fusion with ZZ made the fusion product cleavable exactly at the junction. The promotor and signal peptide of SpA was used for secretion of the fusion protein into the medium. The fusion molecule could then be harvested by affinity chromatography on IgG-Sepharose and after cleavage

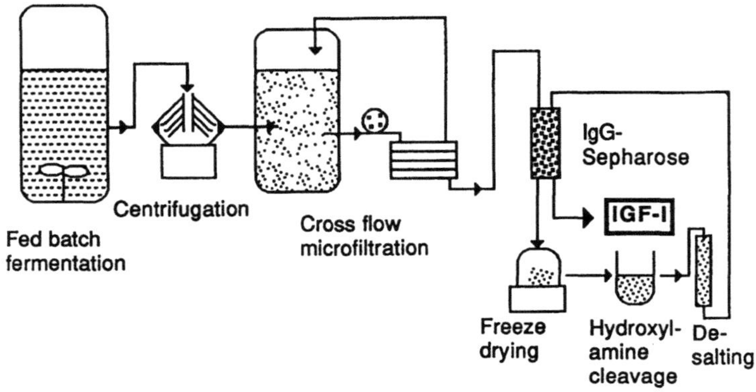

Fig. 4. Purification of extracellular IGF-I from *E. coli* culture medium. Cells are removed by centrifugation and cross flow micro-filtration. The fusion product (ZZ-IGF-I) is purified with IgG-affinity chromatography and freeze dried. After hydroxylamine cleavage and subsequent desalting the hydrolysis products ZZ and IGF-I are separated in a second passage through the IgG column. (Adopted from Ref. 24)

with hydroxylamine the ZZ-ligand could be separated from IGF-I by an additional IgG-affinity chromatography step shown in Fig. 4 [24].

Another versatile affinity system that has been applied for purification of rDNA proteins is the interaction between His and metal ions applied in IMA-chromatography. Histidine in multiples of 2–6 amino acids were fused to the C-terminal of mouse dihydrofolate reductase (DHFR) [25]. Nickel ions were then used as a ligand bound to NTA on a chromatography resin. After elution of the DHFR from the column poly-His chain was removed by digestion with carboxypeptidase A, which has high activity towards aromatic side chains but low activity towards Arg. Since the last few amino acids of the C-terminus were —Arg—Ser—$(His)_n$, the final product had an Arg-terminus.

A similar, but less specific, affinity modification of the product is fusion of a poly-Arg chain to the C-terminal of the product in order to add a surplus of positive charges. It has been applied to the purification of human urogastrone [26]. Cation-exchange chromatography can then be used to enrich the product. After digestion of the C-terminal residues by carboxypeptidase B, which has an affinity for C-terminal basic amino acids, a second chromatography step was used to separate the product from the contaminating proteins.

6.2 Extraction in Aqueous Two-phase Systems

Since fusion of heterologous proteins to β-galactosidase is so common in biotechnology, efficient separation processes for such fusion proteins are of interest. Extraction in aqueous two-phase systems has been applied for this purpose. This principle is based on the very high partition coefficient of *E. coli* β-galactosidase to the top phase of a polyethylene glycol 4000/potassium phosphate system.

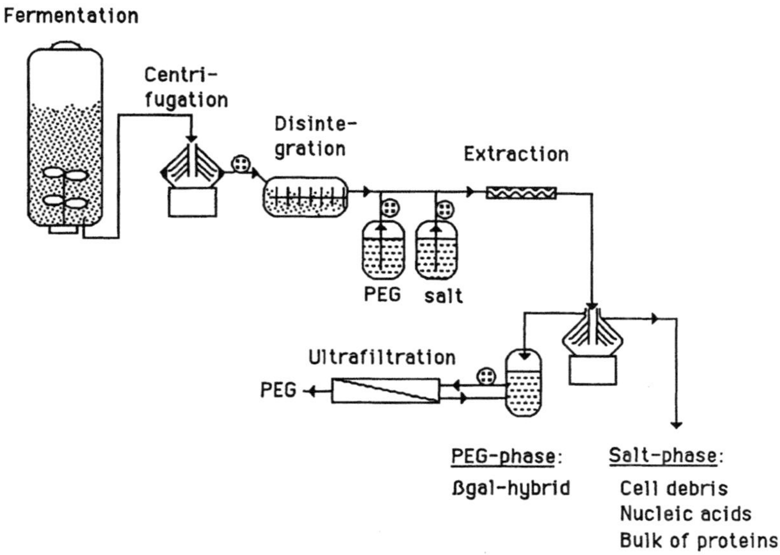

Fig. 5. Flow scheme for continuous extraction of β-galactosidase and β-galactosidase fusion proteins in a PEG-salt two-phase system. The product obtained in the PEG-phase is eventually purified by diafiltration of PEG followed by a concentration step by ultrafiltration

Since in the same system almost all other *E. coli* cell constituents, such as cell debris, nucleic acids and non-β-galactosidase proteins, partition to the bottom phase, it is possible to purify the product in one extraction step [27, 28, 29].

A flow scheme of such a one step purification is shown in Fig. 5. The design of a one-step purification process depends on some key parameters. Fig. 6 shows the binodial, some tie-lines and some iso-volume lines of a PEG 4000/potassium

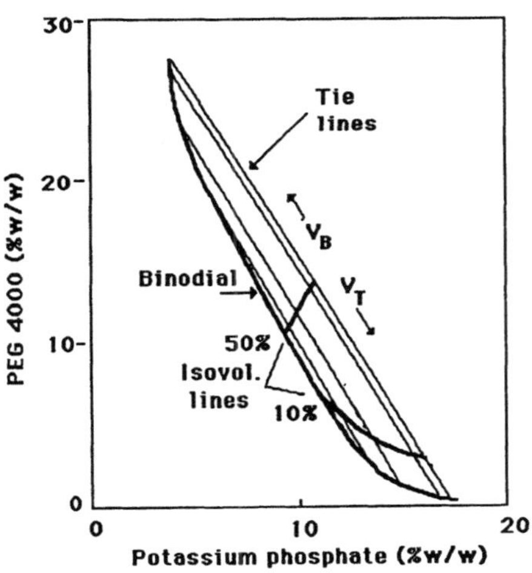

Fig. 6. Phase diagram for PEG 4000 and potassium phosphate including tie lines and iso-volume lines for 10% and 50% top phase volumes. V_T and V_B represent the relationship between the volumes of the top and bottom phases, respectively

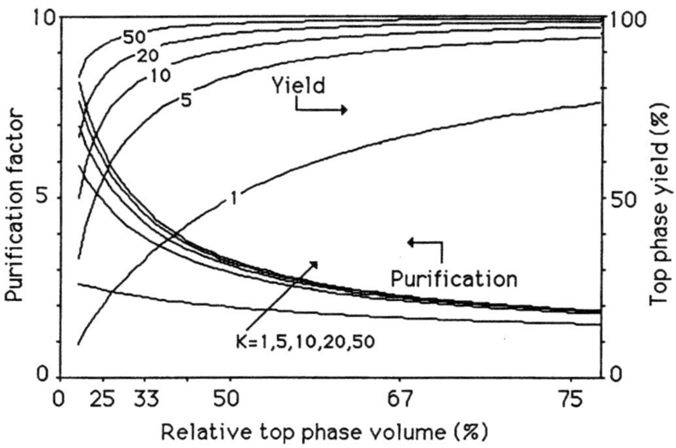

Fig. 7. Influence of relative top phase volume and partition coefficient (*k*) on the top phase yield and purification of a product in one-step extraction to the top phase. Initial product concentration was set as 10% and the partitioning coefficient of non-product proteins was assumed to be 0.3, which is typical for *E. coli* proteins in PEG 4000/potassium phosphate [27]

phosphate system. The relative top phase volume is changed by changing the phase system composition along a tie-line while changes in partition coefficients are obtained through changes of phase system composition along an iso-volume line. The importance of the selection of a proper phase volume ratio is visualized in Fig. 7, which shows a simulation of the top phase yield and purification factor as a function of the top phase volume and partition coefficient. Provided that the initial product concentration is high, as it typically is in production of rDNA proteins in *E. coli*, and that the partition coefficient is high, selection of a low top phase volume system gives the possibility of combining good yield and efficient purification in one step. Furthermore, since the cell debris containing bottom phase is the larger phase, such a system has a high cell load capacity. If the partition coefficient is not high enough, a larger top phase volume must be used to improve the yield at the expense of purification and cell load capacity. An additional advantage of such a system is that it can be applied for continuous extraction of the product in an unpurified cell disintegrate or crude cell extract with only a few minutes residence time. This principe may be important for the reduction of proteolysis during downstream processing.

The applicability of this method depends on the behaviour of β-galactosidase fusions in general. Figure 3 contains partition coefficients of some fusion proteins investigated so far. Since β-glactosidase is a tetrameric protein the fusion product generally contains four additional protein molecules. It is obvious that the properties of β-galactosidase (K = 17) dominate over the properties of protein A (K = 0.7) so that the fusion molecule protein A-β-galactosidase complex, containing 4 molecules of protein A per β-galactosidase molecule, still has a partition coefficient well above 1. According to Albertsson [30] formation of a complex between molecules with little surface contact should give the complex a partition coefficient which is a product of the individual partition coefficients of the molecules. Applied to the fusion between protein A and β-galactosidase this rule should give the fusion molecule a partition coefficient $K_{SpA\text{-}\beta gal} = 17\ (0.7)^4 = 4.1$ The experimental value for the fusion protein was 3.5, which is an indication that the contact surface area between the individual units in the fusion is small.

Biochemical engineers have been trying hard to combine a large number of unit operations to purify a protein but many of the heterologous proteins produced are subject to proteolytic degradation. Genetic engineering now provides a tool that can be used to tailor the protein in such a way that the fusion protein can be purified much more efficiently and it may be stabilized towards proteolysis by conformational changes. A number of purification and stabilization methods can be expected to be included in future biochemical engineering procedures. Depending on the ultimate use of the product additional unit operations may be required to obtain a native protein from a fusion protein.

7 References

1. Cook RA (1988) CRC Critical Reviews in Biotechnology, 8, Issue 3: 159
2. Goldberg AL, Swamy KHS, Chung CH, Larimore FS (1981) Methods Enzymol. 80: 680
3. Larrabee KL, Phillips JO, Williams GJ, Larrabee AR (1980) J. Biol. Chem. 255: 4125
4. Goff SA, Goldberg AL (1985) Cell 41: 587
5. Kitano K, Fujimoto S, Nakao M, Watanabe T, Nakao Y (1987) J. Biotechnol. 5: 77
6. Chesshyre JA, Hipkiss AR (1989) Appl. Microbiol. Biotechnol. 31: 158
7. Goldberg AL, StJohn AC (1976) Ann. Rev. Biochem. 45: 747
8. Groat RG, Matin A (1986) J. Ind. Microbiol. 1: 69
9. Konrad M (1989) TIBTECH 7: 175
10. Marston FAO (1986) Biochem. J. 240: 1
11. Kane JF, Hartley DL (1988) TIBTECH 6: 95
12. Sharma SK (1986) Separation Sci. Technol. 21: 701
13. Shoemaker JM, Brasnett AH, Marston FAO (1985) EMBO J: 4: 775
14. Strandberg L. Hellebust H, Enfors SO (1988) In: Proceedings of biotechnological aspects of protein production in cultured cells, 6–8 July 1988. Prague
15. Strandberg L, Veide A, Enfors SO (1987) J. Biotechnol. 6: 225
16. Hellebust H, Murby M, Abrahmsén L, Uhlén M, Enfors SO (1989) Bio/Technology 7: 165
17. Schein CH, Noteborn MHM (1988) Bio/Technology 6: 291
18. Pietak M, Lane JA, Laird W, Bjorn MJ, Wang A, Williams M (1988) J. Biol. Chem. 263: 4837
19. Mitraki A, King J (1989) Bio/Technology 7: 690
20. Hellebust H, Uhlén M, Enfors SO (1989) J. Biotechnol. 12: 275
21. Hammarberg B, Nygren PÅ, Holmgren E, Elmblad A, Tally M, Hellman U, Moks T, Uhlén M (1989) Proc. Natl. Acad. Sci. USA 86, 4367
22. Moks T, Abrahmsén L, Homgren E, Bilich M, Olsson A, Uhlén M, Pohl G, Sterky C, Hultberg H, Josephson S, Holmgren A, Jörnvall H, Nilsson B (1987) Biochemistry 26: 5239
23. Nilsson B, Moks T, Jansson B, Abrahmsén L, Elmblad A, Holmgrene, Henrichson C, Jones TA, Uhlén M (1987) Protein Eng. 1: 107
24. Moks T, Abrahmsén L, Österlöf B, Josephson S, Östling M, Enfors SO, Persson I, Nilsson B, Uhlén M (1987) Bio/Technology 5: 379
25. Hochuli E, Bannwarth W, Döbeli H, Gentz R, Stüber D (1988) Bio/Technology 6: 13
26. Sassenfeld HM, Brewer SJ (1985) TIBTECH 3: 119

27. Veide A, Lindbäck T, Enfors SO (1984) Enzyme Microb. Technol. 6: 325
28. Veide A, Strandberg L, Enfors SO (1987) Enzyme Microb. Technol. 9: 730
29. Köhler K, von Bonsdorff-Lindeberg L, Enfors S-O (1989) Enzyme Microb. Technol. 11: 730
30. Albertsson PÅ (1981) Meth. Biochem. Anal. 29: 1
31. Szoka PR, Schreiber AB, Chan H, Murthy J (1986) DNA 5: 11
32. Goeddel DV, Kleid DG, Bolivar F, Heyneker HL, Yansura DG, Crea R, Hirose T, Kraszewski A, Itakura K, Riggs AD (1979) Proc. Natl. Acad. Sci. USA 76: 106
33. Nagai K, Perutz MF, Poyart C (1985) Proc. Natl. Acad. Sci. USA 82: 7252

Genetics and Genetic Engineering of the Industrial Yeast *Yarrowia lipolytica*

H. Heslot

Chaire de Génétique Moléculaire et Cellulaire, Institut National Agronomique, 16 Rue Claude Bernard, 75231 PARIS Cédex 05, France

Advances in Biochemical Engineering/
Biotechnology, Vol. 43
Managing Editor: A. Fiechter

Y. lipolytica can be grown on hydrocarbons, and has been used to make single cell proteins and citric acid. The existence of sexuality allowed the performance of inbreeding programs leading to strains that can be utilized for classical genetic analysis. Physiological and genetic studies have been carried out on hydrocarbon utilization, lysine biosynthesis and catabolism, and protein secretion.

Due to the lack of plasmids, transformation of *Y. lipolytica* was first achieved by chromosomal integration, using either homologous or heterologous genes as markers. In a second step, the isolation of ars sequences allowed the building up autoreplicating vectors. A number of heterologous genes have been expressed in *Y. lipolytica*, such as those coding for calf prochymosin, *S. cerevisiae* invertase and porcine alpha-1-interferon. These proteins are secreted.

1 Introduction

Yarrowia lipolytica is an industrial yeast which can be grown on hydrocarbons and has been used for the production of single cell protein and citric acid [1]. It has been utilized to produce a number of metabolites: 2-keto glutaric acid [2] erythritol [3], mannitol [4], isopropyl malic acid [5]. It is also a producer of exocellular proteins and secretes an alkaline protease in fairly high amounts [6].

2 Cell Cycle — Mating Types

Wickerham et al. [7] discovered sexuality in the dimorphic yeast *Candida lipolytica*. Upon sporulation, the asci contained 1 to 4 hat-shaped spores giving rise to haploid colonies of mating types A or B. The strains of *C. lipolytica* showing sexuality were renamed *Saccharomycopsis lipolytica* [8] and then, more recently, *Yarrowia lipolytica* [9]. The life cycle is shown in Fig. 1.

The early genetic studies of *Y. lipolytica* were hampered by the low sporulation frequencies and the low spore viability. For instance, if the fertility of a cross was evaluated as the frequence of asci in a mixture with vegetative cells, low values ranging between 0.5 and 0.8% were found. The germinability of the spores was also low, from 0.1 to 10% [10]. These unfavourable characteristics seemed to exclude the possibility of performing genetic studies. Attempts were made to improve the sporulation media and study factors controlling copulation such as cell density. (10^7 cells ml^{-1} were found to be optimal).

A number of auxotrophs were derived from the two wild type strains W (mating type A) and Z (mating type B), such as W *his1* (A) and Z *lys13* (B). Mixing these strains in minimal medium allowed to isolate prototrophs which were shown to be complementing diploids, on the basis of cell size, DNA content and meiotic segregation [10].

However, the variable number of spores per ascus and the low germinability of the spores precluded the use of microdissection of individual asci. Consequently, ascospores were separated from vegetative cells using paraffin oil. This technique allows one to measure the frequency of spore germination.

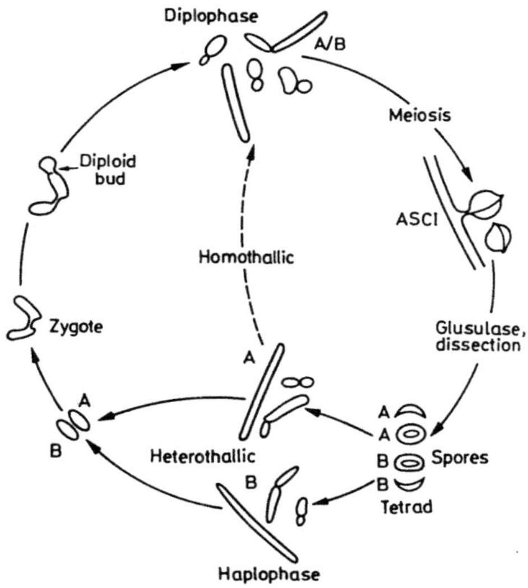

Fig. 1. Life cycle of *Yarrowia lipolytica*. From. Ogrydziak et al. Development of genetic techniques and the genetic map of the yeast *Saccharomycopsis lipolytica* [11]

The effect of brother × sister matings on the behaviour of compatible strains was studied, starting with the diploid D11, which was the result of a cross between W *his1* (A) and Z *lys13* (B). As shown in Table 1, the behaviour of D11 is poor especially with regard to the percentage of spores per ascus and the percentage of spores germinating. Haploid segregants obtained from D11 were checked for mating type and auxotrophy. Then brother × sister matings were performed between compatible pairs. The process was repeated several times. The results were quite striking: 1) the capacity of dipoids to sporulate increased; 2) the mean number of spores per ascus also increased; 3) spore germinability was considerably improved (from 1% to 90%).

Table 1. Effect of brother × sister mating on fertility, mean number of spores per ascus and spore germinability

Compatible strains isolated in		Diploid	% asci[a]	% Spores per ascus[b]			% Spore germination[c]
				2	3	4	
his⁻	*lys*⁻						
W	Z	D 11	35–40	90	7–8	2	1.0
W	D 11	D 1701	30–35	85	12	3	2–3
D 11	D 11	D 1905	40–50	55	30	15	20–40
D 1901	D 1901	D 2801	70–80	10–20	8–10	70–80	80–90
D 2801	D 2801	D 8002	80	5–10	0–5	80–90	90

[a] 1000 to 1500 cells (or asci) were counted.
[b] Established on 150–200 asci.
[c] Counts of spores in paraffin oil were compared with number of colonies formed on YE
From: Gaillardin et al. A study of copulation, sporulation and meiotic segregation in *Candida lipolytica* [10]

The successive diploids were analyzed with respect to meiotic segregation. A considerable unbalance between the two parental types as well as a large excess of prototrophs were found at the beginning. In the course of brother × sister matings, the regularity of segregations was improved. Similar results were obtained by Ogrydziak et al. [11], Kurischko et al. [12].

Barth and Weber [13] used nystatin to develop a new method to select spores of *Y. lipolytica*. At appropriate concentrations, nystatin kills vegetative cells, but has no effect on spores. Moreover, nystatin has no discriminating effects among spores isolates segregating several auxotrophic markers or mating types A and B.

Barth and Weber [14] investigated conditions for mating in order to enhance zygote formation. They were able to increase conjugation frequencies about 25 fold by using late logarithmic growth cells and a yeast extract-malt medium.

In further experiments, Barth and Weber [15] developed a synthetic sporulation medium, containing 1.5% citrate. On such a medium, the sporulation frequency of a diploid was over 80% in the pH range 5,9 to 7.5. The same authors selected, by inbreeding, *Y. lipolytica* strains forming exclusively spherical ascospores. These strains turned out to be more suitable for asci dissection.

Esser and Stahl [16] have studied ascospore formation by cytological and genetical methods. In unselected strains, the number of spores per ascus was very variable. Depending on the strain, instead of the expected 4 spores, up to 90% of the asci contained 2 spores; a few asci contained 3 spores or only one. Diploid cells were plated on a sporulation medium and, after 4–5 days, smear preparations were stained with Giemsa. The asci were checked for the number of spores per ascus and for the number of nuclei in each spore. The conclusion was that in each ascus four nuclei were always present. However, meiotic division and spore wall formation were apparently not correlated, i.e. in 2-spore asci, each spore contained 2 nuclei.

Weber [17] also studied sporulation of diploids, using serial sections of asci examined by electron microscopy. He confirmed that there were 4 nuclei in each ascus, but — in contrast to Esser and Stahl [16] — found that in 1-, 2-, 3- spored asci, unenclosed "naked" nuclei occured additionally to the nuclei incorporated in mature spores. All ascospores derived from asci with different spore numbers were uninuclear.

3 Growth Characteristics

According to the culture medium, the morphology of *Y. lipolytica* can be altered from yeast to mycelium. Ota et al. [18] have found that *Y. lipolytica* grows as true mycelium in a synthetic medium supplemented with olive oil or bovine milk casein. Rodriguez and Dominguez [19] showed that on minimal medium supplemented with either glucose or *n*-hexadecane as the carbon source, the cells remained in the yeast form independent of the age of the culture. On a complex medium (yeast extract) however, a mixture consisting of yeast forms and mycelium was found. The effect of several sugars on the morphology and growth of *Y. lipolytica* was investigated; with *N*-acetylglucosamine, mainly mycelium was formed.

Vega and Dominguez [20] isolated and purified the cell walls of the yeast and mycelial forms of *Y. lipolytica*. Analysis of hydrolysates of both forms revealed the presence of only three sugars: glucose, mannose and *N*-acetylglucosamine. Proteins extractable from the cell walls were separated by polyacrylamide gel electrophoresis. About 50 bands were detected. The cell wall of yeast and mycelial forms of *Y. lipolytica* were qualitatively similar.

Ogrydziak et al. [11] have isolated mutants forming smooth, soft colonies, in contrast to the wild type which produces dry, convoluted colonies containing yeast phase cells as well as branched mycelia. Smooth mutant colonies (ruf) contain almost exclusively yeast phase cells. Fournier et al. [21], in their search for *ars* sequences, have also isolated smooth mutants.

4 Genetic Maps

Ogrýdziak et al. [11] isolated sixty-seven mutations in fifty-eight genes and used them for mapping. Six linkage fragments were established.

Mone recently Ogrydziak et al. [22] reported on mutations in 23 new nuclear genes. Linkage fragments 3 and 4 were shown to be linked. A tentative map of F1 and F3 was presented (Fig. 2). The most recently results have been summarized by Ogrydziak [23].

Kurischko [24] found linkage between markers *metA-lysA* and between *argA-leuA*. The latter was extended to include *MAT* (Mating type locus) which is on linkage fragment 6 [25]. *alk23* was also found to be on fragment 6 Bassel and Mortimer [26], as well as *lys2* and *his1* (De Zeeuw, pers.commun.)

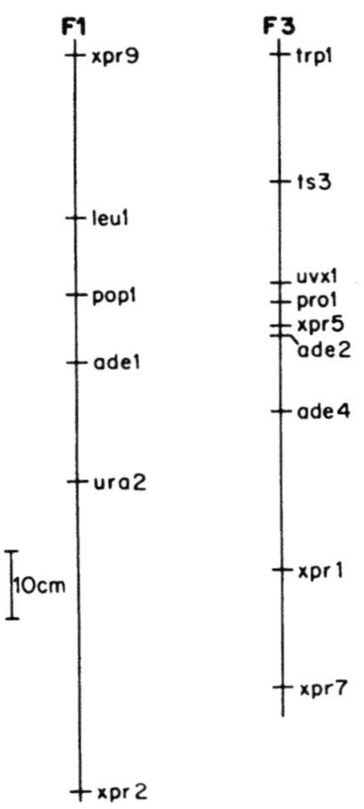

Fig. 2. Tentative genetic map of linkage fragments 1 and 3 of *Yarrowia lipolytica*. From: Ogrydziak et al. Development of the genetic map of the yeast Saccharomycopsis lipolytica [22]

The linkage fragments are now as follows:

Fragment 1: (see Fig. 2). The markers *thr2* and *ts* k were not included in the map.

Fragment 2: *alk1*, *ura1*, *xpr10*, *xpr15*.

Fragment (3 + 4): (see Fig. 2). The markers *alk5*, *phe1* and *ura3* were not included in the map.

Fragment 5: *alk232*, *ts1*.

Fragment 6: *alk2*, *alk4*, *metA*, *alk23*, *argA*, *leuA*, *leu2*, *his1*. It is not known if *leu2* and *leuA* are allelic.

Fragment 7: *metA*, *lysA*.

The genetic map of *Y. lipolytica* is still primitive compared to that of *S. cerevisiae* (Mortimer et al. [27]).

OFAGE (Orthogonal — Field — Alteration Gel Electrophoresis) detects only 2 bands for *Y. lipolytica* [28, 29]. TAFE (Transverse Alternation Field Electrophoresis) detected 4–5 bands (Fournier, pers. commun.) As soon as improved methods allow full separation of *Y. lipolytica* chromosomes, it should become easy to map cloned genes by hybridization techniques.

Useful strains for genetic purposes have been developed in several laboratories in France by Gaillardin et al. [10], in the USA by Ogrydziak et al. [11, 22], and in the GDR by Kurischko et al. [12]. As we have seen, inbreeding has resulted in increased spore viability and the percentage of 4-spored asci. However, when crosses are performed between strains of different origins, for instance French and American, the spore viability drops to 10% or less, making it difficult to incorporate markers obtained in one set of inbred strains into the other set.

5 Protoplast Fusion — Parasexual Cycle

Stahl [30] obtained protoplasts of two auxotrophic strains of *Y. lipopolytica* of identical mating type (*lys*, *ade* and *his*, *arg*) and fused them with PEG. Complementing diploids were selected on minimal medium. Upon treatment of these diploids with the haploidizing agents benomyl or *p*-fluorophenylalanine, 10 out of 12 possible auxotrophic segregants were obtained. In about 3% of the prototrophic fusion products, sexual reproduction was induced by transfer onto a sporulation medium. Asci were isolated and the few spores that were able to germinate gave rise mostly to recombinants, all being of the same mating type as the parents.

Weber et al. [31] have shown that an electric field pulse applied via discharge to protoplasts of *Y. lipolytica* enhanced their fusion considerably (from 10^{-4}–10^{-5} to 10^{-3}–10^{-2}).

Kurischko and Spata [32] performed a comparative genetic analysis of the offspring of an heterozygous diploid, obtained either through meiotic or methylbenzimidazol-2-yl-carbamate (MBC)-induced mitotic segregation. The mitotic segregants were clearly different from those obtained after meiosis. The authors also failed to confirm that diploids homozygous for the mating type locus were able to sporulate.

Kurischko and Weber [33] and Kurischko [25, 34] used special selective conditions to analyse the temporal relationship of diploidization (formation of heterokaryons, karyogamy) and haploidization (meiotic segregation). They showed that, 12 hours after mixing of the parental strains on sporulation media, dissociation products of unstable heterokaryons were selectable. At 16 hours, karyogamy had taken place, as shown by the isolation of spontaneous mitotic recombinants (haploids and aneuploids) whose frequency increased with time. These results showed that in *Y. lipolytica*, in addition to the sexual life cycle, parasexual processes occured, leading to haploidization of the early diploid nuclei.

6 Genome Structure

6.1 DNA Content

Gaillardin et al. [10] have found a total DNA content of 4×10^9 Da per cell and Kurzman et al. [35] have measured a GC content of 49.6%. No plasmid was identified in a survey of 24 strains tested (Treton, unpublished).

6.2 Mitochondrial DNA

Kück et al. [36] have isolated mitochondrial(mit) DNA of *Y. lipolytica*. Its buoyant density in CsCl was found to be 1.687 g cm^{-3}. Its melting point, Tm, of 79.5 °C indicated a GC content of 24.9%. By contrast, nuclear DNA showed a Tm of 90.5 °C (GC content: 51.7%).

Electron microscopy indicated that mit DNA is a circular structure with a contour length of 14.5 μm. This length is much shorter than that of *S. cerevisiae* (25 μm) as shown by Hollenberg and al. [37].

A restriction map of *Y. lipolytica* mit DNA was constructed using the cleavage data of four restriction enzymes. Matsuoba et al. [38] isolated a spontaneous oligomycin-resistant mutant. In protoplast fusion experiments with a drug-sensitive strain, it was found that oligomycin-resistance could be transferred to the sensitive parent without caryogamy. It seems likely that this type of oligomycin resistance is encoded by the mitochondrial genome.

6.3 Ribosomal RNA Genes

The rDNA of eukaryotes is generally organized in one or a few clusters of tandemly repeated units. Each unit is composed of a transcribed unit coding for a 35–45S precursor rRNA and a so-called non-transcribed spacer (NTS). In some species a gene for 5S rRNA is contained within the NTS.

In *Y. lipolytica*, Van Heerikhuizen et al. [39] have showed that there are two major size classes of DNA units (8.7 and 7.7 kb) coding for rRNAs. Both types of units were cloned and it was found that the size difference was mainly in the length of the spacer. No gene for 5S rRNA was observed in the spacer region, showing that the rDNA of *Y. lipolytica* is clearly different from that of *S. cerevisiae*.

Further studies by Fournier et al. [40] used the cloned rDNA units to probe blots of genomic DNA of a number of wild-type and laboratory strains. Each strain displayed a specific pattern and contained two-to-five types of repeated units. When the parents differed in the type (or number) of rDNA repeats, the meiotic segregants showed recombination.

Clare et al. [41], in a less extensive survey of the ribosomal RNA genes of *Y. lipolytica*, confirmed the above findings. They found a repeat unit of 8.9 kb containing the 25S and 18S, but not the 5S rRNA species. The number of copies of these repeat units was estimated to be approx. 50 per haploid genome.

6.4 Virus-like Particles

Some strains of *S. cerevisiae* contain virus-like particles (VLPs), the genome of which consists of one or more double-stranded RNAs (ds RNA) molecules. These VLPs are associated with the killer system (Wickner [42]; Bussey [43]).

Groves et al. [44] have surveyed *Y. lipolytica* for the presence of VLPs. Such particles were indeed isolated from one strain. Their diameter was roughly 50 nm and they contained a double-stranded RNA molecule of MW 3.8×10^6. The major protein component of the VLP had a MW of 75.7 kDa.

Tréton et al. [45] checked for the presence of VLPs in 24 strains of *Y. lipolytica.* These particles were found in four of these strains. Their encapsidated dsRNA was 4.9 kb long. A major VLP polypeptide of MW 80 kDa was observed in all cases, and a second one of MW 77 kDa in three cases. The ds RNA from the VLPs possessing only the large polypeptide showed little homology with the three others. There was no homology between VLP ds RNAs and host DNA or ds RNA from *S. cerevisiae*. No relationship could be found between the presence of VLPs and possible killer behaviour in *Y. lipolytica.*

El-Sherbeini et al. [46] studied VLP possessing two associated polypeptides: P1 (83 kDa) and P2 (77 kDa). These VLPs contained a ds RNA genome designated Ly. Denatured Ly-ds RNA was used to program a cell-free rabbit reticulocyte translation system, resulting in the synthesis of four major proteins P1 (83 kDa), P2 (77 kDa), (74 kDa) and P4 (68 kDa). The in vitro P1 and P2 products were shown to be identical to the in vivo VLP-associated proteins. It was concluded that P1 and P2 were primary translation products of the Ly genome. P3 and P4 are probably fragments of P1 or P2 or both, produced by premature termination of translation in vitro.

Tréton et al. [47] eliminated VLPs from a *Y. lipolytica* strain by UV irradiation. The cured strain was compared with the initial one: it grew slightly slower but did not show other differences.

VLP loss could have been caused either by a direct effect on ds RNA, or through mutation in a chromosomal gene involved in ds RNA maintenance.

The cured strain was crossed with a VLP containing strain, both strains bearing appropriate auxotrophic markers. Segregation of the auxotrophic mutations was compatible with expectation for independent monogenic markers, but all the offspring contained VLPs, an indication that a host mutation was probably not involved in VLPs loss.

7 Citric Acid

Strains of *Y. lipolytica* secrete citric and isocitric acids when grown on *n*-paraffins as carbon sources. Akiyama et al. [48, 49] have investigated strain ATCC 20114 growing on *n*-paraffins. Citric and isocitric acids were accumulated in the ratio 60:40 with a total yield of 130% (w/w). However the development of an improved process required to obtain strains excreting only (or predominantly) citric acid.

In order to decrease the formation of isocitric acid, the effect of monofluoroacetate was studied. A concentration of 0.1% gave an improved ratio of citric to isocitric acid (85:15). This was the result from competitive inhibition of aconitase by monofluorocitrate derived from monofluoroacetate.

From this experiment the conclusion was drawn that mutant strains having low aconitate hydratase activity should produce citric acid without isocitric acid. Cells were mutagenized with *N*-methyl *N'*-nitro-*N*-nitrosoguanidine and a mutant, K20, unable to utilise citrate as carbon source was selected. Strain K-20 was further mutagenized to obtain a fluoroacetate-sensitive strain S-22.

The citric acid productivity from *n*-paraffins by the mutant strains was compared with that of the parent strain (Fig. 3). The citric acid to isocitric acid ratio was changed to 85:15 in strain K-20 and to 97:3 in strain S22. With S22 the yield from *n*-paraffins reached 145% (w/w).

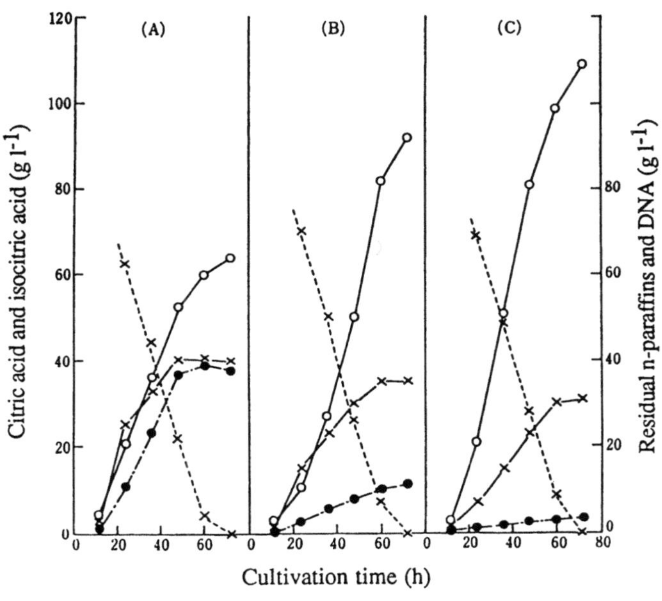

Fig. 3. Comparison of the fermentation time course. ○—○: Citric acid; ●—●: Isocitric acid; ×--×: Residual *n*-paraffins; ×—×: Growth (DNA); (A): Parent strain; (B): Mutant strain K–20 and; (C): Mutant strain S-22. From: Akiyama et al. Induction and citric acid productivity of fluoroacetate-sensitive mutant strains of *Candida lipolytica* [48, 49]

To further investigate the properties of the mutant strains, activities of enzymes involved in the citric acid cycle were measured. The only significant change concerned aconitate hydratase: in strains K-20 and S-22, these activities were about 1/10 and 1/100 of that of the parent strain, respectively.

Treton and Heslot [50] have studied the properties of *Y. lipolytica* aconitate hydratase. Various inhibitors were investigated in vitro, including monofluorocitrate. However, in vivo, the strain was found to be quite resistant to mono fluoroacetate.

Tréton et al. [51] investigated the excretion of citric and isocitric acid in a strain of *Y. lipolytica* grown on either *n*-paraffins, glucose or glycerol. These acids were excreted in the ratio 67:33 on *n*-paraffins and roughly 92:8 on either glucose or glycerol. However, with all carbon sources used, the relative amount of isocitric acid in the intracellular pool remained below 10%. It could be shown that when the cultivation was carried out on glucose or glycerol, neither citric acid nor isocitric acid could be reconsumed after being excreted. It was therefore not surprizing that their relative concentration in the medium reflected the intracellular pool.

During growth on *n*-paraffins, both citric and isocitric acids were excreted but citric acid alone could be reconsumed. This led to an increased proportion of isocitric acid.

Finogenova et al. [52, 53] determined the activity of enzymes of the citrate and glyoxylate cycles in cells of a wild type strain of *Y. lipolytica* which produced a mixture of citric and isocitric acids in a medium containing hexadecane as the carbon source.

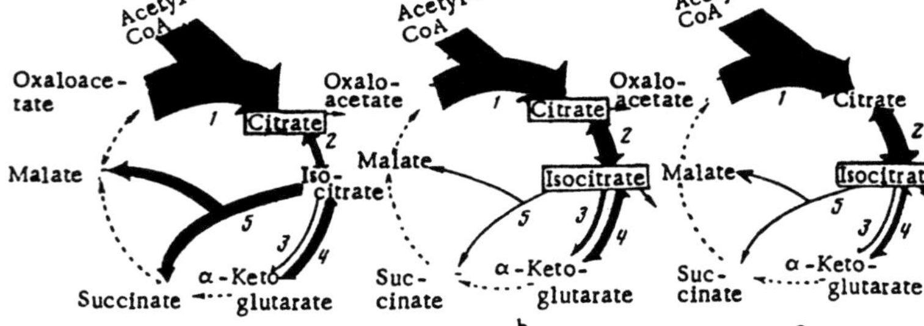

Fig. 4a–c. Sequence of reactions of citrate and glyxylate cycles. The **thick arrows** denote the activity of the enzymes in the yeast cells in the lag phase and in the period of the beginning of excretion of the acids into the medium. **a)** Mutant 1; **b)** natural strain of *C. lipolytica* IBPM U-160; **c)** mutant 2; **1)** citrate synthase **2)** aconitate hydratase; **3)** NAD-isocitrate dehydrogenase; **4)** NADP-isocitrate dehydrogenase; **5)** isocitrate lyase. **Dotted lines**: Acitivities of these enzymes were not determined. **Rectangle**: Metabolite excreted into the medium. From: Finogenova and Glazunova. Activity of enzymes of the citrate and glyoxylate cycles in the synthesis of citric and isocitric acids by various strains of *Candida lipolytica*. Mikrobiologiya (1982) (English translation) 51: 27

These activities were compared with those of two mutants, one producing predominantly citrate (M1), and the other one prodencing predominantly isocitrate (M2). Enzymatic acitivities were determined over three periods: the exponential growth phase, the log phase and the stationary phase. All the strains had high citrate synthase activity. The citrate-producing mutant M1 had high isocitrate lyase and decreased aconitate hydratase activities. In contrast, the isocitrate producer M2 had high aconitate hydrase and low isocitrate dehydrogenase/isocitrate lyase activities (Fig. 4). These alterations in enzymatic activities of key enzymes of the TCA and glyoxylate cycles explain the nature of the secreted products. Unfortunately no information was given concerning the genes that have been mutated in strains M1 and M2.

A citric acid-producing strain DS1 (mating type A) of *Y. lipolytica* was used to induce a number of double auxotrophic mutants through UV mutagenesis [54, 55]. These double mutants showed a considerable decrease in their capacity to secrete citric + isocitric acids (0 to 5% of the original strain DS1).

Two double mutant strains DS-110 (*arg*$^-$ *met*$^-$) and DS-118 (*ade*$^-$ *lys*$^-$), derived from DS1, were used to prepare protoplasts and these were fused giving rise to prototrophs. The DNA content of the latter was found to be approximately double of that of the haploid parents. All of them produced high levels of citric + isocitric acids (80 to 104.8% of DS1).

One of these fusants sporulated and was used to analyze the meiotic segregation pattern by random spore analysis. There was a wide variation of product yield, even among segregants with identical auxotrophic markers; for instance (*arg*$^-$ *met*$^-$) segregants produced citric + isocitric acid in the range 0–60% of DS1.

These experiments may make it possible to identify specific genes affecting citric acid production. They also confirm the previous report [30] that diploid fusants issued from haploids of identical mating types are able to sporulate.

Matsuoka et al. [56] isolated four acetate nonutilizing mutants of *Y. lipolytica* specifically deficient in isocitrate lyase activity. Genetic analysis indicated that these mutations were recessive, non-complementing, and defined a single locus *icl*. This is the structural gene coding for isocitrate lyase, as indicating by the following observations: 1) some revertants of *icl* produced thermostable isocitrate lyase; 2) a dosage effect on isocitrate lyase activity was observed in heterozygous diploids *ICL*/*icl*.

Interestingly, some mutants exhibited temperature-sensitive synthesis, although they produced a thermostable isocitrate lyase, indistinguishalble from that of the wild type.

The purified wild type isocitrate lyase was found to be tetrameric, with a subunit MW of 59000. When induced by acetate at 33 °C, the mutant exhibiting temperature-sensitive synthesis produced inactive isocitrate lyase polypeptides, an indication that the mutation prevented correct maturation.

Hönes [57] has studied the regulation of the enzymes involved in the glyoxylate cycle of *Y. lipolytica*.

Barth [58] isolated *icl* mutants in *Y. lipolytica* to investigate the synthesis of isocitrate lyase. Eighteen mutations were localized in a gene *ICL1*, considered to be the structural gene. A trans-active *ICL2* gene, not linked to *ICL1* was identified.

Another mutation was localized in *ICL3* which was linked to *ICL1*. One mutation in *ICL3* lowered ICL activity in cis position with respect to gene *ICL1*.

Barth and Weber [59] further characterized the *ICL1* locus. Twenty *icl*-alleles were analyzed for their reversion frequency, intraallelic complementation pattern and location within the fine structure map of *ICL1*. Six intragenic complementation groups, A–F, were observed. Mitotic fine mapping was done by the X-ray method of Manney and Mortimer [60].

The distance between the bordering mutations in *ICL1* and the mutation in locus *ICL3*, causing *cis*-dominant inducibility of isocitrate lyase, was estimated. This allowed to infer the direction of transcription.

One revertant expressed a thermosensitive enzyme, which was very unstable at temperatures above 30 °C. This reversion mapped inside the *ICL1* gene.

8 Lysine Metabolism

8.1 Biosynthesis

Lysine biosynthesis in *Y. lipolytica* involves a pathway different from that found in plants and molds.

There are 11 steps leading from acetyl coenzyme A (AcCoA) and α-ketoglutarate (αKG) to L-lysine (Fig. 5). Heslot et al. [61] isolated lysine auxotrophs after mutagenic treatment and tested them for their capacity to grow on intermediates of the pathway. Only α-aminoadipate could be used by some of the mutants and they were classified as utilizers (class 1) and non-utilizers (class 2).

Complementation tests were performed and 11 complementation groups were found, defining a priori 11 genes. Measurement of enzymatic activities of the wild type strain and the mutants allowed some of the blocks to be located: *lys1*

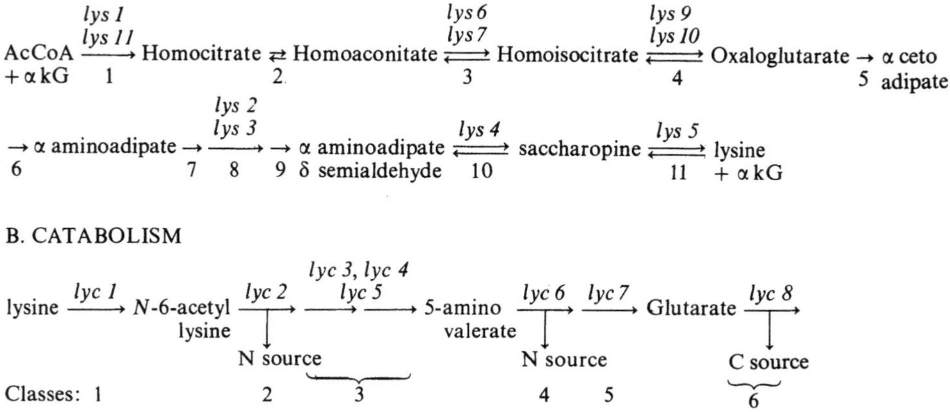

Fig. 5. Biosynthesis and catabolism of lysine in *Yarrowia lipolytica*

and *lys11* lack activity 1; *lys6* and *lys7* have no homoaconitase; *lys9* and *lys10* have no homocitrate dehydrogenase; *lys2* and *lys3* are affected in steps 7, 8 and 9; *lys4* is blocked in step 10; *lys5* has no saccharopine dehydrogenase. There is some intragenic complementation between *lys1* mutants, implicating an oligomeric structure for homocitrate synthetase, which is not surprising, in view of the fact that it is the first enzyme of the pathway.

Homocitrate synthetase is specifically inhibited by lysine, or very close analogs such as cyclohexylalanine and transdehydrolysine.

Homoisocitric dehydrogenase was purified some 500-fold. It consisted of a single protein of MW 48000 [62]. In the presence of homoisocitric acid, a higher molecular weight was observed, suggesting a dimeric structure for the native enzyme. The closely linked *lys9* and *lys10* mutants were devoid of homoisocitric dehydrogenase, and showed partial complementation. It was assumed that the active enzyme was a dimer of identical subunits involved in successive dehydrogenation and decarboxylation steps.

The last enzyme of the pathway, saccharopine dehydrogenase (*lys5*) is apparently repressed two fold by L-lysine in the wild type. This repression disappears completely in mutants (*lyc1*, *lyc2*, *lyc3*, *lyc4*, *lyc5*) which are unable to carry out the first step of the lysine catabolic pathway (see below) i.e. lysine *N*-6-acetyllysine (Fig. 5).

Mutant resistant to transdehydrolysine have been isolated in the wild type strain. Two main types were identified: 1) those altered in the L-lysine transport system; 2) those with an altered homocitrate synthase, desensitized to feedback by L-lysine. In one of the type 2 mutants, the lysine pool was increased five times.

8.2 Catabolism

Wild type strains of *Y. lipolytica* can use lysine as a carbon or nitrogen source. Only *N*-6-acetyllysine and 5-aminovalerate were able to replace lysine as source of carbon or nitrogen [63]. When ^{14}C-lysine was supplied to a cell culture, *N*-6-^{14}C-acetyllysine was detected in the cellular pool, but 5-^{14}C-aminovalerate was only found in very small amounts.

Five phenotypic classes were found among mutants defective in lysine catabolism, comprizing a total of eight complementation groups. According to their growth responses on intermediates, the mutants could be located as indicated on Fig. 5.

Surprisingly, all mutants of class 3 (3 different loci) were found to be unable to use lysine, but they were able to grow on *N*-6-acetyllysine as nitrogen source. The explanation is that these mutants produce *N*-6-acetyllysine only in the presence of 5-aminovalerate which is probably the inducer or activator of the enzyme coded for by *lyc1*.

8.3 Active Transport

Cells of the wild type strain were incubated for 2 min in a minimal growth medium (pH 5.0) containing ^{14}C lysine (range 10–2000 μm). The initial velocity of uptake followed a Michaelis-Menten kinetics with a Km of 1.9×10^{-5} M.

Inhibitors of the energy metabolism, such as 2.4-dinitrophenol and sodium azide inhibited up to 95% of the lysine uptake.

A number of compounds were studied and the following were found to act as competitive inhibitors of uptake: L-arginine, L-ornithine, L-*S*-aminoethylcysteine, and L-4,5-transdehydrolysine. The Ki for arginine (1.8×10^{-3} M is of the same order of magnitude as the Km for lysine, an indication that the permease is probably common to L-lysine and L-arginine [64].

When the lysine uptake was studied at higher L-lysine concentrations (1 to 10 mM), a second transport system, of much lower affinity, was detected. It has a Michaelis Menten kinetics with a Km of 4×10^{-4} M.

Mutants affected in lysine transport systems were isolated by plating mutagenized cells on a mixture of two competitive inhibitors. The resistant mutants were tested for their uptake capacity at 2.5 mM and 50 µM L-lysine. They were affected at one or both of these concentrations.

8.4 Mutants Affected in Lysine and Polyphosphates pools

Sawnor-Korszynska et al. [65] have measured the total pools of free amino acids in *Y. lipolytica*. It varies from 250 to 350 µm/g dry wt and accounts for about 10% of the total amino acid content. 80–90% of the free amino acids were located in the vacuole.

In *S. cerevisiae*, almost all of the polyphosphates are sequestered in the vacuole and it has been suggested that they play a major role in arginine retention Dürr et al. [66]. In order to establish whether or not this was true for lysine in *Y. lipolytica*, Beckerich et al. [67] isolated nine mutants (*ply1* to *ply9*) with low polyphosphates pools. No stochiometric relationship was found between the lysine pool and the polyphosphate pool, excluding that polyphosphates play a role as cationic receptors. On the other hand, the lysine and polyphosphate pools were simultaneously depleted in the mutants, suggesting some kind of relationship between these pools.

Beckerich et al. [68] studied the compartmentation of lysine in vivo by a tracer method and used a mathematical model to calculate the size of the different lysine pools (cytoplasm, vacuole) and lysine fluxes inside the cell. Four strains were studied: a lysine-accumulating mutant and three derived low polyphosphate pool mutants (*ply1*, *ply4*, *ply9*).

8.5 Control of Lysine Metabolism by Mating Type

Beckerich et al. [69] used as starting strain a *LYS1.5 lyc1.5* double mutant with a feedback resistant homocitrate synthase (*LYC1.5*) and a block in the first step of lysine catabolism (*lyc1.5*). The combination of both mutations led to a high intracellular accumulation of free lysine.

This strain was used to isolate low lysine pool mutants. These "revertants" had regained the ability to use lysine as a nitrogen source, i.e. the *lyc1.5* mutation was suppressed. The suppressor was shown to map inside the *LYC1* locus. This revertant, and similar ones, had surprisingly switched the mating type from A to B.

The *lyc1.5* revertant had a modified pattern of lysine storage after nitrogen starvation and its behaviour varied according to the mating type status in haploid (A, B) and diploid (AB, AA, BB) strains.

This phenotype shows similarities with the *ROAM* mutants of *S. cerevisiae* [70] which overproduce an enzyme and are associated with the insertion, in the upstream region of the overexpressed gene, of a Ty1 transposon whose transcription is dependent of the mating type of the cell [71].

Are these findings an indication that transposable elements also exist in *Y. lipolytica*?

9 Overproduction of Methionine

Morzycka et al. [72] have isolated six ethionine-resistant (*Etr*) mutants of *Y. lipolytica*; overproducing methionine. The pool of free methionine in *Etr* mutants was enhanced 1.5 to 18 times. The activity of homocysteine synthase was found to be derepressed in these mutants and insensitive to repression by methionine.

10 Overproduction of Protoporphirine IX

Bassel et al. [73] isolated, after ultraviolet mutagenesis, a mutant (*pop1*) forming red colonies. Spectroscopic analysis of the red pigment indicated that it was protoporphyrin IX. The *pop1* mutant was recessive and segregated in a mendelian way.

Louvel [74] investigated the protoheme biosynthetic pathway in a wild type strain and in the *pop1* mutant. Compared to the wild type strain, the activity of the first enzyme (ALA synthetase) was 12-fold higher in the mutant; that of the two following enzymes (ALA dehydratase and uroporphyrinogen I synthetase) was increased 2.5-fold. However, the activity of the last enzyme of the pathway (ferrochelatase) which has a mitochondrial location was decreased by a factor of 10.

These findings supply a rational explanation for the oversynthesis of protoporphyrin IX, if it is assumed — as is the case for liver cells — that protoheme retroinhibits ALA synthetase.

The ferrochelatase deficiency found in the mutant could result from either: 1) a decrease in the rate of enzyme synthesis; 2) a targeting defect partially pre venting the cytoplasmically synthetized enzyme to enter the mitochondrion; or 3) a decrease in catalytic efficiency as a result of a mutation in the structural gene. Cloning and sequencing of the *pop1* gene would help to clarify the situation.

11 *N*-Alkanes and Fatty Acids Utilization

Markov and Kibarska [75] have isolated mutants of *Y. lipolytica* assimilating *n*-paraffins better than the control strains. The screening procedure used mutagenized colonies growing on kerosene. The wild type strain did not form any

Table 2. Substrate utilization tests of alkane-negative mutants of *S. lipolytica*

Phenotypic designation	Growth on carbon sources						Total mutants
	Alkane	Alcohol	Aldehyde	Fatty acid	Acetate	Glucose	
A	−	+	+	+	+	+	27
B	−	−	+	+	+	+	0
C	−	−	−	+	+	+	7
D	−	−	−	−	+	+	4
E	−	−	−	−	−	+	6
							44

From: Bassel et Mortimer, Genetic and biochemical studies of *N*-alkanes non-utilizing mutants of *Saccharomycopsis lipolytica* [76]

clearing zone around colonies. Some mutants did form such a halo and, when tested quantitatively, showed a better assimilation of kerosene. These mutants were not characterized genetically or biochemically.

Bassel and Mortimer [76] induced alkane- non utilizing mutants of *Y. lipolytica* by ultra violet light. Substrate-utilization tests of the 44 mutants allowed their classification into 5 groups (Table 2).

It is known that the first three enzymatic reactions of *n*-alkane catabolism give rise to fatty acids of the same chain length. Those reactions are catalyzed by a microsomal enzyme aggregate called hydroxylase complex. [77]. The enzymes involved in this complex are *n*-alkane hydroxylase, alcohol dehydrogenase and aldehyde dehydrogenase. The wild type strain showed a one hundred fold increase in hydroxylase complex activity in *n*-decane grown cells, as compared to glucose grown cells.

27 alkane-negative mutants were able to grow on all the intermediates tested (phenotype A). There was no mutant expressing phenotype B, corresponding to a deficiency in alcohol dehydrogenase. 6 mutants were of phenotype C, corresponding to a possible deficiency in aldehyde dehydrogenase. 4 mutants were fatty acid negative (phenotype D) and 6 mutants acetate-negative (phenotype E).

The lack of phenotype B could be explained if *Y. lipolytica* had more than one gene specifying alcohol dehydrogenase.

The 34 mutants of phenotype (A + C) are alkane-negative and fatty acid-positive indicating a defect in *n*-alkane uptake or in hydroxylase complex activity. After eliminating leaky and non-mating mutants, 21 were left and analyzed genetically: all showed a 1:1 pattern of segregation and corresponded to 18 different genes. This is a very complex situation indicating that many genes must be involved in *n*-alkane solubilization and uptake.

Bassel and Mortimer [26] reported the isolation of new alkane non-utilizing, fatty acid-utilizing mutants of *Y. lipolytica*. Including their previous results, the mutants of phenotypes (A + C) corresponded to a total of 26 genetic loci.

Alkane-negative mutants representing the 26 loci were analyzed for their ability to accumulate ^{14}C hexadecane. Sixteen mutants showed significantly reduced *n*-alkane accumulation when compared to wild type, confirming that the alkane uptake process was a complex one.

Cells induced by *n*-decane take up about 100 times more ^{14}C *n*-decane than uninduced, glucose-grown cells. Uptake is inhibited by 2,4 dinitrophenol and KCN, showing that uptake is due to active transport.

Cirigliano and Carman [78, 79] have found that a strain of *Y. lipolytica* produced an inducible extracellular emulsification activity when it was grown on hexadecane. This emulsifier, called liposan, was partially purified from the culture filtrate. The procedure yielded a preparation containing a major band (MW = 27000).

Liposan is composed of approximately 88% carbohydrate and 17% protein. Liposan stabilized oil-in-water emulsion, with a variety of vegetable oils. Nothing is known of the biosynthesis of liposan.

Kamiryo et al. [80] have shown that *Y. lipolytica* possesses two acyl CoA synthetases:

— acyl-CoA synthetase I is found in perxisomes, mitochondria and in the cytoplasm. It is responsible for the production of acyl-CoA to be used for the synthesis of cellular lipids.
— acyl CoA synthetase II is found exclusively in peroxisomes. It provides acyl-CoA to be exclusively degraded via β-oxidation.

Mutants of *Y. lipolyticas* defective in synthetase I cannot incorporate exogenous fatty acids into cellular lipids, but are able to degrade fatty acis via β-oxidation.

Mutants of *Y. lipolytica* defective in synthetase II fail to grow on fatty acids, but are capable of incorporating exogenous fatty acids into cellular lipids.

12 Secretion of Enzymes

12.1 Lipase

Kalle et al. [81] have described two lipases activities in *Y. lipolytica*. One of them is constitutive and not repressed by glucose. The second one is induced by sorbitan monooleate at an optimum concentration of 0.5%. This compound is not a substrate for lipase nor a substrate for growth. The synthesis of the inducible lipase was completely repressed in the presence of glucose.

Dherbomez et al. [82] studied lipase activity in *Y. lipolytica* as a function of medium composition, pH, etc.

Ota et al. [83] found that *Y. lipolytica* grown in a synthetic medium containing no lipid had weak lipase activity. However, when olive oil or long chain fatty acids (such as oleic acid) were added to the medium, the yeast produced a large amount of lipase, most of which being cell-bound. Two kinds of cell-bound lipases were purified that had similar enzymological properties, their molecular weights being respectively 39 and 44 kDa. One was considered as derived from the other by some (unspecified) enzymatic reaction.

An extracellular lipase has also been described by Ota et al. [84, 85]. Its synthesis was stimulated by a protein fraction extracted from soybean.

Nga et al. [86] obtained tributyrin non-utilizing mutants in a *Y. lipolytica* strain subjected to UV mutagenesis, with a frequency of 0.1%. A nystatine selection procedure was used to enhance the efficiency of recovery of tributyrin non-utilizing mutants. Eight such mutants, when assayed for lipase activity, gave values ranging between 0 and 4.2% of the initial strain.

Nga et al. [87] performed a genetic analysis of 14 lipase low-producing mutants (Lip^-) of *Y. lipolytica*. All 14 heterozygous diploids were Lip^+ and Lip^- segregants occured at 1:1 ratio, indicating that the Lip^- mutations were chromosomal and monogenic.

Complementation analysis indicated that 10 of the Lip^- mutants belonged to one complementation group. The four other mutants were distributed among 2 other complementation groups.

In order to establish whether the lipase structural gene had been mutated in some of the Lip^- mutants, reversion studies were conducted. It was found that mutants belonging to the first complementation group gave rise to a number of revertants showing thermosensitive lipase. This was considered as a strong indication that they were affected in the structural gene. This is a first step towards cloning of this gene.

12.2 Extracellular RNase

Cheng and Ogrydziak [88] have investigated the production of RNase by *Y. lipolytica*. RNase production took place during the exponential growth phase. Although several RNases (MW: 45 kDa, 43 kDa and 34 kDa) were detected, it could be demonstrated that only the 45 kDa species is secreted in the pH range 5–7. The other two are degradation products most probably generated through the action of another secreted enzyme, alkaline extracellular protease.

Cheng and Ogrydziak [89] used anti-RNase antiserum to preciptate a 45 KDa protein from ^{35}S methionine-labelled extracts and supernatant medium. A 73 kDa polypeptide was also detected; it is related to the 45 kDa mature RNase but the nature of that relationship is still unclear. It is not sure that the 73 kDa polypeptide is a direct precursor of mature RNase. Cloning of the gene should clarify the matter.

12.3 Extracellular Acid Proteases

Yamada and Ogrydziak [90] showed that *Y. lipolytica* produces at least three extracellular acid proteases, during the exponential phase, on appropriate media. Proteose peptone appeared to be the most efficient nitrogen source. The three proteases were purified and found to have the following characteristics:

— Protease I: MW = 28 kDa; pH optimum = 3.5
— Protease II: MW = 32 kDa; pH optimum = 4.2
— Protease III: MW = 36 kDa; pH optimum = 3.1.

All three proteases were glycoproteins, containing respectively 25, 12 and 1.2% carbohydrate.

12.4 Alkaline Extracellular Protease

Ogrydziak and Mortimer [91] have isolated mutants of *Y. lipolytica* with reduced activity to produce zone of clearing on skim milk agar plates. These mutants (*xpr*) produced reduced levels of an alkaline extracellular protease (AEP). The mutants were recessive and defined 10 or 11 complementation groups, indicating that many genes were involved in AEP secretion. Moreover, some mutants produced reduced levels of extracellular RNase, and these characteristics segregated together, showing there were common steps in the secretion of different extracellular enzymes. These results were confirmed by Mehta and von Borstel [92] and the number of genes involved extended to 16.

Sims and Ogrydziak [93] isolated temperature-sensitive mutants for AEP production. One was found to produce a temperature-sensitive protease and was localized in gene *XPR2*. When it was found that this locus also exhibited a dosage effect by comparing a set of diploids *XPR2/XPR2*, *XPR2/xpr2* and *xpr2/xpr2*, it was concluded that *XPR2* was the structural gene of AEP.

Matoba et al. [94] studied processing and secretion of AEP by pulse-chase and immuno-precipitation experiments. The secreted protease had a molecular weight of 32 kDa but, inside the cell, polypeptides of 55, 52, 44, 36, 32, 20 and 19 kDa, respectively were found. The processing appeared, therefore, to be fairly complex.

Pulse-chase experiments were performed in the presence of tunicamycin, which prevents synthesis of *N*-linked glycoside chains. The various polypeptides observed were also treated with endo-H which hydrolyzes glycosidic-linkages. These experiments indicated that the 55, 52, 44, 20 and 19 kDa proteins contained about 2 kDa of *N*-linked oligosaccharide, but that the 36 kDa and 32 kDa polypeptides had none.

As we shall see later, cloning and sequencing of the AEP structural gene has shed some light on processing of the enzyme during secretion.

13 Genetic Engineering

13.1 Integrative Transformation Systems

Two integrative transformation systems have been described for *Y. lipolytica*:

1. Gaillardin et al. [95] made use of the heterologous *LYS2* gene of *S. cerevisiae* as a selective marker. It is able to complement the corresponding the *lys2* mutation of *Y. lipolytica*.

Eibel and Philippsen [96] had cloned thy *LYS2* gene of *S. cerevisiae* on plasmid YIp333 (Fig. 6). A shot gun of *Y. lipolytica* DNA was inserted at the unique EcoRI site of this plasmid and this gene bank was used to transform spheroplasts of a *lys2* strain of *Y. lipolytica*. The transformation frequency was low, 1–10 transformants per µg of DNA. These transformants were found to contain 10–20 integrated copies of tandem repeats of YIP333 derivatives. The most likely explanation appears to be that several plasmids of the gene bank were cotransformed, gave rise to an hybrid concatemer which was then integrated in the chromosomes

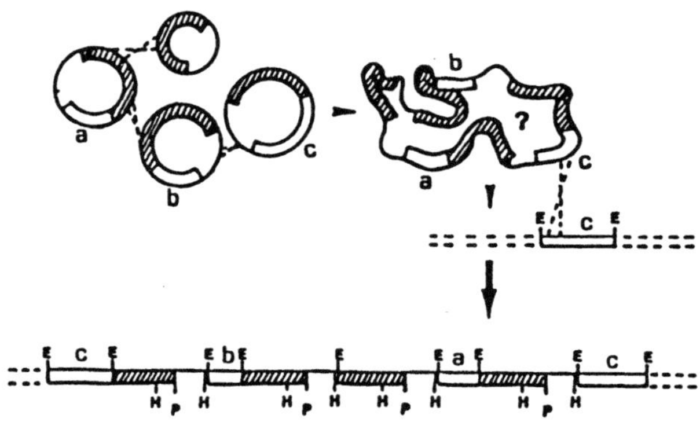

Fig. 6. A possible model of homologous integration of the *Eco*RI shotgun in YIp333 into *Y. lipolytica*. A mixture of YIp333 and YIp333 derivatives carrying different *Y. lipolytica* inserts (labeled *a*, *b*, *c*) recombine by homology to generate tandem repeats of YIp333 separated or not by *Y. lipolytica* inserts. This polymer integrates into the genome by homology between one insert (here insert *c*) and the corresponding chromosomal target. ▨, *S. cerevisiae LYS2* fragment; —, pBR322 DNA; □, *Y. lipolytica* DNA. From: Gaillardin et al. Integrative transformation of the yeast *Yarrowia lipolytica* [95]

by recombination between homologous regions (Fig. 6). This transformation procedure allows gene amplification, but the transformants are unstable to various degrees.

2. Davidow et al. [97] used the homologous *LEU2* gene integrated in a plasmid derived from pBR322. The transformation procedure was that of Ito et al. [98]. Cells entering the late exponential phase were treated with lithium acetate. Then transformation was achieved after PEG treatment and heat shock.

The circular plasmid gave rise to 1–100 transformants per µg of DNA. However, when the plasmid was linearized by an ApaI cut within the *LEU2* gene, there was a 1000 fold increase in transformation frequency and integration occured by recombination with the *LEU2* chromosomal copy.

13.2 Isolation of ars Sequences

In *S. cerevisiae*, high frequency transformation has been achieved with plasmids containing specific chromosomal sequences (ars) which act as origins of replication. Their average frequency is of the order of 1 per 30 to 40 kb of DNA (Beach et al., [99]; Chan and Tye, [100]). These ars plasmids are highly unstable.

In *Y. lipolytica*, Wing and Ogrydziak [101] tested a DNA library in an integrative vector carrying the homologous *LEU2* gene but did not observe any increase in the transformation frequency. Some of the transformants were slow-growing, unstable, and were shown to contain extrachromosomal DNA. However, by retransformation of *E. coli*, no ars sequence could be isolated. The hypothesis was put forward by Fournier et al. [21] that the failure to isolate ars was possibly due to the dimorphic growth of *Y. lipolytica* because, during hyphal growth, only the apical cell divides. In *S. cerevisiae*, Murray and Szostak [102] have demon-

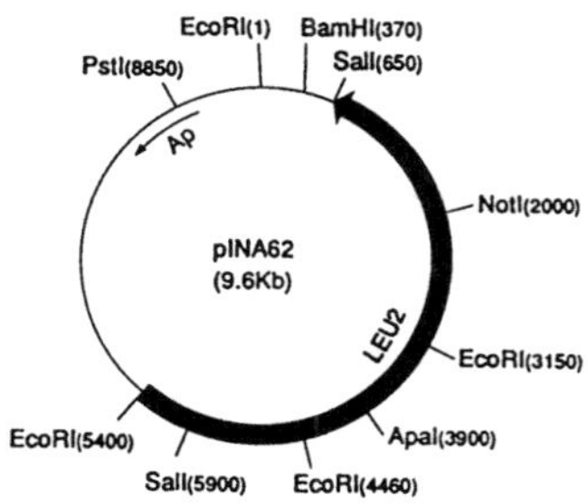

Fig. 7. Structure of the integrative plasmid pINA62

strated that plasmids show a bias during replication and tend to remain associated with the mother cell. They are not irreversibly lost however, because the mother cell has the capacity to form several buds in succession. If such a maternal bias occurs in *Y. lipolytica*, the failure of a plasmid to be transmitted to the apical cell means that it will be lost because the mother cell divides only once.

Morphological mutants unable to form hyphae (Fil^-) were isolated after UV mutagenesis. One of them, forming exclusively budding cells, was chosen for transformation experiments. The integrative plasmid pINA62 (with the *Y. lipolytica LEU2* gene on pBR322) transforms at low frequency (10–100 transf per μg DNA). It was used for the cloning of *Y. lipolytica Bgl* II DNA fragments at its unique *Bam*HI site (Fig. 7). This gene bank was used to transform the Fil^- mutant.

The transformants were checked for their stability on minimal medium (+ leucine); about 11% of them were found to be unstable.

Total uncut DNA of these transformants was probed with plasmid pINA62 which, apart from the chromosomal copy of *LEU2*, detected additional bands corresponding to autonomously replicating plasmids.

Plasmids extracted from a given transformant exhibited a characteristic *Eco*RI pattern with 3 bands, identical to the cloning vector pINA62 and additional band (s) due to the presence of an insert. The Fil^- *Y. lipolytica* strain could be retransformed with these plasmids at a frequency of 10^3–10^4 transformants per μg of DNA. This frequency was much higher than that obtained with the initial pINA62 plasmid.

The restriction map of the five isolated plasmids was established. It turned out that only two ars sequences had been isolated (ars18, ars64). Each of them was shown to be present as a single copy in the genome of *Y. lipolytica.*

The plasmids isolated with the help of a Fil^- mutant, were also able to transform the initial Fil^+ strain at high frequency.

The stability of a number of Fil^- and Fil^+ strains transformed with the same plasmid was checked and the loss rate of the plasmid per generation was calculated. For two different Fil^+ strains, it was 0.4 and 3.6 respectively. The Fil^- strain was distinctly more unstable (rate = 12.6).

The copy number or ars plasmids was found to vary between 2 and 5, a low value compared to that found in *S. cerevisiae.*

The fact that the same ars sequences were repeatedly isolated indicates that there are few sequences able to confer autonomous replication in the *Y. lipolytica* genome.

Retrospectively, the initial inability to isolate ars sequences in Fil^+ strains was most probably due to the high stability of ars plasmids in such strains. The initial hypothesis of preferential maternal inheritance turned out to be wrong. The lesser stability of ars plasmids in the Fil^- mutant remains unexplained.

13.3 Cloning of *Y. lipolytica* genes

13.3.1 Strategy

The *LEU2* [103] and *HIS1* genes [104] were isolated by complementation of *E. coli* auxotrophs.

The availability of an integrative transformation system made cloning possible by direct complementation of *Y. lipolytica* auxtrophs. An integrative *LEU2* vector was used to insert an *Sau*3A digest of DNA into the unique *Bam*HI site (Fig. 7). This gene bank was used to transform *E. coli* and, after amplification, the DNA was reextracted and linearized with *Apa*I which cuts only once in *LEU2*. In a next step the linearized DNA is used to transform an appropriate recipient strain of *Y. lipolytica*, selecting for the *LEU2* marker. Later on, the LEU^+ transformants are checked for complementation of the gene one wants to clone. A difficulty in this method is that an *Apa*I site may be present on the insert; it is therefore advisable to test both complete and partial digestions with *Apa*I.

Xuan et al. [105] have used this technique to clone the *LYS1*, *LYS5* and *ADE1* genes (Table 3). For the *LYS5* and *ADE1* genes, the frequency of LYS^+ or ADE^+ clones within $LEU2^+$ transformants was not changed whether digestion with *Apa*I was complete or partial. This was an indication that there was no *Apa*I site

Table 3. Complementation of *lys5*, *ade1* and *lys1* by transformation with a genomic bank

DNA	Transf per µg of DNA tested	Total nb of clones	Complementation tests			
			*LYS**		*ADE**	
			Nb.	Freq.	Nb.	Freq.
Recipient:	*lys5,*	*leu2, ade1*				
pINA62, C	1.6×10^5	ND				
GB, C	9.3×10^4	4.6×10^4	6	1.3×10^{-4}	8	5×10^{-4}
GB, P	8.9×10^3	8.9×10^3	2	2.2×10^{-4}	2	2.2×10^{-4}
Recipient:	*lys1,*	*leu2*				
pINA62, C	8×10^4	ND				
GB, C	1.2×10^5	2×10^5	0	5×10^{-6}		
GB, P	2×10^4	8×10^4	3	4×10^{-5}		

*: pIN62 = *LEU2* vector; GB = *Sau*3A library into pINA62. All DNAs were digested prior to the transformation with *Apa*I which cuts within the *LEU2* part of pINA62: C = complete digest, *P* = partial digest (J. W. Xuan unpublished).
From: Gaillardin and Heslot, Genetic engineering in *Yarrowia lipolytica* [114]

in the inserts carrying the *LYS5* or *ADE1* genes. Consequently, the linearized plasmids should have been targeted for integration at the *LEU2* site exclusively. This could be confirmed by Southern analysis of the transformants.

With respect to *LYS1* cloning, on the other hand, it is apparent from Table 3 that a complete digestion of the gene bank with *Apa*I does not allow to isolate LYS$^+$ clones among LEU2$^+$ transformants. However, partial digestion gives positive clones. This is due to the presence of one (or several) *Apa*I site (s) within the *LYS1* insert.

When there is no *ApaI* site within the cloned gene, as is the case of *LYS5*, *URA3*, *ADE1* and *BIO* genes, it is easy to recover the plasmic inserted at the *LEU2* locus. A complete digestion of total DNA of the transformant with *Apa*I liberates the plasmid carrying the functional gene. It is then circularized in vitro and used to transform *E. coli*. Fig. 8 shows the recovery process of the *LYS5* gene.

When there is one (or several) *Apa*$^-$ site(s) within the cloned gene, its recovery is more difficult. Using a restriction enzyme that does not cut within the initial vector, it may be possible to recover sequences flanking the vector and containing a complete or partial copy of the gene of interest.

13.3.2 The *LYS5* Gene

The *LYS5* gene has been cloned and sequenced by Xuan et al. [105 and unpublished] revealing an unusual situation: there are two overlapping reading frames ORF1 and ORF2, one on each of the two complementary strands. Promoter and ter-

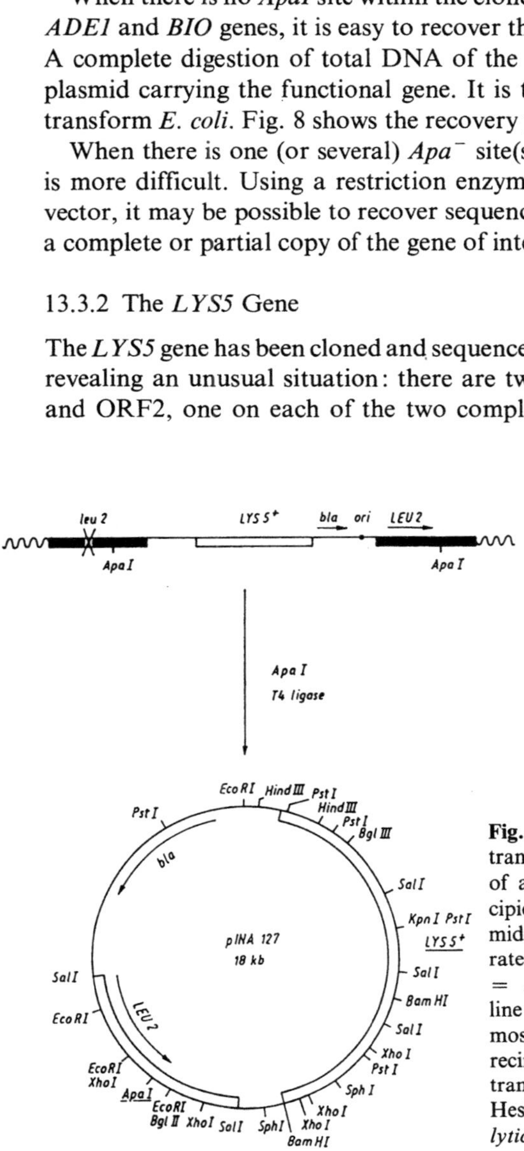

Fig. 8. Cloning of the *LYS5* gene. A Leu$^+$Lys$^+$ transformant was isolated after transformation of a genomic bank into a *lys5*, *leu2*, *ade1* recipient. Southern blotting showed that the plasmid had integrated at the *LEU2* locus to generate the structure depicted on top: black boxes = *LEU2* copies, open boxes = insert, thin line = pBR322 sequences, wawy lines = chromosomal DNA. Digestion of this structure and recircularization regerates pINA127 (i.e. the transformation plasmid). From: Gaillardin and Heslot, Genetic engineering in *Yarrowia lipolytica* [114]

minator sequences are present upstream and downstream of each ORF. Probes detect the presence of mRNAs, but transcription of ORF1 occurs at a very low level. ORF2 is considered to code for saccharopine dehydrogenase; ORF1 could possibly play a regulatory role.

13.3.3 The *XPR2* Gene

This gene codes for the extracellular alkaline protease (AEP). Davidow et al. [106], Matoba et al. [94] and Nicaud et al. [107] have independently cloned *XPR2*. Davidow et al. [106] constructed a gene library using *Sau*3A segments of wild type *Y. lipolytica* DNA inserted into the *Bam*HI site of a *LEU2*-pBR322 vector. *BglII* cuts only in *LEU2* but not in the pBR322 region. The gene library was linearized by partial digestion with *Bgl*II and used to transform a *leu2 xpr2* recipient strain. Only one protease positive transformant was found. The *XPR2*-gene containing plasmid was recovered by partial digestion of total DNA of the transformants with *Bgl*II but had an incomplete copy of the protease gene. This plasmid was used as a probe to screen the original gene library and allowed the isolation of a complete *XPR2* gene.

Nicaud et al. [107] used oligonucleotide probes based on the *N*-terminal aminoacid sequence of the mature AEP molecule. Two probes (51 and 44 bases) were synthetized, using the codon bias of the *Y. lipolytica LEU2* gene [108]. They were used to screen a genomic library and allowed to isolate the complete gene in two successive steps.

Matoba et al. [94] used the same probes to screen a genomic library of *Y. lipolytica* in lambda gt11 and several positive clones were isolated and sequenced.

Although the *XPR2* genes cloned by the three groups came from unrelated wild type strains, they showed a very high degree of homology.

The DNA sequence of *XPR2* and its deduced aminoacid sequence suggest that the enzyme is originally synthetized with an additional pre-proI-proII-proIII NH_2 terminal region (Fig. 9). As already mentioned, the secreted protease had a molecular weight of 32 kDa but that inside the cell, polypeptides of 55, 52, 44, 36, 32, 20 and 19 kDa, were found, indicating a complex processing.

The preregion is a potential secretory signal; it has a basic NH_2-terminal region, followed by a hydrophobic sequence. Starting with Leu14, there is a stretch of 10 X-Ala or X-Pro dipeptides that could be substrates for a dipeptidyl aminopeptidase and give rise to the 52 kDa precursor.

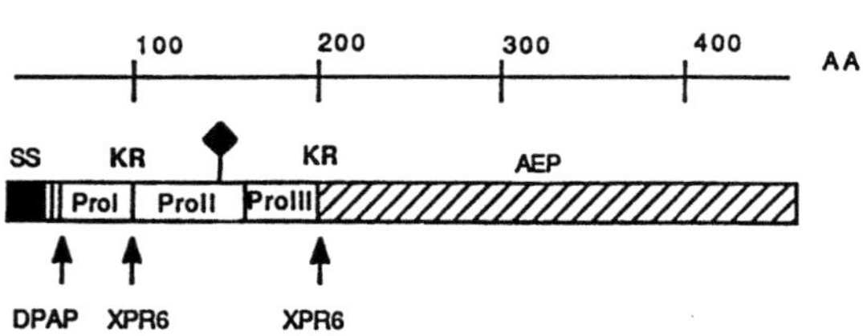

Fig. 9. Structure of the alkaline extracellular protease precursor. Location of processing sites.
DPAP = Dipeptidyl amino peptidase.
XPR6 = KEX2-like peptidase
♦ = Glycosylation site.

At the boundary between proI and proII, there is a Lys59-Arg60 doublet. Between proII and proIII, one finds a doublet Asp128–Lys129 and finally a doublet Lys161-Arg162 between proIII and mature AEP. All these doublets are possible substrates for a KEX2-like endopeptidase. This is presumably encoded by the *XPR6* gene.

There is also an Asn-Ser-Thr sequence, a potential site for *N*-linked glycosylation located in proII (Asn 123). This is in agreement with the fact that the 55, 52 and 44 kDa precursors are glycosylated, whereas the mature AEP (32 kDa) is not.

It appears that the major pathway for AEP processing is 55 kDa → 52 kDa → 32 kDa.

13.3.4 Amplification of *XPR2*

Nicaud et al. [107] have amplified the *XPR2* gene by two approaches: 1) multiple integrations at the XPR2 site; 2) extrachromosomal amplification on a replicative ars vector.

The three selective markers *LEU2*, *LYS5*, and *URA3* were used. The recipient strain JM12 had the genotype *leu2*, *ura3*, *lys5*, *XPR2*. Four plasmids (Fig. 10 and Table 4) were prepared to either disrupt the *XPR2* gene (plasmid pINA157), or to increase its copy numbers (plasmids pINA159 and pINA180).

Before transformation of JM12, the plasmids were linearized by cutting at the unique *Mlu*I site located in *XPR2*. In this way, each plasmid was targeted for integration at the *XPR2* chromosomal site. In one or two successive steps (JM12, JM19, JM83) it was possible to introduce one or two extra copies of *XPR2*.

The second procedure to increase *XPR2* copy number was by using an ars vector. (Fournier et al., 21) bearing one copy of *XPR2*. The transformed strain JM70 was shown to harbor 4–7 copies/cell of the plasmid.

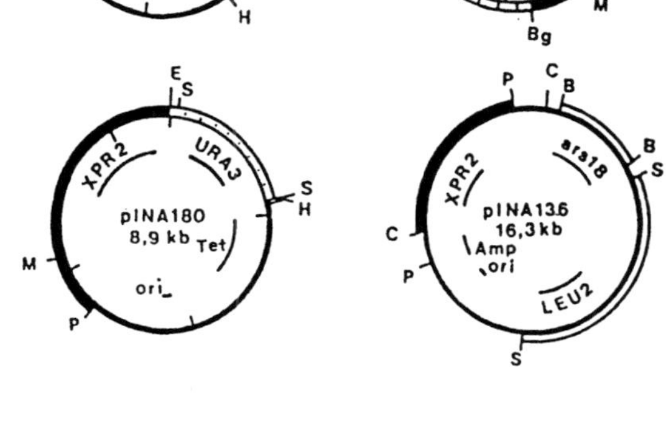

Fig. 10. Plasmids used to inactivate *XPR2* gene or to amplify *XPR2* gene copy number

Table 4. Strains and plasmids used for amplification of *XPR2* gene copy number. Strains JM19 and JM83 derive from JM12 by successive integrations

Strain	Number of *XPR2* copies	Plasmid used	Plasmid structure	Status of plasmid
JM12	1	—	—	—
JM23	1 disrupted	pINA 157	*XPR2::LYS5*	integrated
JM19	2	pINA 159	*XPR2 — LYS5*	integrated
JM83	3	pINA 180	*XPR2 — URA3*	integrated
JM70	4–7	pINA 135	*XPR2-LEU2*-ARS18	ARS vector

AEP production was compared for strains carrying one (JM12), two (JM19) or three (JM83) integrated copies of *XPR2*. There was a linear increase as a function of gene copy number.

13.3.5 Secretion Apparatus — 7S RNA

In mammalian cells, the signal recognition particle (SRP) is required for transport of presecretory proteins across the endoplasmic reticulum (ER) membrane. SRP is composed of six distinct polypeptides of MW 9, 14, 19, 54, 68 and 72 kDa and of a 7S RNA molecule of 300 nucleotides [109].

Despite extensive efforts, it has not been possible to isolate SRP particles or 7S RNA from *S. cerevisiae*, but the search has been successful with the two other yeast species *Schizosaccharomyces pompe* and *Yarrowia lipolytica* [110, 111]. The 7S RNA of the latter possesses 270 nucleotides. Analysis of the genomic clone of this RNA has indicated that it has short conserved sequences that match the A and B boxes defined for polymerase III promoters. It can be folded into a secondary structure very similar to that of eucaryotic 7S RNAs and it shows primary sequence conservation in short regions predicted to be single stranded (Fig. 11). In *Y. lipolytica*, there are two genes encoding the 7S RNA; disruption of both genes is a lethal event [112, 113].

13.3.6 Codon Usage and Gene Structure

The *Y. lipolytica* genes so far sequenced indicate that this yeast uses a codon bias different from that of *S. cerevisiae* [114].

Regarding the 5′ upstream non-translated region, a consensus ATATAA Goldberg-Hogness box is found 40–80 bp downstream of a consensus CAAT box.

3′ untranslated regions of *S. cerevisiae* have a sequence TAG ... TA(T)GT ... TTT which seems important for the termination of transcription [115]. The same sequences are found in all *Y. lipolytica* genes so far investigated.

13.4 Heterologous Expression

The promoters of the *LEU2* [108] and *XPR2* genes [116] have been used to express foreign genes in *Y. lipolytica*.

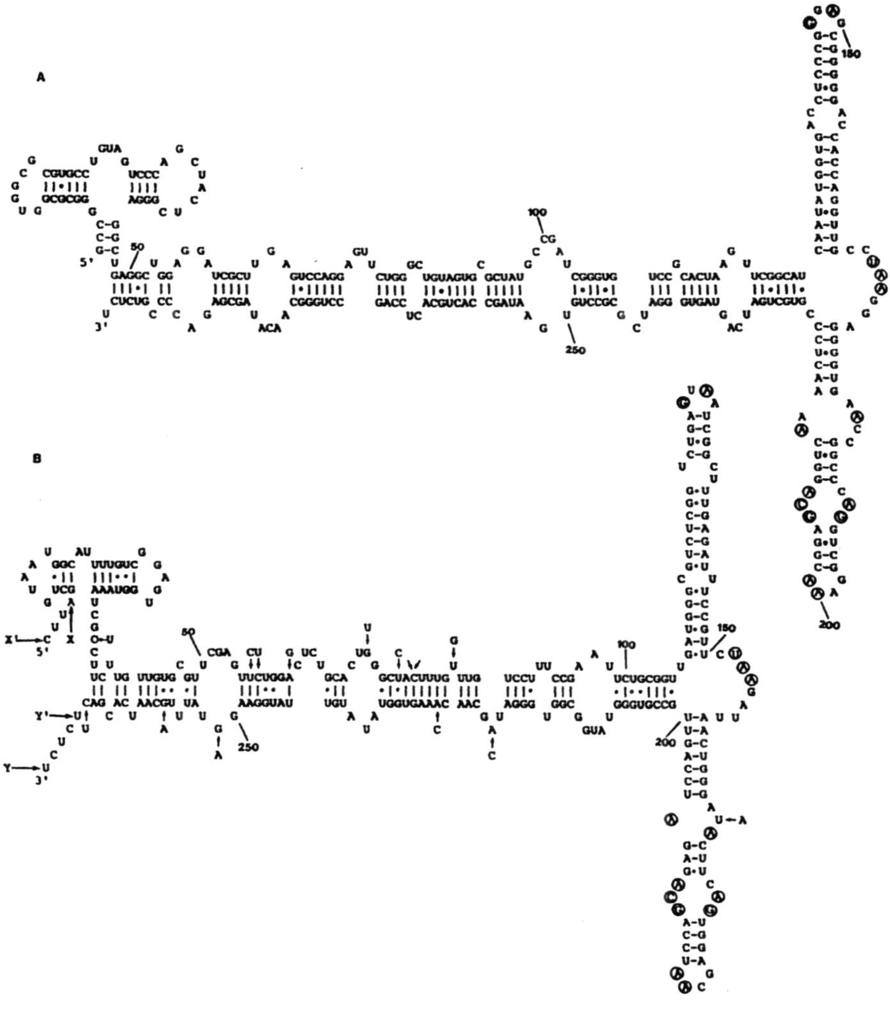

Fig. 11. Comparison of human (A) and *Yarrowia lipolytica* (B) 7SL RNA secondary structures

The *lacZ* gene of *E. coli*, coding for β-glactosidase, has been expressed under control of the *LEU2* promoter. 232 bp are enough for full promoter activity [108]. The same promoter was used to express a *Tn5* gene conferring bleomycin/phleomycin resistance. Integrative transformants of *Y. lipolytica* exhibited an increased phleomycin resistance and direct selection of Ble^R was possible in some conditions.

Nicaud et al. [117] constructed a gene fusion where the *SUC2* gene of *S. cerevisiae*, coding for invertase, was placed under the control of promoter and signal sequences of the *XPR2* gene. *Y. lipolytica* does not make invertase and is unable to grow on sucrose. The *XPR2-SUC2* fusion was put on an integrative plasmid, possessing the *LEU2* marker. By cutting this plasmid at a unique site located within the *XPR2* promoter (*Nhe*I) or within the *LEU2* (*Bgl*II), it was possible to target integration of the plasmid to one or the other of these genes. LEU^+ transformants were obtained at a frequency of 30,000 per μg DNA. They were able to grow on sucrose, in media known to induce AEP. Under appropriate conditions it was possible to select directly for Suc^+ transformants. This means that the *SUC2* gene can be used as a dominant marker.

The porcine pre-α1 interferon has been cloned by La Bonnardière and Laude [118]. Heslot et al. [119] have used the *XPR2* promoter to express this gene and made two types of fusions: preproI-II-III — IFN, or preproI-IFN. After transformation with integrative plasmids, they found that interferon was secreted in the growth medium. When the constructions were put on an ars plasmid, a 6 fold amplification of IFN was observed compared to integrated plasmids.

Chymosin is a milk-clotting enzyme produced by calf stomach. It is secreted as an inactive precursor, prochymosin, which undergoes autocatalytic activation at pH4 to give active chymosin. A42 amino acid NH_2-terminal peptide is cleaved off in this process.

Franke et al. [120] used the *XPR2* and *LEU2* promoters to express prochymosin in *Y. lipolytica.* With *XPR2*, several constructions were made, using various lengths of the prepro region. All these constructions resulted in the secretion of prochymosin in the growth medium. It could be activated at low pH and the NH_2-terminal end of the activated product was found to be identical with that of genuine chymosin. The plasmids were integrated at the *XPR2* locus or at the *BIO* locus, using pBR322 homology in a strain carrying a pBR322:Bio "docking platform" obtained by a previous integrationevent.

14 Prospects

A managable genetic system has been established for *Y. lipolytica*, completed in recent years by genetic engineering technology. One of the interesting characteristics of this yeast is its capacity to secrete high molecular weight proteins. However, compared with the intensive investigations performed on the secretory pathway of *S. cerevisiae* [121] very little is known for *Y. lipolytica.* The recent discovery of a signal recognition particle, and the cloning of its 7S RNA could lead the way to the genetic dissection of the early steps of the secretory pathway.

For expression of heterologous genes, there is clearly a need for new strong promoters that would be easily regulatable.

Chromosomal ars sequences have allowed the building of autonomously replicating vectors. Numerous questions remain however open: what is the nature of these ars sequences? Why do they appear so rarely? Is it be possible to increase the copy number of the corresponding vectors?

Finding proper answers to these questions will require sustained investigations of the *Y. lipolytica* system.

15 References

1. Shah DN, Purokit AP, Sriprakash RS (1982) Enzyme Microbiol Technol 4:69
2. Maldonado P, Gaillardin C (1972) French Patent 41913
3. De Zeeuw J, Tynan EJ (1973) USA Patent 3736229
4. De Zeeuw J, Tynan EJ (1973) USA Patent 3736917
5. De Zeeuw J, Stasko I (1983) USA Patent 4407953
6. Tobe ST, Takami T, Ikeda S, Mitsugi K (1976) Agric Biol Chem 40: 1087
7. Wickerham LJ, Kurtzman CP, Herman AI (1970) Science 167: 1141
8. Yarrow D (1972) Ant Van Leeuwenh 36: 357
9. Van der Walt JP, Arx JA (1980) Ant van Leeuwenh 46: 517
10. Gaillardin CM, Charoy V, Heslot H (1973) Arch Microbiol 92: 69
11. Ogrydziak D, Bassel J, Contopoulou R, Mortimer R (1978) Molec Gen Genet 163: 229
12. Kurischko C, Inge-Vechtomov SG, Weber H (1983) Zeitsch Allg. Mikrobiol 23: 513
13. Barth G, Weber H (1984) Zeitsch Allg Mikrobiol 24: 125
14. Barth G, Weber H (1984) Zeitsch Allg Mikrobiol 24: 403
15. Barth G, Weber H (1985) Ant van Leeuwenh 51: 167
16. Esser K, Stahl U (1976) Molec Gen Genet 146: 101
17. Weber H (1978) Zeitsch Allg Mikrobiol 19: 283
18. Ota Y, Oikawa S, Morimoto Y, Minoda Y (1984) Agri Biol Chem 48: 1933
19. Rodriguez C, Dominguez A (1984) Can J Microbiol 30: 605
20. Vega R, Dominguez A (1986) Arch Microbiol 144: 124
21. Fournier P, Guyaneux l, Chasles M, Gaillardin C (submitted for publ) Yeast
22. Ogrydziak D, Bassel J, Mortimer R (1982) Mol Gen Genet 188: 179
23. Ogrydziak D (1988) J Basic Microbiol 28: 185
24. Kurischko C (1984) Zeitsch Allg Mikrobiol 24: 545
25. Kurischko C (1986) J Basic Mikrobiol 26: 33
26. Bassel JB Mortimer RK (1985) Curr Genet 9: 579
27. Mortimer RK, Tavares FC (1976) Genetic mapping in yeast In: Microbiology ASM. Washington DC, pp 572–574
28. De Jonge P, de Jonge FCM, Meijers R, Steensha HY, Scheffers WA (1986) Yeast 2: 193
29. Johnston JR, Mortimer RK (1986) Int J Syst Bacteriol 36: 569
30. Stahl U (1978) Molec Gen Genet 160: 111
31. Weber H, Förster W, Jacob HE, Berg H (1980) 5th Intern Sympos Yeasts July 20–25, 1980, pp 219–224
32. Kurischko C, Spata L (1984) Zeitsch Allg Mikrobiol 24: 551
33. Kurischko C, Weber H (1985) J Basic Microbiol 26: 137
34. Kurischko C (1986) Curr Genet 10: 709
35. Kurtzman CP, Phaff HJ, Meyer SA (1983) In: Spencer JTF, Spencer DM, Smith ARW, (eds) Yeast genetics. Springer, Berlin Heidelberg New York, p 139
36. Kück V, Stahl U, Lhermitte A, Esser K (1980) Curr Genet 2: 97
37. Hollenberg CP, Borst P, Van Bruggen EFJ (1970) Biochem Biophys Acta 209: 1
38. Matsuoka M, Uchia K, Aiba S (1982) J Bacteriol 152: 530
39. Van Heerikhuizen H, Ykema A, Klootwijk G, Gaillardin C, Ballas C, Fournier P (1985) Gene 39: 213
40. Fournier P, Gaillardin C, Persuy PA, Klootwijk J, van Heerikhuize H (1986) Gene 42: 273
41. Clare JJ, Davidow LS, Gardner DC, Oliver SG (1986) Curr Genet 10: 449
42. Wickner RB (1979) Plasmid 2: 303
43. Bussey H (1981) Adv Microb Physiol 22: 93
44. Groves DP, Clare JJ, Oliver SG (1983) Curr Genet 7: 185
45. Tréton BY, Le Dall MT, Heslot H (1985) Curr Genet 9: 279
46. El-Sherbeini M, Bostian KA, Levitre J, Mitchell D (1987) Curr Genet 11: 483
47. Tréton BY, Le Dall MT, Heslot H (1987) Curr Genet 12: 37
48. Akiyama SI, Suzuki T, Sumino Y, Nakao Y, Fukuda H (1973) Agr Biol Chem 37: 879
49. Akiyama SI, Suzuki T, Sumino Y, Nakao Y, Fukuda H (1973) Agri Biol Chem 37: 885
50. Tréton B, Heslot H (1977) Agri Biol Chem 42: 1201

51. Tréton B, Le Dall MT, Heslot H (1978) Europ J Appl Microbiol Biotech 6: 67
52. Finogenova TV, Glazunova LM (1982) Mikrobiologiya 51: 27
53. Finogenova TV, Shishkanova NV, Ermakova IT Kataeva IA (1986) Appl Microbiol Biotechnol 23: 378
54. Shah DN, Purohit AP, Sriprakash KS (1982) Enzym Microb Techn. 4: 116
55. Shah DN, Sriprakash KS, Chattoo BB (1989) J Biotechn 12: 211
56. Matsuoka M, Himeno T, Aiba S (1984) J Bacteriol 157: 899
57. Hönes I (1984) Zeitsch Allg Mikrobiol 24: 599
58. Barth G (1985) Curr Genet 10: 119
59. Barth G, Weber H (1987) Yeast 3: 255
60. Manney TR, Mortimer RK (1964) Science 143: 581
61. Heslot H, Gaillardin C, Beckerich JM, Fournier P (1979) In: Sebek O, Laskin A (eds) Genetics-Industrial microorganisms. Americ Soc Microbiol, Washington DC, p 54
62. Gaillardin C, Ribet AM, Heslot H (1982) Eur J Biochem 128: 489
63. Gaillardin C, Fournier P, Sylvestre G, Heslot H (1976) J Bacteriol 125: 48
64. Beckerich JM, Heslot H (1978) J Bacteriol 133: 492
65. Sawnor-Korszynska D, Morzycka E, Zaborowska D, Ryczynska-Bojunowska K (1977) Acta Biochim Polonica 24: 75
66. Dürr M, Urech K, Boller T, Wiemken A, Schwenke J, Nagy M (1979) Arch Microbiol 121: 169
67. Beckerich JM, Lambert M, Heslot H (1981) Biochem Biophys Res Comm 100: 1292
68. Beckerich JM, Pommies E, Faivre C, Lambert M, Heslot H (1986) Biochimie 68: 517
69. Beckerich JM, Colonna Ceccaldi B, Lambert M, Heslot H (1984) Curr Genet 8: 531
70. Errede B, Cardillo TS, Sherman F, Deschamps J, Wiame JM (1980) Cell 22: 427
71. Elder RT, St John TP, Stinchcomb DT, Daviw Rw (1980) Cold Spring Harbor Sympos Quant Biol 45: 581
72. Morzycka E, Sawnor-Korzynska D, Paszawski A, Grabski J, Raczynska-Bojanowska K (1976) Appl Envir Microbiol 32: 125
73. Bassel J, Hambright P, Mortimer R, Bearden AJ (1975) J Bacteriol 123: 118
74. Louvel L (1979) Contribution à l'étude de la biosynthèse de l'hème chez *Saccharomycopsis lipolytica*. Thèse Institut National Agronomique Paris
75. Markov KI, Kibarska T (1971) CR Acad Sci Agri Bulgarie 4: 413
76. Bassel JB, Mortimer RK (1982) Curr Genet 5: 77
77. Yorifugi T (1978) *n*-Alkane oxidation by *Saccharomycopsis lipolytica*. Ph D Thesis Massachusetts Institute of Technology
78. Cirigliano MC, Carman GM (1984) Appl. Environm Microbiol 48: 747
79. Cirigliano MC, Carman GM (1985) Appl Environm Microbiol 50: 846
80. Kamiryo T, Nishikawa Y, Mishina M, Terao M, Numa S (1979) Proc Natl Acad Sci USA 76: 4390
81. Kalle GP, Gadkari SV, Deshpande SY (1972) Indian J Biochem Biophys 9: 171
82. Dherbomez M, Lacrampe JL, Larrouquere J (1975) Rev Franc Corps gras 22: 147
83. Ota Y, Gomi K, Kato S, Sugiura T, Minoda Y (1982) Agri Biol Chem 46: 2885
84. Ota Y, Nakamiya T, Yamada K (1970) Agri Biol Chem 34: 1368
85. Ota Y, Morimoto Y, Sugiura T, Minoda Y (1978) Agric Biol Chem 42: 1937
86. Nga BH, Heslot H, Gaillardin CM, Fournier P, Chan K, Chan YN, Lim EW, Nai PC (1988) J Biotechnol 7: 83
87. Nga BH, Gaillardin CM, Fournier P, Heslot H (1989) J Gen Microbiol 135: 2439
88. Cheng SC, Ogrydziak DM (1986) J Bacteriol 168: 581
89. Cheng SC, Ogrydziak DM (1987) J Bacteriol 169: 1433
90. Yamada T, Ogrydziak DM (1983) J Bacteriol 154: 23
91. Ogrydziak DM, Mortimer RK (1977) Genetics 87: 621
92. Mehta RD, von Borstel RC (1979) Genetics 91: 80
93. Simms PC, Ogrydziak DM (1981) J Bacteriol 145: 404
94. Matoba S, Ogrydziak DM (1989) J Biol Chem 264: 6037
95. Gaillardin CM, Ribet A, Heslot H (1985) Curr Genet 10: 49
96. Eibel H, Philippsen P (1983) Mol Gen Genet 191: 66

97. Davidow LS, Apostolakos D, O'Connel MM, Proctor AA, Ogrydziak DM, Wing RA, Stasko I, De Zeeuw JR (1985) Curr Genet 10: 39
98. Ito H, Fukuda Y, Murata K, Kimura A (1983) J Bacteriol 153: 163
99. Beach D, Piper M, Shall S (1980) Nature 284: 185
100. Chan CSM, Tye BK (1980) Proc Natl Acad Sci USA 77: 6329
101. Wing RA, Ogrydziak DM (1985) In: Timberlake WE (ed) Molecular genetics of filamentous fungi, AR Liss, New York, p 367
102. Murray AW, Szostak JM (1983) Cell 34: 961
103. Davidow LS, Kaczmarek FS, De Zeeuw JR, Coulon SW, Lath MR, Pereira DA, Franke AE (1987) Curr Genet 11: 377
103. Davidow LS, Kaczmarek FS, De Zeeuw JR, Coulon SW, Lauth MR, Pereira DA, Franke AE (1987) Curr Genet 11: 377
104. Davidow LS, De Zeeuw JR (1985) European Patent 138 508
105. Xuan JW, Fournier P, Gaillardin C (1988) Curr Genet 14: 15
106. Davidow LS, O'Donnell MM, Kaczmarek FS, Pereira DA, De Zeeuw JR, Franke AE (1987) J Bacteriol 169: 4621
107. Nicaud JM, Fabre E, Beckerich JM, Fournier P, Gaillardin C (1989) J Biotechn 12: 285
108. Gaillardin C, Ribet AM (1987) Curr Genet 11: 369
109. Siegel V, Walter P (1988) TIBS 13: 314
110. Brennenwald P, Liao XB, Holm K, Poter G, Wise JA (1988) Molec Cell Biol 8: 1580
111. Poritz MA, Siegel V, Hansen W, Walter P (1988) Proc Natl. Acad Sci USA 85: 4315
112. He F, Beckerich JM, Ribes V, Tollervey D, Gaillardin C (1989) Curr Genet 16: 347
113. He F, Yaver D, Beckerich JM, Ogrydziak DM, Gaillardin CM (1990) Curr Genet (accepted)
114. Gaillardin C and Heslot H (1988) J Basic Microbiol 28: 161
115. Zaret KS, Sherman F (1982) Cell 28: 563
116. Davidow LS, Franke AE, De Zeeuw JR (1987) European Patent 220 864
117. Nicaud JM, Fabre E, Gaillardin C (1989) Curr Genet 16: 253
118. La Bonnardière C, Laude H (1981) Infect Immun 32:28
119. Heslot H, Nicaud JM, Fabre E, Beckerich JM, Fournier P, Gaillardin C (1989) Proc 2nd SSM Intern Congress Microbiology Singapore Oct 31–Nov 3, 1989 (in press)
120. Franke AE, Kaczmarek FS, Eisenhard ME, Geoghegan KF, Danley DE, De Zeeuw JR, O'Donnell MM, Gollaher MG, Davidow LS (1988) in Develop Industr Mikrobiol 29: 43
121. Shekman R, Novick P (1982) In: Strathern J, Jones E, Broach J (eds) The molecular biology of the yeast *Saccharomyces*: Metabolism and gene expression. Cold Spring Harbor, p 361

Transfer and Expression of Heterologous Genes in Yeasts Other Than *Saccharomyces cerevisiae*

Jakob Reiser, Virpi Glumoff, Markus Kälin and Urs Ochsner
Institut für Biotechnologie, ETH-Hönggerberg, CH-8093 Zürich, Switzerland

In the past few years, yeasts other than those belonging to the genus *Saccharomyces* have become increasingly important for industrial applications. Species such as *Pichia pastoris*, *Hansenula polymorpha*, *Schizosaccharomyces pombe*, *Yarrowia lipolytica* and *Kluyveromyces lactis* have been modified genetically and used for the production of heterologous proteins.

For a number of additional yeasts such as *Schwanniomyces occidentalis*, *Zygosaccharomyces rouxii*, *Trichosporon cutaneum*, *Pachysolen tannophilus*, *Pichia guilliermondii* and members of the genus *Candida* genetic transformation systems have been worked out. Transformation was achieved using either dominant selection markers based on antibiotic resistance genes or auxotrophic markers in conjunction with cloned biosynthetic genes involved in amino acid or nucleotide metabolism.

Advances in Biochemical Engineering/
Biotechnology, Vol. 43
Managing Editor: A. Fiechter

1 Introduction

Eukaryotic microbial systems such as yeasts and filamentous fungi have been applied successfully for the production of heterologous proteins during the past few years (see Ref. [1–7] for some recent reviews).

For the production of heterologous eukaryotic proteins, yeasts and filamentous fungi offer certain advantages over bacteria. First, since yeasts are eukaryotes, their intracellular environment is likely to be more suitable for the correct folding and assembly of eukaryotic proteins (see e.g. Ref. [8]). Second, yeasts such as *Saccharomyces cerevisiae* have the ability to glycosylate (see Ref. [9] for a recent review), to fatty acylate (see Ref. [10] for a review) and to phosphorylate proteins [11] and to modify them in yet additional ways that alter their electrophoretic mobility [12]. *S. cerevisiae* has also been reported to be able to add farnesyl moieties to certain proteins [13]. These post-translational modifications are important for maintaining the structural integrity, solubility, biological activity and cellular location of such proteins. Third, since yeasts can secrete proteins efficiently (see Ref. [14] for a review), the separation of a desired recombinant product from cellular constituents is facilitated.

S. cerevisiae was the initial choice as a host due to the accumulation of information concerning molecular and genetic aspects of this organism. Since the first report on heterologous gene expression in *S. cerevisiae* [15], a number of groups have reported success in producing small quantities of foreign proteins on a laboratory scale using this organism. Several problems have been encountered, however, especially during attempts to adapt laboratory-scale production schemes to an industrial scale [3]. One additional problem in conjunction with the *S. cerevisiae* system concerns aerobic ethanol fermentation, which is a characteristic feature of its metabolism [16] leading to reduced biomass yields and consequently to a reduced biomass-dependent production of host proteins. Furthermore, *S. cerevisiae* tends to hyperglycosylate its proteins [17]. This could conceivably lead to undesired immunological reactions if such proteins are to be used for therapeutic purposes.

Disadvantages such as these have promoted a search for alternative hosts, and research was initiated, aiming at developing bettter host-vector systems based on yeasts other than *S. cerevisiae*. To date, a number of alternatives are available [18]. In addition to *S. cerevisiae* and other related species such as *S. uvarum* [19], the list includes the methylotrophic yeasts *Pichia pastoris* and *Hansenula polymorpha* [20–23], the hydrocarbon-utilizing yeasts *Candida maltosa* [24–28], *Pichia guilliermondii* [25], and *Yarrowia lipolytica* [29–31], and the industrial yeasts *Kluyveromyces lactis* [32–34], *Kluyveromyces fragilis* [35], *Schizosaccharomyces pombe* [36–43] *Schwanniomyces occidentalis* [44, 45], *Trichosporon cutaneum* [46], *Rhodosporidium toruloides* [47], *Pachysolen tannophilus* [48] and *Zygosaccharomyces rouxii* [49].

A number of reviews dealing with gene transfer systems in fungi have recently been published [50, 6, 51, 52]. In this review we summarize some of the recent results concerning transfer and expression of heterologous genes in non-*Saccharomyces* yeasts.

2 Gene Transfer Systems in Yeasts

Hinnen et al. [53] first succeeded in complementing a *leu2* mutation in haploid laboratory strains of *S. cerevisiae* with a cloned yeast *LEU2* gene. Since then a number of yeasts, as outlined above, have been transformed. Two types of selectable markers have been used to transform yeast cells: positive (dominant) markers and complementing markers. The following summarizes the principles of the two systems.

2.1 Gene Transfer Systems Based on Positive Selection Markers

2.1.1 Drug Resistance Markers

The positive or dominant marker system frequently involves an antitiotic to which the yeast is sensitive, as well as a gene which confers resistance to the drug. The genes encoding resistance to chloramphenicol (chloramphenicol acetyltransferase [54–56]), G418/geneticin (aminoglycoside phosphotransferase [57–60]), hygromycin B (hygromycin B phosphotransferase [61, 62]), and phleomycin (DNA scission [63]), originate from sequences which are carried by *E. coli* transposons. More recently, a bleomycin-resistance gene conferring resistance to bleomycin and phleomycin has been isolated from *Streptoalloteichus hindustanus* [64] and applied successfully for the construction of transformation vectors [65]. Since bacterial promoters, as expected, operate inefficiently in yeasts they were made to function efficiently as selectable markers with yeast promoter and terminator signals. Resistance to methotrexate was accomplished by modifying the mouse dihydrofolate resistance gene (*dhfr*) for expression in yeast [66–68].

A general procedure for identifying enzymes inactivated by inhibitors relies on an increased copy number of the corresponding genes. By using minimal growth limiting concentrations of the inhibitors tunicamycin, compactin and ethionine in a haploid strain of *S. cerevisiae*, Rine et al. [69] were able to isolate, from a *S. cerevisiae* episomal plasmid (YEp) library, wild type genes encoding the three different enzymes inhibited by these compounds including UDP-*N*-acetyl-glucosamine-1-P-transferase, 3-hydroxy-3-methylglutaryl-CoA reductase and Ado Met synthetase, respectively. The same approach has been used to isolate the *S. cerevisiae* lanosterol 14α-demethylase gene encoding cytochrome P450 by virtue of the fact that this enzyme is inhibited by ketoconazole [70].

Multicopy plasmids have also been used to isolate resistant allelic variants of three ribosomal proteins. The recessive alleles *tcm1*, encoding a trichodermin-resistant ribosomal protein L3 [71, 72], *cyh2*, encoding a cycloheximide-resistant ribosomal protein L29 [73], and *cry1*, encoding a cryptoleurine-resistant ribosomal protein 59 [74], result in resistance to trichodermin, cycloheximide and cryptoleurine, respectively. The same methodology was shown to be applicable to the selection of methylglyoxal resistance that was subsequently expressed in bakers, brewery and sake yeasts [75]. The sequence responsible for methylglyoxal resistance was also shown to confer resistance to $CdCl_2$, $CoCl_2$, tetramethylthiuram disulfide, iodoacetamide, $ZnCl_2$, $NiCl_2$, $CuCl_2$, phenylglyoxal, *N*-ethylmaleimide and 8-hydroxyquinoline.

A further application of this approach was used to isolate a sulfometuron methyl (SM) resistance gene [76]. Deletions of the SM resistance gene eliminated acetolactate synthase (ALS), the *ILV2* gene product, leading to a requirement for isoleucine and valine [76]. Mutations within the plasmid borne *ILV2* gene were selected on high concentrations of SM resulting in SM-resistant ALS activity. These *ILV2* resistance alleles were redesignated as *SMR1* [76]. The *SMR* allele was transformed into different industrial *Saccharomyces* strains [77] including strains from the ale, lager, baking, distillery, wine and sake industries.

Single gene inherited resistance to multiple unrelated inhibitors has been reported for the *pdr1* locus of *S. cerevisiae*. Different alleles of this gene [78] have resulted in altered resistance to more than 20 different cellular inhibitors [79]. Resistance results from an alteration in plasma membrane permeability [80]. In view of the diverse resistance phenotype of *pdr1* mutations, the cloned gene [81] will be invaluable for the construction of gene transfer vectors.

2.1.2 Markers Conferring Heavy Metal Tolerance

For the isolation of the *S. cerevisiae* copper-resistance gene, a copper sensitive, *trp1* haploid strain of *S. cerevisiae* was transformed by a genomic library cloned in a high copy number yeast replicating plasmid (YRp) vector containing *TRP1* as a selectable marker. A 3.1 kb insert was shown to confer resistance to copper. The *CUP1* gene, which encodes a metallothionein, was found to be located on chromosome VIII [82]. Resistance to copper required the presence of up to 15 tandem nuclear copies of *CUP1* [82]. The *CUP1* gene has been used for transformation purposes in a number of industrial yeast strains [83, 84].

2.1.3 Genes Encoding Amylases, α-Galactosidase, β-Galactosidase and Invertase as Selectable Markers

The absence of degradative amylases prevents starch utilization in most commercial *Saccharomyces* species [85]. The availability of cloned amylase genes permits their transfer to and expression in deficient strains as a means of selection and strain improvement. The glucoamylase genes from *S. diastaticus* have previously been cloned [86–88]. Since glucoamylse is an extracellular enzyme that releases glucose by sequential hydrolysis of α-1,4-linked sugars from the nonreducing end, its activity is conveniently monitored by a clear halo in starch bromocresol medium. Transformants from a *S. diastaticus* genomic library were selected in haploid *S. cerevisiae* strains by expression of a prototrophic marker. Halo forming colonies were used to identify clones carrying two different glucoamylase genes (*STA1* and *STA3* [86, 87]). An independently isolated glucoamylase gene, *DEX1*, was obtained by screening prototrophic transformants on medium containing 2% dextrin [88]. The same authors reported that *DEX1* could be used as a selectable marker in haploid laboratory strains by allowing plasmid treated protoplasts to regenerate on glucose-sorbitol medium prior to selection on dextrin medium.

Melibiose is produced by invertase cleavage of raffinose; however, further utilization of melibiose requires the *MEL1* gene product, encoding α-galactosidase (melibiase), which is absent in most industrial *Saccharomyces* strains except *S. carlsbergensis* [89]. *MEL1* has been independently cloned [90, 91] and sequenced [91–93] and shown to be under the control of the *GAL* regulatory system [90]. Gendre and Guerineau [94] constructed a vector carrying *MEL1* and the G418 resistance marker. By initially selecting on melibiose and then screening on G418 medium they were able to increase the transformation frequency by 25-fold. In a practical application Xiao and Rank [95] have used the yeast *SMR1* gene as a dominant resistance marker in the construction of recombinant industrial *Saccharomyces* strains carrying a *MEL1* gene devoid of bacterial plasmid DNA. The absence of bacterial DNA in the integrant structure facilitates the commercial application of recombinant DNA technology in the food and beverage industry.

Saccharomyces yeasts are unable to use lactose as the sole carbon source. By transferring to *S. cerevisiae* a plasmid carrying a 13 kb region of the *Kluyveromyces lactis* genome, including the *LAC4* gene encoding β-galactosidase, strains capable of growth on lactose were obtained. Several lines of evidence indicate that the LAC^+ phenotype in the *S. cerevisiae* transformants was due to the simultaneous expression of a *K. lactis* lactose permease and a β-galactosidase gene [96].

The yeast *P. pastoris* cannot use sucrose as the sole carbon source. The introduction of the invertase gene of *S. cerevisiae* rendered *P. pastoris* SUC^+, allowing it to grow in sucrose-based media [97]. The same principle has been used to construct sucrose-utilizing strains of *Y. lipolytica*. In this case the *S. cerevisiae SUC2* gene was placed under the control of the promoter and signal sequence of the *Y. lipolytica* gene encoding alkaline extracellular protease [98]. This system should be applicable to other yeasts which are unable to use sucrose as the sole carbon source.

2.1.4 Transfer of Killer Toxin Immunity

M1 double-stranded RNA (dsRNA) found in killer yeast strains codes for an extracellular toxin and an immunity factor that protects infected cells from the toxin; sensitive cells that do not contain dsRNA are killed by the killer toxin. Bussey and Meaden [99] developed a transformation protocol for expression of a cDNA of M1 dsRNA on medium containing killer toxin. Two industrial yeasts transformed by this procedure [99] were shown to have stable inheritance of the killer and immunity phenotype.

Examples of dominant marker genes which have been applied for selection purposes in *Saccharomyces* yeasts are shown in Table 1 and transformation systems for non-*Saccharomyces yeasts* based on positive selection markers are presented in Table 2.

2.2 Gene Transfer Systems Based on Complementing Auxotrophic Markers

For the development of this kind of marker system an auxotrophic mutant with a specific enzymatic defect and a complementing biosynthetic gene are needed.

Table 1. Positive selection markers for gene transfer in *Saccharomyces* yeasts

Marker gene encoding	Source	Selection principle	Ref.
Chloramphenicol acetyltransferase	*E. coli*	Chloramphenicol^r	[54–56]
Aminoglycoside phosphotransferase	transposons Tn5/Tn903	G418/Geneticin^r	[57–60]
Hygromycin phosphotransferase	*E. coli*	Hygromycin B^r	[61], [62]
Phleomycin resistance	transposon Tn 5	Phleomycin^r	[63]
Phleomycin resistance	*Streptoalloteichus hindustanus*	Phleomycin^r	[64], [65]
Dihydrofolate reductase	*E. coli*	Methotrexate^r	[66–68]
UDP-N-acetyl-glucoseamine-1-P-transferase	*S. cerevisiae*	Tunicamycin^r	[69]
3-hydroxy-3-methylglutaryl-CoA reductase	*S. cerevisiae*	Compactin^r	[69]
Ado Met synthetase	*S. cerevisiae*	Ehionine^r	[69]
Lanosterol 14α-demethylase	*S. cerevisiae*	Ketoconazole^r	[70]
Ribosomal protein L3	*S. cerevisiae*	Trichodermin^r	[71], [72]
Ribosomal protein L29	*S. cerevisiae*	Cycloheximide^r	[73]
Ribosomal protein 59	*S. cerevisiae*	Cryptoleurine^r	[74]
Methylglyoxal reductase	*S. cerevisiae*	Methylglyoxal^r	[75]
Acetolactate synthase	*S. cerevisiae*	Sulfometuron methyl^r	[76], [77]
Metallothionein	*S. cerevisiae*	Copper^r	[83], [84]
Thymidine kinase	Herpes simplex virus	Sulfanilamide^r Amethopterin^r	[100]
Ribosome modification ativity	*Candida maltosa*	Cycloheximide^r	[101]
Glucoamylase	*S. diastaticus*	Starch utilization	[86–88]
α-galactosidase	*S. carlsbergensis*	Melibiose utilization	[94–95]
β-galactosidase	*K. lactis*	Lactose utilization	[96]
Killer toxin	*S. cerevisiae*	Killer toxin immunity	[99]

Table 2. Transformation systems for non-Saccharomyces yeasts based on positive selection markers

Organism	Marker gene	Ref.
Kluyveromyces lactis	G418^r	[32], [33], [180]
Kluyveromyces fragilis	G418^r	[35]
Pichia pastoris	G418^r	[23]
Pichia pastoris	Invertase	[97]
Yarrowia lipolytica	Phleomycin^r	[102]
Schizosaccharomyces pombe	G418^r	[103]
Trichosporon cutaneum	Phleomycin^r, Hygromycin^r	[46]
Pachysolen tannophilus	Hexokinase	(48]
Yarrowia lipolytica	Invertase	[98]

To isolate specific types of auxotrophic mutants of yeasts a number of positive mutant-selection and enrichment schemes have been described [104–106]. Ureido-succinic acid has been used as a drug to select pyrimidine auxotrophs of *S. cerevisiae* [104] and α-aminoadipic acid was instrumental in obtaining *lys2* and *lys5* mutants [105]. Mutants of *S. cerevisiae* which are resistant to 5-fluoro orotic acid are often defective in their orotidine-5′-P decarboxylase or orotidine-5′-P pyrophosphorylase (*URA3* gene product and *URA5* gene product, respectively). Nystatin has been used in a number of systems such as *S. cerevisiae*, *H. polymorpha*, and *Y. lipolytica* for mutant enrichment [107–110] and inositol enrichment techniques have been applied for the isolation of *S. cerevisia* and *Y. lipolytica* mutants [111, 112]. For many of the non-*Saccharomyces* yeasts for which transformation systems are available auxotrophic mutants had been isolated without the aid of a selection technique, and subsequently transformed using heterologous or homologous marker genes (Table 3).

Table 3. Transformation systems for non-*Saccharomyces* yeasts based on nutritional markers

Organism	Marker gene	Origin	Ref.
Kluyveromyces lactis	*LAC4*	*K. lactis*	[32]
Kluyveromyces lactis	*URA3*, *TRP1*	*S. cerevisiae*	[113], [114]
Kluyveromyces lactis	*TRP1*	*K. lactis*	[115]
Yarrowia lipolytica	*LEU2*	*Y. lipolytica*	[30]
Yarrowia lipolytica	*LYS2*	*S. cerevisiae*	[29]
Trichosporon cutaneum	*ARGB*	*Aspergillus nidulans*	[116]
Trichosporon cutaneum	*ARGB*	*T. cutaneum*	[116]
Pichia guilliermondii	*ARG4*	*S. cerevisiae*	[25]
Candida maltosa	*LEU2*, *LYS2*, *ARG4*	*S. cerevisiae*	[24–27]
Candida maltosa	*HIS5*	*C. maltosa*	[240]
Candida albicans	*ADE2*	*C. albicans*	[117], [118]
Pichia pastoris	*HIS4*	*P. pastoris*	[20], [23]
Pichia pastoris	*HIS4*	*S. cerevisiae*	[20], [23]
Hansenula polymorpha	*LEU2*, *URA3*	*S. cerevisiae*	[21], [22], [119]
Schizosaccharomyces pombe	*lys1*	*S. pombe*	[120]
	his2	*S. pombe*	(6)
	ade5, *ade6*	*S. pombe*	[121]
	ura1	*S. pombe*	[40]
	ura4	*S. pombe*	[122]
	arg7	*S. pombe*	[123]
	leu1	*S. pombe*	[124]
Schizosaccharomyces pombe	*LEU2*, *URA3*	*S. cerevisiae*	[36], [125]
Schwanniomyces occidentalis	*ADE*	*S. occidentalis*	[44]
Schwanniomyces occidentalis	*TRP5*	*S. cerevisiae*	[45]
Rhodosporidium toruloides	*PAL*	*R. toruloides*	[47]
Rhodosporidium toruloides	*LEU2*	*S. cerevisiae*	[47]
Zygosaccharomyces rouxii	*LEU2*	*S. cerevisiae*	[49]

3 Production of Heterologous Proteins in Non-*Saccharomyces* Yeasts

3.1 Heterologous Protein Production in the Methylotrophic Yeasts *Pichia pastoris* and *Hansenula polymorpha*

A few representatives of the yeast genera *Candida*, *Hansenula*, *Pichia*, and *Torulopsis* are capable of growth on methanol. Of these the ascomycetous yeasts *P. pastoris* and *H. polymorpha* have been studied extensively in terms of their methanol metabolism, genetics and molecular biology (for recent reviews see references [126] and [23]).

Many of the enzymes involved in the methanol-utilization pathway appear to be induced and are present at substantially higher levels in methanol-grown cells than in cells grown on other carbon sources. In *P. pastoris*, alcohol oxidase (AOX), the first enzyme in the methanol utilization pathway, is not present in glucose-grown cells, but comprises as much as 30% of the total protein from methanol-grown cells [127] being concentrated primarily in the peroxisomes. For *P. pastoris*, three of these methanol-regulated genes have been isolated, including two highly homologous *AOX* genes (*AOX1* and *AOX2*; [128, 129]) and a dihydroxyacetone synthase (DAS)-encoding gene [128]. Synthesis of AOX appears to be regulated primarily at the transcriptional level as measured using *lacZ* fusions [130]. In cells grown on methanol a massive proliferation of peroxisomes occurs, occupying up to 80% of the cytoplasmic volume.

Strains of *P. pastoris* have been isolated for growth at very high cell density. Concentrations of as much as 130 g (dry weight) per liter could be harvested in a continuous cultivation process [131], [132]. Thus, the utilization of methanol-regulated promoters in conjunction with the bioreactor technology developed for this yeast is attractive as a production system for the synthesis of valuable foreign proteins.

The *H. polymorpha* methanol oxidase (MOX) has been studied in detail. Under fully induced conditions, up to 35% of the cellular protein consisted of MOX [133]. The corresponding *MOX* gene and its flanking sequences have been cloned and sequenced [134]. The gene encodes a 664 amino acids long protein and contains no intervening sequences. The gene encoding the *H. polymorpha* dihydroxyacetone synthase (DHAS) has been isolated by differential plaque hybridization and the sequence of the 2106 bp structural gene and its flanking sequences reported [135].

The regulatory sequences of the *MOX* and *DHAS* genes show several similarities possibly reflecting similarities in their regulation by methanol. The regulation of MOX production in *H. polymorpha* has been studied in continuous cultures using a mixture of glucose and methanol as carbon sources [136]. The study focussed on the identification of stages in the biosynthesis affecting the formation of active MOX in glucose-methanol grown cells. The levels of *MOX* mRNA, MOX protein in monomeric and octameric form, the FAD/MOX ratio, and the actual MOX activity have been quantified as functions of the dilution rate D. Hybridization studies with *MOX* mRNA probes showed an induction of *MOX* mRNA formation up to D = 0.29 h^{-1}. The induction of MOX protein synthesis (up to 37% of the cellular protein) at low D values appeared to be determined at the transcriptional level. The MOX activity at high D values, on the other hand, appeared to be tuned by the incorporation of FAD and (post) translation events. Despite the high levels of *MOX* mRNA, decreasing levels of MOX activity and MOX protein were found at D values ranging from 0.14 to 0.29 h^{-1}. The maximal FAD/MOX ratio was determined at D = 0.1 h^{-1}, which correlated with the maximal specific activity of MOX. In glucose-methanol media, both, the protein levels and the MOX activity were repressed by increasing levels of residual glucose at high D values.

Transformation in methylotrophic yeasts has been reported in the cases of *P. pastoris* [20] and *H. polymorpha* [21], [22], [119]. Mutants corresponding to certain of the known auxotrophic mutants of *S. cerevisiae* were isolated. They could either be complemented by existing vectors carrying the corresponding *Saccharomyces* gene or by the homologous sequence. Plasmid replication could be sustained either by a fortuitous ARS activity present in *S. cerevisiae* sequences or by the isolation of homologous ARS sequences. Cregg et al. [20] developed the system for *P. pastoris* based upon the isolation of mutants defective in histidinol dehydrogenase, the enzyme defective in *his4* mutants of *S. cerevisiae*. The *Pichia* gene was isolated by complementation of *S. cerevisiae his4* mutants with a *Pichia* gene library. Transformation of *Pichia* cells was achieved using protoplasts which had been induced following treatment with Zymolyase, and it was found that complementation would occur using either the homologous or heterologous sequences. Two autonomous replicating (ARS) sequences isolated from a *Pichia* library, PARS1 and PARS2, respectively, allowed autonomous replication for over 50 generations of plasmids carrying either the homologous or the heterologous *HIS4* genes.

S. cerevisiae genes were also used for the development of *H. polymorpha* transformation systems by complementing *ura3* and *leu2* auxotrophic mutants of *H. polymorpha* [22, 21, 119]. Transformation was possible using either the lithium procedure of Ito et al. [137] or the PEG procedure of Klebe et al. [138] or by inducing protoplasts using glucuronidase or Zymolyase. The transforming plasmids were found to integrate at random sites within the *H. polymorpha* genome and were kept as unstable, episomal elements provided that a *S. cerevisiae* ARS sequence was present. Roggenkamp et al. [22] isolated two ARS elements (HARSI and HARSII) from *Hansenula* chromosomal fragments. These HARS elements increased the transformation frequency some two to seven fold. A general conclusion from both the *Pichia* and *Hansenula* studies is that auxotrophic mutations can be complemented by either the homologous or by heterologous genes. In the absence of ARS activity, the homologous gene transforms with a low frequency and quickly integrates.

The attractiveness of the *AOX* and *DHAS* gene promotors stems from the fact that they can be used to allow the high-level production of a heterologous gene product in a regulated fashion. The feasibility of this approach has been demonstrated both in *Hansenula* and in *Pichia*. Tschopp et al. [130] constructed fusions of the *E. coli lacZ* gene to both the *AOX* and *DAS* gene promoters of *P. pastoris*.

The hybrid constructs were integrated into the chromosomal *HIS4* gene. When read from the *AOX* promoter, β-galactosidase activity increased from 0 to 17 U mg^{-1} protein upon glucose derepression, and methanol caused a further increase to 5–13 × 10^3 U mg^{-1} protein. Expression from the *DAS* gene promoter paralleled physiological finding in that it required both derepression and methanol induction.

The *SUC2* gene of *S. cerevisiae* was transformed into *P. pastoris* and used to study the synthesis, post-translational processing, and secretion of invertase [139]. Expression of the *SUC2* gene under the control of the *AOX* gene promoter resulted in efficient secretion of biologically active invertase into the periplasmic space and growth medium of *P. pastoris*. Invertase produced and secreted reached levels of up to 2.5 g l^{-1} of growth medium. The N-terminus of invertase secreted from *P. pastoris* was homogeneous and found to be identical to invertase secreted from *S. cerevisiae*. Invertase secreted from *P. pastoris* was glycosylated but contained significantly less carbohydrate than proteins secreted by *S. cerevisiae*.

In a separate study Grinna and Tschopp [140] have investigated the size distribution and general structural features of N-linked oligosaccharides in *P. pastoris* proteins. The largest oligosaccharides isolated were significantly shorter than the hypermannosylated structures typical of *S. cerevisiae* proteins, indicating that the factors which influence the processing of N-linked oligosaccharides in *P. pastoris* are different from those which influence processing in *S. cerevisiae*. The smaller N-linked oligosaccharides resembled high-mannose oligosaccharides synthesized by animal cells.

In a commercial application the hepatitis B virus surface antigen (HBsAg) was produced in *P. pastoris* under the control of the *AOX* promoter system [233]. A cassette was constructed which consisted of the HBsAg structural gene sandwiched in between the *AOX* promoter and terminator, along with the *HIS4* gene. The linear cassette was targeted into the *AOX1* chromosomal locus by gene replacement in such a way that the *AOX1* gene was deleted. *P. pastoris* encodes a second *AOX* gene (*AOX2*) which provides a much less efficient source of alcohol oxidase [129]; however, enough alcohol oxidase is synthesized from the *AOX2* gene transcripts to allow slow growth of *Pichia* on methanol. Expression of HBsAg was undetectable during the growth phase on glycerol and was induced to 3–4% of soluble cell protein upon addition of methanol. The non-secreted, unglycosylated product was apparently identical to the unglycosylated product recovered from human serum. A high proportion of monomers were found that spontaneously assembled into 22 nm particles, which are necessary for immunogenicity. The process was successfully scaled up to 240 liters, using two stage batch cultures, which yielded 90 g of product. This system achieves high-level expression from a single integrated copy.

Shen et al. [141] have expressed the hepatitis B virus pre S2-HBsAg in *H. polymorpha* under the control of its *MOX* promoter. The synthesized pre S2-HBsAg protein was found to be secreted outside the cell membrane into the periplasm and further excreted into the culture medium following permeabilization of the cell wall with β-glucanase. Cell cultures treated in this way were able to continuously synthesize and secrete 22 nm particles of pre S2-HBsAg into the medium for several days.

Janowicz et al. [142] have reported on a systematic analysis of the *H. polymorpha* expression system. A number of foreign genes were expressed under the control of the *MOX*, *DHAS* and formate dehydrogenase gene (*FMDH*) promoters using expression cassettes which had been stably integrated in the *H. polymorpha* genome. The gene encoding bacterial β-lactamase (lacking the signal sequence) was expressed very efficiently, and was stringently regulated by glucose catabolite repression/derepression mechanisms. Methanol had a 2–3 fold induction effect. The expressed β-lactamase constituted some 10% to 20% of the total cell protein under derepressed (glycerol) and induced (methanol) growth conditions, respectively. The β-lactamase gene was also fused to the prepro segment of the *S. cerevisiae* alpha factor and expressed under the control of the *MOX* promoter. More than 90% of the foreign protein was secreted into the medium. Secreted proteins, like alpha amylase from *Schwanniomyces castellii* and invertase from *S. cerevisiae* were also efficiently produced and secreted, indicating that *H. polymorpha* can recognise foreign secretory signals. The gene encoding the HBsAg was expressed under the control of the *FMDH* promoter. In the transformants grown in complete media containing methanol or glycerol as a carbon source, HBsAg protein constituted approximately 3–5% of the total soluble protein in the form of 22 nm particles.

A cDNA for bovine lysozyme c2, a novel lysozyme characteristic of the cow stomach was expressed in *P. pastoris* and the protein product was secreted, using its native signal sequence [143]. A semi-continuous cultivation process which was used for the production resulted in extremely high cell densities and product concentrations. During fed-batch cultivation, bovine lysozyme concentrations exceeded 550 mg l^{-1} and cell densities exceeded 120 g dry cells l^{-1}. In continuous culture, bovine lysozyme and cell concentrations remained constant at approximately 350 mg l^{-1} and 100 g dry cells l^{-1}, respectively. The signal sequence of the pre-lysozyme was accurately processed.

Streptokinase, which is a potent plasminogen activator with widespread clinical use as a thrombolytic agent is naturally secreted by several strains of hemolytic *Streptococci*. The structural gene encoding streptokinase, minus the bacterial signal sequence, was fused to the *AOX* promoter of *P. pastoris* [144]. Upon induction on methanol, strains of *P. pastoris* transformed with the recombinant plasmid produced a 47 kDa protein (the size of native streptokinase). Lysates of transformed cells induced on methanol activated plasminogen, as determined by casein degradation on agar plates. During continuous cultivation on methanol, the streptokinase gene was stably maintained and expressed and no adverse effects on cell growth were observed.

3.2 Heterologous Protein Production in *Schizosaccharomyces pombe*

The fission yeast *Schizosaccharomyces pombe* has received little attention as an organism for technical applications so far [123] although its ability to metabolize malic acid which is used for deacidification of grape must [145], [146] and its ability to produce ethanol are well established [147].

S. pombe is an unicellular eukaryote showing a vegetative reproduction by splitting. It has a rather narrow range of carbon sources and produces alcohol when growing under oxygen limited conditions. Due to its property of forming asci with one to four smooth, oval or round ascospores this yeast is equally amenable to genetic manipulation as is *S. cerevisiae*. The genome consists of three electrophoretically separable chromosomes with sizes of 5.7, 4.6 and 3.5 megabase pairs, respectively [148].

Transformation procedures for *S. pombe* using protoplasts have been described [36, 37, 40, 38, 41, 149, 42, 37, 43]. The efficiency was $1–5 \times 10^4$ transformants per μg of plasmid DNA using protoplasts [36] and an efficiency of $4–9 \times 10^3$ transformants per μg of plasmid DNA was obtained with the lithium acetate procedure [150]. Transformation of *S. pombe* by electroporation yielding between $1–8 \times 10^3$ colonies per μg of plasmid DNA has recently been achieved [151].

S. pombe vectors typically contain a yeast selectable marker gene and an autonomous replication sequence [43] besides a bacterial origin of replication and a selectable marker for *E. coli*. The cloned *S. cerevisiae LEU2* and *URA3* genes can complement the *S. pombe leu1* and *ura4* mutations and thus have been used in the design of transformation vectors for *S. pombe*. The homologous *ura4* marker gene [152] has recently been used for gene disruption and replacement in *S. pombe* [122, 153]. The *S. pombe sup 3–5* marker has been used as well. This marker is an opal nonsense suppressing tRNA gene which suppresses the *ade6-704* mutation in *S. pombe* [154].

Vectors for heterologous gene expression containing different promoters have been developed. *S. cerevisiae* promoters gave weak expression levels in *S. pombe* [155] because of inherent differences in transcription initiation. The analysis of five different *S. pombe* genes showed that the TATA box region in *S. pombe* is located some 35–45 nucleotides upstream of the transcription start site [156, 157]. In *S. cerevisiae* this distance varies between 35 and 120 nucleotides. A number of additional promoters including the SV40 early promoter [158] and the cauliflower mosaic virus 35 S promoter have been shown to be functional in *S. pombe* [103, 159].

The *S. pombe* alcohol dehydrogenase gene has been isolated and characterized [160] and its promoter has been used for expressing foreign genes in *S. pombe* [161, 162] and a transcriptionally regulated expression vector for *S. pombe* has recently been described [163]. This vector was designed to facilitate the construction of transcriptional fusions to the promoter of the *S. pombe* fructose biphosphatase gene [164]. Transcription from this promoter is regulated by glucose repression over a range of greater than 100-fold. Intermediate levels of expression could be induced by growth on different carbon sources.

For the isolation of *S. pombe* promoters, random *S. pombe* genomic fragments were cloned upstream of the promoterless *E. coli lacZ* gene and tested in vivo for their capacity to promote expression of the β-galactosidase gene [165]. An efficient *S. pombe* promoter called 54/1 was isolated in this way and shown to drive up to 5% of the total protein synthesis as β-galactosidase.

A number of heterologous proteins have been expressed in *S. pombe* using homologous and heterologous promoters. A *Saccharomyces diastaticus* glucoamylase gene was introduced into *S. pombe* and low levels of expression from its own promoter could be detected [166]. Kuranda and Robbins [167] have successfully expressed the *S. cerevisiae* α-mannosidase, exoglucanase and endochitinase genes in *S. pombe* using the homologous *S. cerevisiae* promoters.

The polyoma virus middle-T antigen gene, lacking its intron, was inserted into a expression plasmid containing the *S. cerevisiae* phosphoglycerate kinase promoter and introduced into *S. cerevisiae* and *S. pombe*. Compared to the level of expression in *S. cerevisiae*, expression of middle T-antigen in *S. pombe* was significantly lower [155] possibly indicating that the *S. cerevisiae* phosphoglycerate kinase promotor is used inefficiently in *S. pombe*. Recombinant plasmids containing a human antithrombin III cDNA under the control of the *S. cerevisiae* alcohol dehydrogenease promoter were constructed and introduced into *S. pombe* [237]. The signal sequence of antithrombin III was recognized and antithrombin III was secreted into the medium in a biologically active form.

A human liver epoxide hydrolase cDNA was inserted into the yeast expression vector pEVP11 containing the *S. pombe* alcohol dehydrogenase (*adh*) promoter and introduced into *S. pombe*. The epoxide hydrolase protein and enzymic activity were subsequently produced and identified in the 105,000 g pellet after centrifugal fractionation of homogenized yeast cells [168]. Likewise, the *Halobacterium halobium* gene encoding bacterio-opsin was also expressed from the *adh* promoter. Western blot analyses revealed processing of the precursor protein and addition of the chromophore retinal to the culture medium resulted in a slight purple colour of the yeast cells indicating the in vivo regeneration of bacteriorhodopsin from bacterio-opsin and its incorporation into membranes [169].

A general expression vector for *S. pombe* was constructed [170]. In this vector the expression of the heterologous gene is driven by the *S. pombe adh* promoter. Transcription termination signals were isolated from the *S. pombe* actin gene and the vector carries a *S. cerevisiae URA3* gene, which complements the *S. pombe ura4* mutation. The stability of this plasmid is conferred by the *S. pombe* ars and stb elements respectively, which were derived from plasmid pFL20 [125]. The functionality of this vector system was shown by the production of the human blood coagulation factor XIIIa [170].

S. pombe can directly utilize D-xylose as a sole carbon source after it is transformed with plasmids containing the xylose isomerase gene from *E. coli* [171]. In these transformants the low xylose isomerization activity was the limiting step for D-xylose utilization since yeast proteases decreased the xylose isomerase activity [172].

3.3 Heterologous Protein Production in *Kluyveromyces lactis*

Members of the budding yeast genus *Kluyveromyces* have been used industrially for many years for the production of β-galactosidase (lactase). The species *K. lactis* and *K. fragilis* have the ability to use lactose as a sole carbon source [173] and whey can be used as a culture medium without any additions. Thus, the production of heterologous proteins, using a cheap carbon source such as whey, is feasible with *Kluyveromyces* yeasts.

Lactose utilization is accomplished by inducing both lactose transport and an extracellular β-galactosidase by more than 100-fold above a moderate basal level [174, 175]. Galactose is catabolized by inducing enzymes of the Leloir pathway: galactokinase, transferase, and epimerase. These enzymes are encoded by the *GAL1*, *GAL7* and *GAL10* genes, respectively [176]. Induction of the Leloir pathway in *K. lactis* occurs at the level of transcription and the organization and regulation of the *GAL* gene cluster in *K. lactis* is closely related to the homologous gene cluster in *Saccharomyces* [176]. The *K. lactis GAL4* gene encoding β-galactosidase has been cloned into *E. coli* by complementation of lactose-non utilizing mutants [177].

A high frequency transformation system for *K. lactis* based on vectors containing a bacterial kanamycin resistance gene and the *K. lactis LAC4* gene has been worked out [32]. For the selection of the transformants, *lac4* mutant strains were used. Selection was on G418-containing media first, followed by testing of the resistant colonies for their ability to grow on lactose as a sole carbon source. Vectors containing one of a series of *K. lactis* DNA fragments that presumably carried ARS sequences (called KARS) could transform *K. lactis* with high frequencies ranging from 10^3 to 3×10^4 colonies per μg DNA. Under non-selective conditions the plasmid sequences were unstable indicating that they existed in an episomal state. Vectors containing *S. cerevisiae* ARS sequences only led to integrative transformation of *K. lactis*. Using a similar strategy, Sreekrishna et al. [33] have transformed *K. lactis* with plasmids containing a Tn903-derived kanamycin resistance gene in conjunction with *K. lactis* ARS sequences. For the related yeast, *K. fragilis*, Das et al. [35] have constructed a vector containing a bacterial kanamycin resistance gene, the *TRP1* gene of *S. cerevisiae* and a KARS sequence from *K. lactis*. By utilizing the method based on treatment with alkali cations and with the kanamycin resistance gene as the selective marker, G418 resistant colonies were obtained. In the transformants the plasmid replicated autonomously.

The *K. lactis TRP1* gene has recently been cloned from a genomic library by complementation of a *S. cerevisiae trp1* mutation [115]. Transposon mutagenesis and DNA sequencing studies have indicated that the *TRP1* gene encodes a 210-amino acid polypeptide which shows 53% identity to the homologous *S. cerevisiae* protein. The *K. lactis TRP1* gene has been disrupted by substituting the *S. cerevisiae URA3* gene for a large part of the *TRP1* coding sequence. Replacement of the chromosomal *TRP1* locus with this construction has enabled the production of non-reverting *trp1*$^-$ strains of *K. lactis*. DNA sequencing has also shown that the *K. lactis TRP1* sequence is flanked by genes encoding inorganic pyrophosphatase and histone H3.

Some strains of *K. lactis* contain two linear double-stranded DNA plasmids, k1 and k2. The presence of the two plasmids confer on the cells a "killer" character, due to the production of a toxin that kills the sensitive cells [178]. Hybrid plasmids containing parts of the k1 plasmid and the *URA3* gene of *S. cerevisiae* have been constructed and used to transform a uracil-requiring strain of *K. lactis* to a prototrophic form [113]. The transforming plasmids contained the sequence of one or both ends of the linear k1 DNA integrated into a circular molecule.

A circular plasmid, pKD1 (or 1.6 μm DNA) has recently been isolated from *K. drosophilarum* [179]. This plasmid appears to have a functional organization analogous to that of the 2 μm DNA of *S. cerevisiae* although the respective nucleotide sequences show little homology. pKD1 can be transferred to *K. lactis* where it is replicated stably. Using recombinant molecules derived from pKD1 carrying the *S. cerevisiae URA3* gene or the Tn903-derived kanamycin resistance gene, a practical transformation system has been developed for *K. lactis*, with an efficiency and stability comparable to the 2 μm-based *S. cerevisiae* transformation system. The transforming plasmids based on this pKD1-plasmid was stably maintained under non-selective conditions [180]. Several species belonging to the genus *Kluyveromyces* could be transformed to G418 resistance with the pKD1 plasmid containing the kanamycin/G418 resistance gene from Tn903 with different kinds of frequencies ranging from 33 to 8060 transformants per μg of DNA and stabilities of 21% to 99.9% [181]. Thus, this broad host range plasmid has properties which are lacking in the *S. cerevisiae* 2 μm plasmid.

Kluyveromyces as a host organism for heterologous protein production was found to be superior to *S. cerevisiae* both for internal and extracellular production of proteins [182]. Intracellular expression of prochymosin and of fusion proteins obtained with vectors carrying the 5′-end of the *K. lactis* lactase gene linked to the calf prochymosin cDNA has been achieved [34]. Moreover, it was found, when using the *Kluyveromyces* alpha-factor signal sequence fused to prochymosin, that more than 85% of the produced prochymosin was secreted in *K. lactis*. Authentic prochymosin found in the culture medium was converted into active chymosin by a pH shift. These results have led to the development of a process at the 41,000 liter scale [34]. Under these large scale conditions the stability of the recombinant plasmid was excellent.

3.4 Heterologous Protein Production in *Yarrowia lipolytica*

The genetics and genetic engineering of the hydrocarbon degrading yeast *Yarrowia lipolytica* are summerized by Heslot in a separate chapter of this issue of Advances in Biochemical Engineering/Biotechnology and will therefore not be dealt with in this review.

4 Gene Transfer Systems for Other non-*Saccharomyces* Yeasts

4.1 Gene Transfer in *Schwanniomyces occidentalis*

The genus *Schwanniomyces* was initially described as containing four species, i.e. *Schwanniomyces alluvius*, *Schwanniomyces castellii*, *Schwanniomyces persoonii* and

Schwanniomyces occidentalis [183]. A recent study of DNA reassociation between species of *Schwanniomyces* has led to a single species designation, *Schwanniomyces occidentalis* [184]. This yeast is of great interest because of its superior ability to produce amylolytic enzymes, i.e. α-amylase and glucoamylase and because it is able to degrade starch completely and to convert starch and inulin to fuel-grade ethanol [185]. *S. occidentalis* can secrete proteins greater than 140 kDa efficiently into the culture medium, does not hyperglycosylate secreted proteins, and does not secrete measurable quantities of proteases [186]. In addition the genes for the secreted proteins are substrate induced and catabolite repressed [187] and the organism is capable of growth on inexpensive substrates [185].

The enzymes glucoamylase and α-amylase were found in the cell free supernatant of *S. occidentalis* on a variety of carbon sources and both amylolytic enzymes were purified from culture media containing either starch, maltose or cellobiose. Unlike extracellular enzymes from *S. cerevisiae*, these enzymes were found not to be hyperglycosylated [186].

The *S. occidentalis* glucoamylase gene (*GAM1*) was isolated using synthetic oligonucleotid as probes [188]. The *GAM1* gene contains an open reading frame of 2874 nucleotides coding for 958 amino acids. The identity of the glucoamylase gene was confirmed by disruption and displacement of the *GAM1* gene in *S. occidentalis* leading to its inability to grow efficiently on starch. These *gam1* strains still exhibited low glucoamylase activity suggesting that at least one more, weakly expressed glucoamylase gene (*GAM2*), must be present in *S. occidentalis*. The *GAM1* gene was not expressed in *S. cerevisiae* under the control of its 5′ non-coding sequences.

For the isolation of the α-amylase gene (*AMY1*), a *S. occidentalis* cosmid gene library was transformed into *S. cerevisiae* and the gene was identified by expression screening on starch agar plates after staining with iodine [189]. The positive clone contained a DNA fragment harbouring an open reading frame of 1536 nucleotides coding for a 512-amino acid polypeptide with a calculated molecular weight of 56.5 kDa. The deduced amino acid sequence revealed significant similarities to the sequence of the *Saccharomycopsis fibuligera* and *Aspergillus oryzae* α-amylases. A similar strategy was followed by Abarca et al. [190] to clone an α-amylase gene from *S. castellii*. The *AMY1* gene was found to be expressed from its original promotor in *S. cerevisiae*, *K. lactis* and *S. pombe*, leading to an active secreted gene product and thus enabling the different yeast transformants to grow on starch as a sole carbon source.

For the purpose of studying the sequence and organization of *Schwanniomyces* secretory signals and to provide materials essential for the analysis of gene regulation and protein secretion in *S. occidentalis*, the gene encoding invertase (*INV*) has been cloned from *S. occidentalis*. The enzyme consists of 533 amino acids, 8 potential glycosylation sites and has 45% identity with the invertase from *S. cerevisiae*. The proenzyme has a 22 amino acid signal sequence that has a high α-helical transmembrane potential which differs significantly from that predicted for the *S. cerevisiae* enzyme [191].

For the design of suitable cloning vectors, the *S. occidentalis* orotidine-5′-phosphate decarboxylase (*ODC*) gene has been isolated by complementation of the *ura3* mutation of *S. cerevisiae* [192] showing that the *ODC* gene promoter, like the one of the α-amylase gene of *S. occidentalis* are functional in *S. cerevisiae*.

An efficient transformation system for *S. occidentalis* based on *ade2* mutants of *S. occidentalis* deficient for phosphoribosylaminoimidazole carboxylase has been worked out by Klein and Favreau [44]. As a selectable marker the *S. occidentalis* *ADE2* gene was isolated by complementation in an *ade2* strain of *S. cerevisiae*. Intact plasmids could be recovered in *E. coli* from whole-cell lysates of ADE^+ transformants, indicating that plasmids were replicating autonomously. High-molecular-mass species of the transforming plasmid were found by Southern hybridization analysis of intact genomic preparations. Subclones of such plasmids were shown to complement *ade2* mutants of *S. occidentalis* with high efficiencies (> 1000 colonies per μg DNA).

Dohmen et al. [45] have achieved high frequency transformation of a *S. occidentalis* mutant defective in the last step of tryptophan synthesis with plasmids containing the tryptophan synthetase gene (*TRP5*) of *S. cerevisiae* and an autonomous replication sequence (SwARS1) from *S. occidentalis*. The SwARS1 fragment was found to be active in *S. cerevisiae* as well. The average copy number of the plasmids in both yeast species was 5 to 10 per cell under selective conditions. *S. occidentalis* cells that were transformed with an autonomously replicating plasmid carrying the cloned α-amylase gene from *S. occidentalis* secreted about five times more α-amylase than cells without additional copies of the α-amylase gene. Thus, this transformation system provides a possibility of improving starch degradation by *S. occidentalis*.

4.2 Gene Transfer in *Candida Yeasts*

4.2.1 Genc Transfer in *Candida maltosa*

Among the large number of known *Candida* species only a few members have been studied extensively for biotechnological applications. These yeasts have received considerable interest because of their ability to metabolize a broad range of different carbon sources, such as methanol, hydrocarbons, fatty acids and pentoses [18].

Since *Candida* yeasts are non-sporulating microbes, classical genetic manipulations are somewhat limited. However, genetic analysis has been carried out in the past by protoplast fusion, mitotic recombination through the parasexual cycle, and by mutagenesis of the desired strains. Recombinant DNA techniques improve the possibility for genetic manipulation of these industrially and medically important yeasts and until now most of the investigations have been carried out using *Candida maltosa* and *Candida albicans* [193].

Candida maltosa is an *n*-alkane utilizing asporogenic yeast, which has been used in large scale bioprocesses for single cell protein production in the past. For the purpose of setting up a genetic transformation system for this yeast a *LEU* gene from *C. maltosa* was cloned, and found to complement leu^- mutations in *E. coli* and *S. cerevisiae* [28]. The nucleotide sequence analysis of the *C. maltosa*

LEU gene revealed 72% similarity with the *S. cerevisiae LEU2* gene at the nucleotide level and 76% at the deduced amino acid sequence level [194].

To construct a host-vector system for *C. maltosa*, ARS sequences from the *C. maltosa* genome were isolated by transforming *C. maltosa* with a gene library prepared by using a *S. cerevisiae* YEp vector which contained the *S. cerevisiae LEU2* gene, and LEU^{+} transformants were obtained with a high frequency [24]. Such transformants contained plasmids with *C. maltosa* ARS sequences. A 3.8 kb fragment with ARS activity was subcloned and designated TRA (transformation ability region). Transformation of *C. maltosa* with such TRA plasmids was possible using both spheroplast and lithium acetate methods yielding between 330 and 1650 colonies per μg of transforming DNA. Southern blot analyses suggested that the copy number of these plasmids was between 10 and 20, and that they were stably maintained during growth without selective pressure in the medium. The *C. maltosa* ARS sequence also functioned in *S. cerevisiae*. More recently, an improved host-vector system for *C. maltosa* using the homologous *HIS5* gene together with the TRA region was described [240].

To develop a transformation system in *C. maltosa*, *arg4* mutants deficient in argininosuccinate lyase and *lys2* mutants deficient in α-aminoadipate reductase were isolated [25, 26, 27]. Genetic transformation of the arginine auxotrophic mutants was attained using the *S. cerevisiae* hybrid plasmid pYe(*ARG4*) [195]. In all transformants argininosuccinate lyase as well as β-lactamase could be detected and the ARG^{+} phenotype was mitotically unstable. Closed circular pYe(*ARG4*) DNA was detected in DNA preparations from the transformants by hybridization to pBR322 DNA and by transformation of *E. coli* to ampicillin resistance, indicating that the plasmid had be maintained extrachromosomally [25]. The *C. maltosa lys*-auxotrophic strain could be transformed by the plasmids pDP12 and pDP13, which contain the *LYS2*-encoding gene of *S. cerevisiae* [196]. In these transformants plasmid maintenance, involving site-directed integrative transformation into, and aberrant excision from the yeast genome was observed, generating rearranged plasmids which could be recovered by transformation into *E. coli* [26, 27].

The *POX2* and *POX4* genes encoding the peroxisomal acyl coenzyme A oxidase subunits PXP-2 and PXP-4, respectively, of peroxisomal fatty acyl-coenzyme A oxidase of *Candida tropicalis* were introduced into *C. maltosa* [197]. The fact that *C. tropicalis* promoters were recognized and the gene products were targeted into the *C. maltosa* peroxisomes indicates that gene expression and targeting signals and mechanisms in these two yeasts are similar.

4.2.2 Gene Transfer in *Candida albicans*

Candida albicans is a dimorphic, opportunistic pathogen, causing disease in immunocompromised patients. Therefore, its use as a potential host for heterologous gene expression will be limited. Because of its diploid genome [198] and the lack of a known sexual cycle [199], genetic analysis relied on protoplast fusion experiments [200–202].

A transformation system for *C. albicans* was described by Kurtz et al. [117]. They complemented an *ade2* mutant of *C. albicans* with the homologous *ADE2* gene by integrative spheroplast transformation at frequencies of 0.5 to 5 transformants per μg of DNA. The development of autonomously replicating plasmids rised the frequency up to 200–1000 ADE^{+} transformants per μg DNA [118]. A method for introducing specific mutations into diploid *C. albicans* cells by a one-step gene disruption and a subsequent UV-induced recombination procedure was developed [203]. The cloned *C. albicans URA3* gene [204] was disrupted with the *C. albicans ADE2* gene [117] and the linearized DNA was used for transformation of two *ade2* mutants. Both, an insertional inactivation of the *URA3* gene, and a disruption resulting in a 4 kb deletion, were made. The transformants, heterozygous for the *ura3* mutation, were rendered homozygous by UV-induced recombination.

4.3 Gene Transfer in *Pichia guilliermondii*

Arginine-auxotrophic mutants of the *n*-alkane utilizing yeast *P. guilliermondii* have been transformed [25] using the *S. cerevisiae* plasmid pYe(*ARG4*)411, carrying the *S. cerevisiae ARG4* gene [195]. The ARG^{+} phenotype was mitotically unstable and the transforming plasmid could be rescued into *E. coli*. This indicates that plasmid pYe(*ARG4*) 411 was able to replicate autonomously in *P. guilliermondii*.

4.4 Gene Transfer in *Trichosporon cutaneum*

Our own interest deals with the filamentous *Trichosporon* yeasts and with *T. cucaneum* in particular. This yeast has a purely oxidative metabolism and des not form ethanol even under oxygen limitation [205]. It shows a high biomass yield of around 50 to 55% on glucose, and its behaviour in bioreactors has been studied extensively [206]. In addition, *Trichosporon* yeasts have the potential to use a very large variety of different carbon sources [207] and *T. cutaneum* has been shown to grown on various disaccharides, such as cellobiose, maltose, lactose, sucrose, melibiose, and trehalose [208], on xylans [209–211], starch [212], and on pectic acid, carboxymethylcellulose and locust-bean gum [213]. *T. cutaneum* appears to be very versatile in metabolizing numerous aromatic compounds [214–221] and grows on ethylamine [222]. These properties indicate the extraordinary potential of *T. cutaneum* for the efficient conversion of various cheap carbon sources into biomass [223].

Realizing its potential as a host organism for foreign gene expression, we have worked out a versatile transformation system for this yeast [46]. Transformation of *T. cutaneum* was obtained with plasmids carrying either the *E. coli* hygromycin B phosphotransferase-encoding gene [61] or the *Streptoalloteichus hindustanus* phleomycin-resistance gene [64], as dominant selection markers. Expression of both resistance-conferring genes was controlled by the glyceraldehyde-3-phosphate dehydrogenase promoter and the *trpC* gene terminator, from *Aspergillus nidulans* [224, 65]. The transforming DNA was found to consist of multiple tandem plasmid copies of high molecular weight. This polymeric structure, in nonselective media, was mitotically unstable, possibly indicating that it existed in an episomal state.

More recently, we have isolated stable auxotrophic mutants of *T. cutaneum* using UV mutagenesis [116]. Some of the arginine-auxotrophic mutants obtained could be transformed using the *A. nidulans* ornithine transcarbamoylase (*OTC*) gene. For the purpose of setting up a homologous transformation system, cosmids carrying the *T. cutaneum OTC* gene were isolated using the *A. nidulans OTC* gene as a heterologous hybridization probe and by a sib-selection strategy [116]. The nucleotide sequence of part of the *T. cutaneum OTC* coding region showed 68% homology with the *A. nidulans* and *A. niger OTC* genes, respectively, but only 52% homology with the corresponding gene from *S. cerevisiae*. The cloned *T. cutaneum OTC* gene transformed *T. cutaneum argB* mutants with an efficiency of roughly 100 colonies per µg DNA [116].

4.5 Gene Transfer in *Rhodosporidium toruloides*

The basidiomycetous yeast *Rhodosporidium toruloides* has been investigated in the past with regard to its capacity of producing phenylalanine ammonia lyase (PAL) [225]. This enzyme is of interest to clinicians as a possible oral treatment for phenylketonuria. PAL catalyzes the deamination of phenylalanine to *trans*-cinnamic acid and ammonia and thus has industrial applications in the synthesis of L-phenylalanin from *trans*-cinnamic acid. In *R. toruloides* phenylalanine can act as the sole source of carbon, nitrogen, and energy. As PAL catalyzes the initial reaction in the catabolism of the amino acid, the enzyme plays a key role in regulating phenylalanine metabolism. In *R. toruloides* PAL is induced by the presence of L-phenylalanine and L-tyrosine [226].

R. toruloides protoplasts were transformed, in the presence of polyethylene glycol, at frequencies of approximately 1000 transformants per µg of DNA [47]. The plasmid used contained the *PAL* gene of *R. toruloides* and was able to replicate in the yeast as an unstable element, or able to integrate at the *PAL* locus to give stable transformants. Plasmids that function in *R. toruloides* were constructed using either the *PAL* gene or the *LEU2* gene of *S. cerevisiae* as selection markers. It has also been shown that the 2 µm replicon of *S. cerevisiae* does not function in *R. toruloides*; neither was the *PAL* gene expressed in *S. cerevisiae* [47].

4.6 Gene Transfer in *Pachysolen tannophilus*

Pachysolen tannophilus is a D-xylose-utilizing yeast which has originally been isolated from wood sulfite liquor [227]. This yeast has the ability to convert hexose and pentose sugars to ethanol. It has, therefore, potential use in the production of ethanol from lignocellulose hydrolysates which are comprised of a mixture of hexose and pentose sugars, and, ultimately in providing an efficient method of converting biomass into energy. Although this organism has been studied extensively with regard to its ability to convert xylose to ethanol, little is known about its genetic organization and gene expression. This yeast is a strongly homothallic organism in which the haplophase is predominant. Typically, diploidy is a consequence of fusion of two mitotic products within a specialized conjugant cell [228]. The diploid nucleus appears to proceed directly to meiosis thereby producing a four-spored ascus. A technique for isolating both homozygous and

Table 4. Fate of transforming plasmid DNA in non-*Saccharomyces* yeasts

Organism	Site of homologous Integration	Heterologous Integration	Autonomous replication sequence	Ref.
P. pastoris	*AOX1* gene			[23]
	AOX2 gene			[23]
	HIS4 gene			[130]
			PARS1	[20], [23]
			PARS2	
			PARS3	
H. polymorpha		yes		[22]
			HARS1	[22]
			HARS2	
S. pombe	*ura4* gene			[122]
		yes		[37], [153]
			ars, stb	[43]
K. lactis	*TRP1* gene			[115]
		yes		[32]
			KARS	[32]
			1.6 μm DNA	[180]
Y. lipolytica	*LEU2* gene			[29], [30]
	XPR2 gene			[239]
		yes		[29]
			ars	[232]
S. occidentalis			yes	[44]
			SwARS1	[45]
C. maltosa			TRA	[24]
		yes		[25], [26], [27], [240]
C. albicans	*ADE2* gene			[117]
			ARS	[118]
T. cutaneum			ARS	[46], [116]
R. toruloides	*PAL* gene			[47]
			ARS	[47]

heterozygous diploids has been devised. Auxotrophic mutants are easily obtained, and the yeast is amenable to tetrad analysis [228]. It is possible that *P. tannophilus* is adequate for commercial purposes as it occurs in nature, but it is reasonable to suppose that more competent strains can be produced through genetic engineering.

Utilization of D-xylose, the major pentose in lignocellulose hydrolysates, by *P. tannophilus* is subject to hexose catabolite repression. Hexose negative mutants of *P. tannophilus*, unable to utilise D-glucose, D-fructose, and D-mannose, have been isolated in an attempt to alleviate hexose sugar catabolite repression of D-xylose utilisation [229]. Lithium chloride treated cells of such hexose negative mutants have recently been transformed using the cloned hexokinase PII structural gene of *S. cerevisiae* [48] as a selectable marker. This opens up the possibility for more directed strain improvement programs in this yeast.

Table 5. Heterologous protein production in non-*Saccharomyces* yeasts

Organism	Product	Ref.
P. pastoris	Bovine lysozyme	[143]
	Hepatitis B virus HBs-antigen	[233]
	E. coli β-galactosidase	[130]
	S. cerevisiae invertase	[139]
	Streptokinase	[144]
	Human tumor necrosis factor	[238]
	Human epidermal growth factor	[241]
	Human tissue plasminogen activator	[241]
H. polymorpha	Hepatitis B virus pre S2-HBs-antigen	[141]
	Bacterial β-lactamase	[142]
	Schwanniomyces. castellii α-amylase	[142]
	S. cerevisiae invertase	[142]
	Hepatitis B virus HBs-antigen	[142]
S. pombe	*S. diastaticus* glucoamylase	[166]
	Human α-1 antitrypsin	[234]
	Polyoma virus middle T-antigen	[155]
	S. cerevisiae α-mannosidase	[167]
	S. cerevisiae exoglucanase	[167]
	S. cerevisiae endochitinase	[167]
	Human antithrombin III	[237]
	Liver epoxide hydroxylase	[168]
	Human blood coagulation factor XIIIa	[170]
	E. coli xylose isomerase	[172]
	E. coli β-galactosidase	[165]
	E. coli β-glucuronidase	[159]
	Halobacterium halobium bacteriorhodopsin	[169]
K. lactis	Bovine chymosin	[34]
Y. lipolytica	*E. coli* β-galactosidase	[102]
	S. cerevisiae invertase	[98]
	Porcine α-1 interferon	[235]
	Calf prochymosin	[236]

4.7 Gene Transfer in *Zygosaccharomyces rouxii*

A number of plasmids resembling structurally the *S. cerevisiae* 2 μm DNA have been isolated from osmophilic haploid yeasts of the genus *Zygosaccharomyces* [230]. Although structurally and functionally related, none of them have homology with the *S. cerevisiae* 2 μm DNA. Among these plasmids, pSR1 has been most intensively investigated [231]. A transformation system based on a leucine auxotrophic strain of *Zygosaccharomyces rouxii* and plasmid pSR1 carrying a *S. cerevisiae* *LEU2* gene has recently been published [49]. pSR1 can replicate in *S. cerevisiae* as well as in its native host and exists in two isomeric forms generated by intramolecular recombination events [231].

The fate of the transforming plasmid DNA in the various yeasts described above is summarized in Table 4, and Table 5 lists some of the heterologous proteins which have been produced using non-*Saccharomyces* yeasts.

5 Acknowledgements

Our project dealing with *Trichosporon* yeasts is being supported by the Swiss National Science Foundation, grant No. 3.006–0.87. We thank Dr. C. P. Hollenberg and Dr. R. Klein for providing us with preprints of their unpublished work.

6 References

1. Kingsman SM, Kingsman AJ, Dobson MJ, Mellor J, Roberts NA (1985) Biotechnol Genet Engineering Reviews 3: 377
2. Van Brunt J (1986) Bio/Technology 4: 1057
3. Kingsman SM, Kingsman AJ, Mellor J (1987) Tibtech 5: 53
4. Rambosek J, Leach J (1987) CRC Crit Rev Biotechnol 6: 357
5. Bitter GA (1987) Heterologous gene expression in yeast. In: Berger SL, Kimmel AR (eds) Meth in Enzymol 152: 673
6. Esser K, Kämper J (1988) Proc Biochem 23: 36
7. Saunders G, Picknett TM, Tuite MF, Ward M (1989) Tibtech 7: 283
8. Valenzuela P, Medina A, Rutter WJ, Ammerer G, Hall BD (1982) Nature 298: 347
9. Kukuruzinska MA, Bergh MLE, Jackson BJ (1987) Ann Rev Biochem 56: 915
10. Towler DA, Gordon JI, Adams SP, Glaser L (1988) Ann Rev Biochem 57: 69
11. Miyamoto C, Chizzonite R, Crowl R, Rupprecht K, Kramer R, Schaber M, Kumar G, Poonian M, Ju G (1985) Proc Natl Acad Sci USA 82: 7232
12. Sambucetti LC, Schaber M, Kramer R, Crowl R, Curran T (1986) Gene 43: 69
13. Lowy DR, Willumsen BM (1989) Nature 314: 384
14. Smith RA, Duncan MJ, Moir DT (1985) Science 229: 1219
15. Hitzeman RA, Hagie FE, Levine HL, Goeddel DV, Ammerer G, Hall BD (1981) Nature 293: 717
16. Käppeli O (1986) Adv Microb Phys 28: 183
17. Van Arsdell JN, Kwok S, Schweickart VL, Ladner MB, Gelfand DH, Innis MA (1987) Bio/Technology 5: 60
18. Spencer JFT, Spencer DM, Reynolds N (1988) J Basic Microb 28: 321
19. Holmberg S (1984) Tibtech 2: 98
20. Cregg JM, Barringer KJ, Hessler AY, Madden KR (1985) Mol Cell Biol 5: 3376
21. Gleeson MA, Ortori SA, Sudbery PE (1986) J Gen Microb 132: 3459
22. Roggenkamp R, Hansen H, Eckart M, Janowicz Z, Hollenberg CP (1986) Mol Gen Genet 202: 302

23. Cregg JM, Madden KR (1987) Development of yeast transformation systems and construction of methanol-utilization-defective mutants of *Pichia pastoris* by gene disruption. In: Stewart GG, Russell I, Klein RD, Hiebsch RR (eds) Biological research on industrial yeasts, vol II. CRC Press, Boca Raton, Florida, pp 1–18
24. Takagi M, Kawai S, Chung Chang M, Shibuya I, Yano K (1986) J Bact 167: 551
25. Kunze G, Petzoldt C, Bode R, Samsonova I, Hecker M, Birnbaum D (1985) Curr Genet 9: 205
26. Kunze G, Bode R, Schmidt I, Samsonova I, Birnbaum D (1987a) Curr Genet 11:385
27. Kunze G, Bode R, and Birnbaum D (1987b) Autonomous and integrative transformation of the yeast *Candida maltosa*. In: Proc 4th Europ Congr of Biotechnol., Vol 1 (Neijssel OM, van der Meer RR, Luyben KCAM, eds) Elsevier, Amsterdam, pp 398–401
28. Kawamura M, Takagi M, Yano K (1983) Gene 24: 157
29. Gaillardin C, Ribet AM, Heslot H (1985) Curr Genet 10: 49
30. Davidow LS, Apostolakos D, O'Donnell MM, Proctor AR, Ogrydziak DM, Wing RA, Stasko I, DeZeeuw JR (1985) Curr Genet 10: 39
31. Wing RA, and Ogrydziak DM (1985) Development of the genetics of the dimorphic yeast *Yarrowia lipolytica* In: Timberlake WE (ed) Molecular genetics of filamentous fungi. A. Liss, New York, pp 367–381
32. Das S, Hollenberg CP (1982) Curr Genet 6: 123
33. Sreekrishna K, Webster TD, Dickson RC (1984) Gene 28: 73
34. van den Berg JA, van der Laken KJ, van Ooyen AJJ, Renniers TCHM, Rietveld K, Schaap A, Brake AJ, Bishop RJ, Schultz K, Moyer D, Richman M, Shuster JR (1990) Bio/Technology 8: 135
35. Das S, Kellermann E, Hollenberg CP (1984) J Bact 158: 1165
36. Beach D, Nurse P (1981) Nature 290: 140
37. Wright APH, Maundrell K, Shall S (1986) Curr Genet 10: 503
38. Fournier P, Gaillardin C, de Louvencourt L, Heslot H, Lang BF, Kaudewitz F (1982) Curr Genet 6: 31
39. Beach D, Piper M, Nurse P (1982) Mol Gen Genet 187: 326
40. Sakaguchi J, Yamamoto M (1982) Proc Natl Acad Sci USA 79: 7819
41. Gaillardin C, Fournier P, Budar F, Kudla B, Gerbaud C, Heslot H (1983) Curr Genet 7: 245
42. Toda T, Nakasako Y, Niwa O, Yanagida M (1984) Curr Genet 8: 93
43. Heyer WD, Sipiczki M, Kohli J (1986) Mol Cell Biol 6: 80
44. Klein RD, Favreau MA (1988) J Bact 170: 5572
45. Dohmen RJ, Strasser AWM, Zitomer RS, Hollenberg CP (1989) Curr Genet 15: 319
46. Glumoff V, Käppeli O, Fiechter A, Reiser J (1989) Gene 84: 311
47. Tully M, and Gilbert HJ (1985) Gene 36: 235
48. Wedlock DN, Thornton RJ (1989) Biotechnol. Lett. 2: 601
49. Jearnpipatkul A, Araki H, Oshima Y (1987) Mol Gen Genet 206: 88
50. Cantwell BA, McConnell DJ (1987) Transformation of brewing yeast strains. In: Stewart GG, Russell I, Klein RD, Hiebsch RR (eds) Biological research on industrial yeasts, vol II. CRC Press, Boca Raton, Florida, pp 77–89
51. Rank GH, Casey G, Xiao W (1988) Food Biotechnol 2: 1
52. Fincham JRS (1989) Microb Rev 53: 148
53. Hinnen A, Hicks JB, Fink GR (1978) Proc Natl Acad Sci USA 75: 1929
54. Cohen JD, Eccleshall TR, Needleman RB, Fedoroff H, Buchferer BA, Marmur J (1980) Proc Natl Acad Sci USA 77: 1078
55. Hadfield C, Cashmore AM, Meacock PA (1986) Gene 45: 149
56. Hadfield C, Cashmore AM, Meacock PA (1987) Gene 52: 59
57. Jimenez A, Davies J (1980) Nature 287: 869
58. Webster TD, Dickson RC (1983) Gene 26: 243
59. Sakai K, Yamamoto M (1986) Agric Biol Chem 50: 1177
60. Lang-Hinrichs C, Berndorff D, Seefeldt C, Stahl U (1989) Appl Microb Biotech 30: 388
61. Gritz L, Davies J (1983) Gene 25: 179
62. Kaster KR, Burgett SG, Ingolia TD (1984) Curr Genet 8: 353
63. Gatignol A, Baron M, Tiraby G (1987) Mol Gen Genet 207: 342

64. Durand H, Baron M, Calmels T, and Tiraby G (1988) Classical and molecular genetics applied to *Trichoderma reesei* for the selection of improved cellulolytic industrial strains. In: Aubert JP, Beguin P, Millet J (eds) Biochemistry and genetics of cellulose degradation. Academic, London, pp 135–151
65. Mattern IE, Punt P (1988) Fungal Genet Newslett 35: 25
66. Miyajima A, Miyajima I, Arai K, Arai N (1984) Mol Cell Biol 4: 407
67. Zhu J, Contreras R, Gheysen D, Ernst J, Fiers W (1985) Bio/Technology 3: 451
68. Zhu J, Contreras R, Fiers W (1986) Gene 50: 225
69. Rine J, Hansen W, Hardeman E, Davis RW (1983) Proc Natl Acad Sci USA 80: 6750
70. Kalb VP, Loper JC, Dey CR, Woods CW, Sutter TR (1986) Gene 45: 237
71. Fried MF, Warner JR (1981) Proc Natl Acad Sci USA 78: 238
72. Schultz LD, Friesen JD (1983) J Bact 155: 8
73. Fried MF, Warner JR (1982) Nucl Acids Res 10: 3133
74. Larkin JC, Woolford JL (1983) Nucl Acids Res 11: 403
75. Murata K, Fukuda Y, Shimosaka M, Watanabe K, Saikusa T, Kimura A (1985) Appl Environ Microb 50: 1200
76. Falco SC, Dumas KS (1985) Genetics 109: 21
77. Casey GP, Xiao W, Rank GH (1988) J Inst Brew 94: 93
78. Saunders GW, Rank GH (1982) Can J Genet Cytol 24: 493
79. Rank GH, Sheard JW (1979) Molec Gen Genet 167: 309
80. Rank GH, Robertson A, Philips K (1975) J Bact 122: 359
81. Balzi E, Weining C, Stanislaw U, Capieaux E, Goffeau A (1987) J Biol Chem 35: 16871
82. Fogel F, Welch JW (1982) Proc Natl Acad Sci USA 79: 5342
83. Henderson RCA, Cox BC, Tubb R (1985) Curr Genet 9: 133
84. Hincliffe E, Daubney CJ (1985) Am Soc Brw Chem 44: 98
85. McCann AK, Barnett JA (1986) Yeast 2: 109
86. Yamashita I, Fukui S (1983) Agric Biol Chem 47: 2689
87. Yamashita I, Maemura T, Hatano T, Fukui S (1985) J Bact 161: 574
88. Meaden P, Ogden K, Bussey H, Tubb RS (1985) Gene 34: 325
89. Tubb RS, Liljestrom PL (1986) J Inst Brew 92: 588
90. Post-Beittenmiller MA, Hamilton RW, Hopper JE (1984) Mol Cell Biol 4: 1238
91. Ruohola H, Liljestrom PL, Torkkeli T, Kopu H, Lehtinen P, Kalkinen N, Korhola M (1986) FEMS Microb Lett 34: 179
92. Liljestrom PL (1985) Nucl Acids Res 13: 7257
93. Summer-Smith M, Bozzato RP, Skipper N, Davies RW, Hopper JE (1985) Gene 36: 333
94. Gendre F, Guerineau M (1986) Yeast 2: S127
95. Xiao W, and Rank GH (1989) Gene 76: 99
96. Sreekrishna K, Dickson RC (1985) Proc. Natl. Acad. Sci. USA 82: 7909
97. Sreekrishna K, Tschopp JF, Fuke M (1987) Gene 59: 115
98. Nicaud JM, Fabre E, Gaillardin C (1989) Curr Genet 16: 253
99. Bussey H, Meaden P (1985) Curr Genet 9: 285
100. Goodey AR, Doel SM, Piggott JR, Watson MEE, Zealey GR, Cafferkey R, Carter BLA (1986) Mol Gen Genet 204: 505
101. Takagi M, Kawai S, Shibuya I, Miyazaki M, Yano K (1986) J Bact 168: 417
102. Gaillardin C, Ribet AM (1987) Curr Genet 11: 369
103. Gmünder H, Kohli J (1989) Mol Gen Genet 220: 95
104. Bach M-L, LaCroute F (1972) Mol Gen Genet 115: 126
105. Chattoo BB, Sherman F, Azubalis DA, Fjellstedt TA, Mehvert D, and Ogur M (1979) Genetics 93: 51
106. Boeke JD, LaCroute F, Fink GR (1984) Mol Gen Genet 197: 345
107. Snow R (1966) Nature 211: 206
108. Sanchez S, Demain AL (1977) Europ J Appl Microbiol 4: 45
109. Bystrykh LV, Aminova LR, Trotsenko YA (1988) FEMS Microb Lett 51: 89
110. Nga BH, Heslot H, Gaillardin CM, Fournier P, Chan K, Chan YN, Lim EW, Nai PC (1988) J Biotech 7: 83
111. Henry SA, Donohue TF, Culbertson MR (1975) Mol Gen Genet 143: 5

112. Bigelis R, Black KA (1989) J Ind Microb 4: 435
113. de Louvencourt L, Fukuhara H, Heslot H, Wesolowski M (1983) J Bact 154: 737
114. Yamakawa M, Hishinuma F, Gunge N (1985) Agri Biol Chem 49: 1537
115. Stark MJR, Milner JS (1989) Yeast 5: 35
116. Ochsner U, Glumoff V, Kälin M, Fiechter A, Reiser J (unpublished)
117. Kurtz MB, Cortelyou MW, Kirsch DR (1986) Mol Cell Biol 6: 142
118. Kurtz MB, Cortelyou MW, Miller SM, Lai M, Kirsch DR (1987) Mol Cell Biol 7: 209
119. Tikhomirova LP, Ikonomova RN, Kuznetsova EN (1986) Curr Genet 10: 741
120. Ye Z-H, Bhattacharjee JK (1988) Yeast 4: S179
121. Niwa O, Matsumoto T, Yanagida M (1986) Mol Gen Genet 203: 397
122. Grimm C, Kohli J, Murray J, Maundrell K (1988) Mol Gen Genet 215: 81
123. Yamamoto M (1989) Fission Yeat. In: Barr PJ, Brake AJ, Valenzuela P (eds) Yeast Genetic Engineering. Butterworths, Boston, pp 53–64
124. Kikuchi Y, Kitazawa Y, Shimatake H, Yamamoto M (1988) Curr Genet 14: 375
125. Losson R, Lacroute F (1983) Cell 32: 371
126. Gleeson MA, Sudbery PE (1988) Yeast 4: 1
127. Couderc R, Baratti J (1980) Agric Biol Chem 44: 2279
128. Ellis, SB, Brust PJ, Koutz PJ, Waters AF; Harpold MM, Gingeras TR (1985) Mol Cell Biol 5: 1111
129. Kroutz P, Davis GR, Stillman C, Barringer K, Cregg J, Thill G (1989) Yeast 5: 167
130. Tschopp JF, Brust PF, Cregg JM, Stillman CA, Gingeras TR (1987) Nucl Acids Res 15: 3859
131. Hitzman DO, Wegner GH, Shay LK (1987) Applications of ultrahigh cell density yeast fermentations. In: Stewart GG, Russell I, Klein RD, Hiebsch RR (eds) Biological research on industrial yeasts, vol I. CRC Press, Boca Raton Florida, pp 21–25
132. Siegel RS, Brierley RA (1989) Biotechnol Bioeng 34: 403
133. Ledeboer AM, Zoetmulder MCM, Veale R, Sudbery PE (1987) In: Neijssel OM, van der Meer RR, Luyben KCAM (eds) Proc 4th Europ Congr on Biotechnol, vol 1, Elsevier, Amsterdam, p 510
134. Ledeboer AM, Edens L, Maat J, Visser C, Bos JW, Verrips CT, Janowicz Z, Eckart M, Roggenkamp R, Hollenberg CP (1985) Nucl Acids Res 13: 3063
135. Janowicz ZA, Eckart MR, Drewke C, Roggenkamp RO, Hollenberg CP, Maat J, Ledeboer AM, Visser C, Verrips CT (1985) Nucl Acids Res 13: 3043
136. Giuseppin MLF, van Eijk HMJ, Bes BCM (1988) Biotechnol Bioengin 32: 577
137. Ito H, Fukuda Y, Murata K, Kimura A (1983) J Bacteriol 153: 163
138. Klebe RJ, Harries JV, Sharp ZD, Douglas MG (1983) Gene 25: 333
139. Tschopp JF, Sverlow G, Kosson R, Craig W, Grinna L (1987) Bio/Technology 5: 1305
140. Grinna LS, Tschopp JF (1989) Yeast 5: 107
141. Shen S-H, Bastien L, Nguyen T, Fung M, Slilaty SN (1989) Gene 84: 303
142. Janowicz ZA, Merckelbach A, Eckart M, Weydemann U, Roggenkamp R, Hollenberg CP (1988) Yeast 4: S155
143. Digan ME, Lair SV, Brierley RA, Siegel RS, Williams ME, Ellis SB, Kellaris PA. Provow, SA, Craig WS, Velicelebi G, Harpold MM, Thill GP (1989) Bio/Technology 7: 160
144. Hagenson MJ, Holden KA, Parker KA, Wood PJ, Cruze JA, Fuke M, Hopkins TR, Stroman DW (1989) Enzyme Microb. Technol. 11: 650
145. Rankine BC (1966) J Sci Food Agric 17: 312
146. Gallander JF (1977) Am J Enol Vitic 28: 65
147. Kunkee RE, Amerine MA (1970) Yeasts in Wine-making. In: Rose AH, Harrison JS (eds) The Yeasts, vol 3, Academic, pp 5–71
148. Smith CL, Matsumoto T, Niwa O, Klco S, Fan JB, Yanagida M, Cantor CR (1987) Nucl Acids Res 15: 4481
149. Sakai K, Sakaguchi J, Yamamoto M (1984) Mol Cell Biol 4: 651
150. Bröker M (1987) Biotechniques 5: 516
151. Hood MT, Stachow C (1990) Nucl Acids Res 18: 688
152. Bach M-L (1987) Curr Genet 12: 527
153. Grimm C, Kohli J (1988) Mol Gen Genet 215: 87

154. Hottinger H, Pearson D, Yamao F, Gamulin V, Cooley L, Cooper T, Söll D (1982) Mol Gen Genet 188: 219
155. Belsham GJ, Barker ADG, Smith AE (1986) Eur J Bioch 156: 413
156. Russell PR (1985) Gene 40: 125
157. Losson R, Fuchs RPP, Lacroute F (1985) J Mol Biol 185: 65
158. Jones RH, Moreno S, Nurse P, Jones NC (1988) Cell 53: 659
159. Pobjecky N, Rosenberg GH, Dinter-Gottlied G, Käufer NF (1990) Mol Gen Genet 220: 314
160. Russell PR, Hall BD (1983) J Biol Chem 258: 143
161. Russell P, Nurse P (1986) Cell 45: 145
162. McLeod M, Stein M, Beach D (1987) EMBO J 6: 729
163. Hoffman CS, Winston F (1989) Gene 84: 473
164. Vassarotti A, Friesen JD (1985) J Biol Chem 260: 6348
165. Kudla B, Persuy M-A, Gaillardin C, Heslot H (1988) Nucl Acids Res. 16: 8603
166. Erratt JA, Nasim A (1986) J Bact 166: 484
167. Kuranda MJ, Robbins PW (1987) Proc Natl Acad Sci USA 84: 2585
168. Jackson MR, Burchell B (1988) Bioch J 251: 931
169. Hildebrandt V, Ramezani-Rad M, Sida U, Wrede P, Grzesiek S, Primke M, Büldt G (1989) FEBS Lett. 243: 137
170. Bröker M, Bäuml O (1989) FEBS Lett 248: 105
171. Chan EC, Ueng PP, Chen LF (1986) Biotechnol Lett 8: 231
172. Chan EC, Ueng PP, Chen LF (1989) Appl Microbiol Biotechnol 31: 524
173. Phaff HJ (1985) Biology of yeasts other than *Saccharomyces*. In: Demain AL, Solomon NA, (eds) Biology of industrial microorganisms Benjamin/Cummings, Menlo Park, pp 537–562
174. Dickson RC, Barr K (1983) J Bact 154: 1245
175. Dickson RC, Markin JS (1980) J Bact 142: 777
176. Webster TD, Dickson RC (1988) Nucl Acids Res 16: 8011
177. Dickson RC, Markin JS (1978) Cell 15: 123
178. Gunge N, Sakaguchi K (1981) J Bact 147: 155
179. Falcone C, Saliola M, Chen XJ, Frontali L, Fukuhara H (1986) Plasmid 15: 248
180. Bianchi MM, Falcone C, Jie CX, Weslowski-Louvel M, Frontali L, Fukuhara H (1987) Curr Genet 12: 185
181. Chen XJ, Bianchi MM, Suda K, Fukuhara H (1989) Curr Genet 16: 95
182. Rietveld K, Bakhuis JG, Jansen in de Wal NJ, van Leen RW, Noordermeer ACM, van Ooyen AJJ, Schaap A, van den Berg JA (1988) Yeast 4: S163
183. Phaff HJ (1970) *Schwanniomyces* Klöcker. In: Lodder J (ed) The yeasts- a taxonomic study, North-Holland, Amsterdam, pp 756–766
184. Price CW, Fuson GB, Phaff HJ (1978) Microbiol Rev 42: 161
185. Ingledew WM (1987) Crit Rev Biotechnol 5: 159
186. Deibel MR, Hiebsch RR, Klein RD (1988) Prep Bioch 18: 77
187. Wilson JJ, Ingledew WM (1982) Appl Environ Microb 44: 301
188. Dohmen RJ, Strasser AWM, Dahlems UM, Seeboth PG, Hollenberg CP (1988) Yeast 4: S145
189. Strasser AWM, Selk R, Dohmen RJ, Niermann T, Bielefeld M, Seeboth P, Tu G, Hollenberg CP (1989) Eur J Bioch 184: 699
190. Abarca D, Fernandez-Lobato M, Gonzalo Claros M, Jimenez A (1989) FEBS Lett. 255: 455
191. Klein RD, Poorman RA, Favreau MA, Shea MH, Hatzenbuhler NT, Nulf SC (1989) Curr Genet 16: 145
192. Klein RD, Roof LL (1988) Curr Genet 13: 29
193. Magee BB, D'Souza TM, Magee PT (1987) J Bact 169: 1639
194. Takagi M, Kobayashi N, Sugimoto M, Fujii T, Watari J, Yano K (1987) Curr Genet 11: 451
195. Tschumper G, Carbon J (1982) J Mol Biol 156: 293
196. Fleig UN, Pridmore RD, Philippsen P (1986) Gene 46: 237
197. Kamiroyo T, Sakasegawa Y, Tan H (1989) Agric Biol Chem 53: 179
198. Olaiya AF, Sogin SJ (1979) J Bact 140: 1043
199. Whelan WL, Beneke ES, Rogers AL, Sol DR (1981) Antimicrob Agents Chemother 19: 1078

200. Poulter R, Hanrahan V, Jeffry K, Markle D, Shepherd MG, Sullivan PA (1982) J Bact 152: 969
201. Nombela C, Pomés R, Gil C (1987) CRC Critical Rev in Microb 15: 79
202. Gil C, Pomés R, Nombela C (1988) J Gen Microb 134: 1587
203. Kelly R, Miller SM, Kurtz MB, Kirsch DR (1987) Mol Cell Biol 7: 199
204. Gillum AM, Tsay YE, Kirsch DR (1984) Mol Gen Genet 198: 179
205. Käppeli O, Fiechter A (1982) Biotechn Bioeng 24: 2519
206. Fiechter A, Käppeli O, Meussdoerffer F (1987) Batch and continuous culture. In: Rose AH, Harrison JS (eds), The yeasts, vol 2. Academic, London, pp 99–129
207. Barnett JA, Payne RW, Yarrow D (1983) Yeasts: Characteristics and identification. Cambridge University Press, pp 512–514
208. Mörtberg M, Neujahr HY (1986) J Bact 168: 734
209. Stüttgen E, Sahm H (1982) Europ J Appl Microb Biotechnol 15: 93
210. Hrmova M, Biely P, Vrsanska M, Petrakova E (1984) Arch Microb 138: 371
211. Lee H, To RJB, Latta RK, Biely P, Schneider H (1987) Appl Env Microb 53: 2831
212. De Mot R, Demeersman M, Verachtert H (1984) System Appl Microb 5: 421
213. Zimmermann M, Emeis CC (1988) Yeast 5: S131
214. Anderson JJ, Dagley S (1980) J Bact 141: 534
215. Anderson JJ, Dagley S (1981) J Bact 146: 291
216. Gaal A, Neujahr HY (1979) J Bact 137: 13
217. Powlowski JB, Dagley S (1985) J Bact 163: 1126
218. Sparnins VL, Anderson JJ, Omans J, Dagley S (1978) J Bact 136: 449
219. Sparnins VL, Burbee DG, Dagley S (1979) J Bact 138: 425
220. Sze IS-Y, Dagley S (1984) J Bact 159: 353
221. Varga JM, Neujahr HY (1970) Eur. J. Bioch. 12: 427
222. Veenhuis M, van der Klei IJ, Harder W (1986) Arch Microb 145: 39
223. Neujahr HA (1989) Yeasts in biodegradation and biodeterioration processes. In: Verachtert H, and De Mot R (eds) Yeast, biotechnology and biocatalysis Marcel Dekker, New York, pp 321–348
224. Punt PJ, Oliver RP, Dingemanse MA, Pouwels PH, van den Hondel CAMJJ (1987) Gene 56: 117
225. Gilbert HJ, Tully M (1982) J Bact 150: 498
226. Marusich WC, Jensen RA, Zamir LO (1981) J Bact 146: 1013
227. Wickerham LJ (1970) Genus 14. *Pachysolen* Boidin et Adzet. In: Lodder J (ed) The yeasts, a taxonomic study, 2nd edn. North-Holland, Amsterdam, pp 448–454
228. James AP, Zahab DM (1982) J Gen Microb 128: 2297
229. Wedlock DN, James AP, Thornton RJ (1989) J Gen Microb 135: 2019
230. Toh-e A, Tada S, Oshima Y (1982) J Bact 151: 1380
231. Araki H, Jearnpipatkul A, Tatsumi H, Sakurai T, Ushio K, Muta T, Oshima Y (1985) J Mol Biol 182: 191
232. Fournier P, Kudla B, Chasles M, Gaillardin C (1988) Yeast 4: S127
233. Cregg JM, Tschopp JF, Stillmann C, Siegel R, Akong M, Craig WS, Buckholz RG, Madden KR, Kellaris PA, Davis GR, Smiley BL, Cruze J, Torregrossa R, Velicelebi G, Thill GP (1987) Bio/Technology 5: 479
234. Russell PR (1984) European patent application 84305571.6
235. Heslot H, Nicaud JM, Fabre E, Beckerich JM, Fournier P, Gaillardin C (1989) Proc 2nd SSM Intern Congress Microbiology Singapore (in press)
236. Franke AE, Kaczmarek FS, Eisenhard ME, Geoghegan KF, Danley DE, De Zeeuw JR, O'Donnell MM, Gollaher MG, Dawidow LS (1988) Develop Industr Microbiol 29: 43
237. Bröker M, Ragg H, Karges HE (1987) Biochim Biophys Acta 908: 203
238. Sreekrishna K, Potenz RHB, Cruze JA, McCombie WR, Parker KA, Nelles L, Mazzaferro PK, Holden KA, Harrison RG, Wood PJ, Phelps DA, Hubbard CE, Fuke M (1988) J Basic Microb 28: 265
239. Nicaud JM, Fabre E, Beckerich JM, Fournier P, Gaillardin C (1989) J Biotech 12: 285
240. Hikiji T, Ohkuma M, Takagi M, Yano K (1989) Curr Genet 16: 261
241. Ratner M (1989) Bio/Technology 7: 1129

Author Index Volumes 1–43

GPSR Compliance

The European Union's (EU) General Product Safety Regulation (GPSR) is a set of rules that requires consumer products to be safe and our obligations to ensure this.

If you have any concerns about our products, you can contact us on ProductSafety@springernature.com

In case Publisher is established outside the EU, the EU authorized representative is:

Springer Nature Customer Service Center GmbH
Europaplatz 3
69115 Heidelberg, Germany

Batch number: 10620141

Printed by Printforce, the Netherlands